Foundations of
Optical Waveguides

Foundations of Optical Waveguides

Gilbert H. Owyang
Professor of Electrical Engineering
Worcester Polytechnic Institute

Elsevier
New York • Oxford

Elsevier North Holland, Inc.
52 Vanderbilt Avenue, New York, New York 10017

Distributors outside the United States of America and Canada:

Edward Arnold (Publishers), Ltd.
41 Bedford Square
London WC1B 3DQ England

Library of Congress Cataloging in Publication Data

Owyang, Gilbert H.
Foundations of dielectric optical waveguides.

Includes bibliographies and index.
1. Optical wave guides. I. Title.
TA1800.098 621.36′92 81-5490
ISBN 0-444-00560-9 AACR2

Desk Editor John Haber
Design Edmée Froment
Design Editor Virginia Kudlak
Art rendered by Vantage Art, Inc.
Cover Design by José García and Michele Maldeau
Production Manager Joanne Jay
Compositor Science Typographers, Inc.
Printer Haddon Craftsmen

Manufactured in the United States of America

To my wife

Lily

and the boys

Kevin and Colin

To the memory of
my parents

Contents

Preface

As newer processes are developed for the manufacture of optical fibers, better quality is increasingly available while production costs quickly become more feasible. Consequently, communication in the optical frequencies is among the most rapidly developing areas of technology.

Although a number of books treating this material are presently available, they are generally designed for research scientists. This text is the result of notes developed for undergraduate and graduate courses in optical waveguides—including classes for practical engineers. It is intended to serve readers with some background in electromagnetic theory, but a thorough grounding is not assumed. Advanced undergraduates as well as practicing engineering scientists in the field of communications could find it useful. A basic understanding of wave propagation in bounded and unbounded regions should be sufficient prerequisite.

The text investigates from the field point of view the behavior of waves propagating in planar and cylindrical waveguides. Owing to the complexity of the problem, the analysis is mathematical in nature, but the physical interpretation of the theoretical results is emphasized throughout.

Beginning with a brief review of electromagnetic theory, wave propagation in free space, and guided waves in homogeneous media, the book analyzes the basic dielectric sheet, imperfect, and inhomogeneous waveguides. The more practical cladded cylindrical and inhomogeneous circular waveguides are also considered. Mathematical techniques involving eigenfunctions and Green's function are discussed in detail, as are methods of approximation.

One of the goals of this treatment is to enable readers to grasp fully the material in current research papers. Consequently, many techniques usually found only in higher-level mathematics textbooks are presented, eliminating the annoying and frustrating task of locating suitable references. While the text should remain accessible to those wishing to focus on the highlights of the mathematical development, it is intended to make further progress possible.

I would like to thank those reviewers whose criticisms and suggestions helped refine the manuscript. I also wish to extend my appreciation to John Haber of Elsevier North Holland, Inc., to Harit Majmudar, head of the Department of Electrical Engineering at Worcester Polytechnic Institute, for his support and encouragement, and to Geri Hicky for typing the manuscript.

Foundations of
Optical Waveguides

1

Electromagnetic Theory

This chapter provides a survey of the background material that will be the basis of the entire book. The chapter begins with a review of the field relations and their mathematical solution. In order to analyze the wave equation, the method of separation of variables is introduced. The one-dimensional wave equation is investigated in detail and the results generalized to describe wave propagation in an arbitrary direction. The formulation of problems in unbounded nonhomogeneous media follows. Finally, wave propagation in a guiding structure filled with a homogeneous medium will be studied.

1.1. Maxwell's Equations

The *field* of a quantity is defined as the mathematical function that describes the variation of the quantity in a region. The electromagnetic field obeys Maxwell's equations,

$$\nabla \times \tilde{\mathbf{E}}(\mathbf{r}, t) = -\frac{\partial \tilde{\mathbf{B}}(\mathbf{r}, t)}{\partial t} \qquad \text{(Faraday's law)}, \tag{1}$$

$$\nabla \times \tilde{\mathbf{H}}(\mathbf{r}, t) = \tilde{\mathbf{J}}(\mathbf{r}, t) + \frac{\partial \tilde{\mathbf{D}}(\mathbf{r}, t)}{\partial t} \qquad \text{(Ampere's circuital law)}, \tag{2}$$

$$\nabla \cdot \tilde{\mathbf{B}}(\mathbf{r}, t) = 0, \tag{3}$$

$$\nabla \cdot \tilde{\mathbf{D}}(\mathbf{r}, t) = \tilde{\rho}(\mathbf{r}, t) \qquad \text{(Gauss's law)}, \tag{4}$$

where tildes label functions of position *and* time, and

$\tilde{\mathbf{E}}$ = electric field intensity in volts per meter,

$\tilde{\mathbf{H}}$ = magnetic field intensity in amperes per meter,

$\tilde{\mathbf{D}}$ = electric flux density in coulombs per square meter,

$\tilde{\mathbf{B}}$ = magnetic flux density in webers per square meter,

$\tilde{\mathbf{J}}$ = electric current density in amperes per square meter
$= \tilde{\mathcal{J}}(\mathbf{r}, t) + \tilde{\mathbf{J}}_c(\mathbf{r}, t)$,

$\tilde{\mathcal{J}}$ = the source current density,

$\tilde{\mathbf{J}}_c$ = the conduction current density,

$\tilde{\rho}$ = electric charge density in coulombs per cubic meter, and

$\mathbf{r}$ = the position vector of the field point.

Maxwell's equations are supplemented by the following *constitutive relations*, which characterize the properties of the medium:

$$\tilde{\mathbf{D}} = \epsilon \tilde{\mathbf{E}}, \tag{5}$$

$$\tilde{\mathbf{B}} = \mu \tilde{\mathbf{H}}, \tag{6}$$

$$\tilde{\mathbf{J}}_c = \sigma \tilde{\mathbf{E}}, \tag{7}$$

where

ϵ = the permittivity in farads per meter,

μ = the permeability in henry per meter, and

σ = the conductivity in mhos per meter.

Equations (5)–(7) are valid for linear isotropic media. A medium is linear if these relations hold independent of the magnitude of the field. A medium is isotropic if these relations hold independent of the direction of the field. The *constitutive parameters* (ϵ, μ, or σ) can be functions of position and time.

1.2. Periodic Time Dependent Fields

Maxwell's equations are partial differential equations in which the independent variables are the spatial coordinates and time. Consider the *simple-harmonic time varying field*, which can be expressed in the exponential form

$$\tilde{\mathbf{E}}(\mathbf{r}, t) = \mathbf{E}(\mathbf{r}) e^{j\omega t} \equiv \mathbf{E} e^{j\omega t}, \qquad \mathbf{E} \equiv \mathbf{E}(\mathbf{r}), \tag{1}$$

$$\tilde{\mathbf{H}}(\mathbf{r}, t) = \mathbf{H}(\mathbf{r}) e^{j\omega t} \equiv \mathbf{H} e^{j\omega t}, \qquad \mathbf{H} \equiv \mathbf{H}(\mathbf{r}). \tag{2}$$

Taking simple-harmonic time varying fields in the time domain does not restrict the applicability of the results. This may be readily understood when one recalls that any function of time can be represented as a Fourier series or Fourier integral of simple-harmonic functions. Once the solution of a simple-harmonic time varying field is known, the principle of superposition may be used to find the total field.

For simple-harmonic time varying fields, Maxwell's equations take the following form:

$$\nabla\times\mathbf{E}=-j\omega\mu\mathbf{H}, \tag{3}$$

$$\nabla\times\mathbf{H}=\mathbf{J}+j\omega\epsilon\mathbf{E}, \tag{4}$$

$$\nabla\cdot\mathbf{B}=\nabla\cdot(\mu\mathbf{H})=\mathbf{H}\cdot\nabla\mu+\mu\nabla\cdot\mathbf{H}=0, \tag{5}$$

$$\nabla\cdot\mathbf{D}=\nabla\cdot(\epsilon\mathbf{E})=\mathbf{E}\cdot\nabla\epsilon+\epsilon\nabla\cdot\mathbf{E}=\rho. \tag{6}$$

In a region excluding sources, that is, where $\mathcal{J}=0$ and $\rho=0$, Eqs. (4) and (6) become

$$\nabla\times\mathbf{H}=\mathbf{J}_c+j\omega\epsilon\mathbf{E}=(\sigma+j\omega\epsilon)\mathbf{E}=j\omega\varepsilon\mathbf{E}, \tag{7}$$

$$\varepsilon\equiv\epsilon(1+\sigma/j\omega\epsilon), \tag{8}$$

$$\nabla\cdot\mathbf{D}=\mathbf{E}\cdot\nabla\epsilon+\epsilon\nabla\cdot\mathbf{E}=0. \tag{9}$$

The complex permittivity, which will be written ε, is a convenient parameter, enabling Eq. (7) to have the same form whether the medium is a perfect conductor, a perfect insulator, or somewhere in between. Note that all field quantities in Eqs. (3)–(9) are functions of position only, the time dependency of these functions having been removed through Eqs. (1) and (2). The problem is thus simplified to finding the fields as function of position, that is, to solving the above equations. The complete expression of the fields as function of both position *and* time can then be obtained by factoring in $e^{j\omega t}$.

1.3. Generalized Wave Equations

Consider fields in a region which contains no sources, that is, where $\mathcal{J}=0$ and $\rho=0$. Maxwell's equations are then

$$\nabla\times\mathbf{E}=-j\omega\mu\mathbf{H}, \tag{1}$$

$$\nabla\times\mathbf{H}=j\omega\varepsilon\mathbf{E},\qquad \varepsilon\equiv\epsilon(1+\sigma/j\omega\epsilon). \tag{2}$$

The divergence of Eqs. (1) and (2) yields

$$\mathbf{H}\cdot\nabla\mu+\mu\nabla\cdot\mathbf{H}=0, \tag{3}$$

$$\mathbf{E}\cdot\nabla\varepsilon+\varepsilon\nabla\cdot\mathbf{E}=0. \tag{4}$$

These equations involve two field quantities, **E** and **H**, and can be reduced to a single equation for one field. The **H** field is first obtained from Eq. (1) and then substituted into Eq. (2):

$$\mathbf{H}=\frac{1}{-j\omega\mu}\nabla\times\mathbf{E},$$

$$\begin{aligned}\nabla\times\mathbf{H}&=\nabla\times\left(\frac{1}{-j\omega\mu}\nabla\times\mathbf{E}\right),\\ &=\nabla\left(\frac{1}{-j\omega\mu}\right)\times(\nabla\times\mathbf{E})+\frac{1}{-j\omega\mu}\nabla\times(\nabla\times\mathbf{E}),\\ &=\frac{1}{j\omega\mu^2}\nabla\mu\times(\nabla\times\mathbf{E})+\frac{1}{-j\omega\mu}\nabla\times(\nabla\times\mathbf{E}).\end{aligned}\tag{5}$$

Substitution of Eq. (5) into Eq. (2) yields

$$\nabla\times\nabla\times\mathbf{E}-\frac{1}{\mu}\nabla\mu\times(\nabla\times\mathbf{E})=-\gamma^2\mathbf{E},\tag{6}$$

where

$$\gamma^2\equiv j\omega\mu\, j\omega\varepsilon=-\omega^2\mu\varepsilon\tag{7}$$

and γ is the *wave number* of the medium.

Similarly, if $\mathbf{E}$ is solved from Eq. (2) and then substituted into Eq. (1), one obtains

$$\nabla\times\nabla\times\mathbf{H}-\frac{1}{\varepsilon}\nabla\varepsilon\times(\nabla\times\mathbf{H})=-\gamma^2\mathbf{H}.\tag{8}$$

With use of the vector identity

$$\nabla\times\nabla\times\mathbf{g}=\nabla\nabla\cdot\mathbf{g}-\nabla^2\mathbf{g}\tag{9}$$

and Eq. (4), Eq. (6) then takes the following form:

$$\begin{aligned}&\nabla\nabla\cdot\mathbf{E}-\nabla^2\mathbf{E}-\frac{1}{\mu}\nabla\mu\times(\nabla\times\mathbf{E})=-\gamma^2\mathbf{E},\\ &\nabla\left(\frac{-1}{\varepsilon}\nabla\varepsilon\cdot\mathbf{E}\right)-\nabla^2\mathbf{E}-\frac{1}{\mu}\nabla\mu\times(\nabla\times\mathbf{E})=-\gamma^2\mathbf{E},\\ &\nabla^2\mathbf{E}+\nabla\left(\frac{1}{\varepsilon}\nabla\varepsilon\cdot\mathbf{E}\right)+\frac{1}{\mu}\nabla\mu\times(\nabla\times\mathbf{E})-\gamma^2\mathbf{E}=0.\end{aligned}\tag{10}$$

Similarly, the use of Eqs. (3) and (9) in Eq. (8) yields

$$\nabla^2\mathbf{H}+\nabla\left(\frac{1}{\mu}\nabla\mu\cdot\mathbf{H}\right)+\frac{1}{\varepsilon}\nabla\varepsilon\times(\nabla\times\mathbf{H})-\gamma^2\mathbf{H}=0.\tag{11}$$

In the case where the constitutive parameters μ and ε are not functions of spatial variables, their gradients vanish and Eqs. (10) and (11) are simplified.

$$\nabla^2\mathbf{E}-\gamma^2\mathbf{E}=0,\tag{12}$$

$$\nabla^2\mathbf{H}-\gamma^2\mathbf{H}=0.\tag{13}$$

These are the *vector wave equations* and are valid for each field component in rectangular coordinates.

$$\nabla^2 E_i - \gamma^2 E_i = 0, \tag{14}$$

$$\nabla^2 H_i - \gamma^2 H_i = 0. \tag{15}$$

Here the subscript indicates the i component, where i stands for either x, y, or z. Equations (14) and (15) are known as the *scalar wave equations*, or *Helmholtz equations*.

1.4. One-Dimensional Wave Equation

It has been shown that both the electric and magnetic fields satisfy the wave equation,

$$\nabla^2 \mathbf{G} - \gamma^2 \mathbf{G} = 0, \qquad \gamma^2 \equiv -\omega^2 \mu \varepsilon, \tag{1}$$

where $\mathbf{G} \equiv \mathbf{G}(\mathbf{r})$ stands for either **E** or **H**. When **G** is a function of a single variable, say z [i.e., $\mathbf{G} \equiv \mathbf{G}(z)$], Eq. (1) becomes

$$\frac{d^2 \mathbf{G}}{dz^2} - \gamma^2 \mathbf{G} = 0 \tag{2}$$

and each component of **G** satisfies the scalar wave equation

$$\frac{d^2 G_i}{dz^2} - \gamma^2 G_i = 0, \qquad i = x, y, \text{ or } z. \tag{3}$$

This is the *one-dimensional wave equation*. Let the trial solution of Eq. (3) be

$$G_i(z) = G_0 e^{pz}, \tag{4}$$

where G_0 is a constant and the quantity p is still to be determined. Then substitution of Eq. (4) into Eq. (3) yields

$$(p^2 - \gamma^2) G_0 e^{pz} = 0,$$

and therefore

$$p^2 - \gamma^2 = 0, \tag{5}$$

$$p = \pm \gamma. \tag{6}$$

Equation (5) is known as the *characteristic equation* of the problem, and its roots, given by Eq. (6), are the *characteristic values* (or *eigenvalues*) of the problem. The independent solutions corresponding to each characteristic value are known as the *characteristic functions* (or *eigenfunctions*) of the problem.

According to the theory of differential equations, the general solution of Eq. (3) is a linear combination of the eigenfunctions,

$$G_i(z) = G_i^+ e^{-\gamma z} + G_i^- e^{\gamma z}, \tag{7}$$

where G_i^+ and G_i^- are two arbitrary constants of integration for the second-order differential equation and can be determined by the specified boundary conditions. G_i^+ (G_i^-) is the amplitude of the positive (negative) traveling wave, as will be explained.

The sum of all three spatial components gives the complete vector expression of the field. In rectangular coordinates

$$\begin{aligned}\mathbf{G}(z) &= \hat{\mathbf{x}}(G_x^+ e^{-\gamma z} + G_x^- e^{\gamma z}) + \hat{\mathbf{y}}(G_y^+ e^{-\gamma z} + G_y^- e^{\gamma z}) \\ &\quad + \hat{\mathbf{z}}(G_z^+ e^{-\gamma z} + G_z^- e^{\gamma z}) \\ &= \mathbf{G}^+ e^{-\gamma z} + \mathbf{G}^- e^{\gamma z},\end{aligned} \tag{8}$$

where $\mathbf{G}^{\pm} = \hat{\mathbf{x}}G^{\pm} + \hat{\mathbf{y}}G_y^{\pm} + \hat{\mathbf{z}}G_z^{\pm}$ and $\hat{\mathbf{x}}$ is a unit vector in the x direction. The complete expression of the field in spatial and time domains is then given by

$$\tilde{\mathbf{G}}(z,t) \equiv \mathbf{G}(z)e^{j\omega t} = \mathbf{G}^+ e^{j\omega t - \gamma z} + \mathbf{G}^- e^{j\omega t + \gamma z}. \tag{9}$$

The *wave number* or *propagation constant* γ is a complex quantity in general. From its definition, Eq. (1),

$$\begin{aligned}\gamma &= j\omega\sqrt{\mu\varepsilon} \\ &= j\omega\sqrt{\mu\epsilon(1+\sigma/j\omega\epsilon)} \equiv \alpha + j\beta.\end{aligned} \tag{10}$$

Then

$$e^{-\gamma z} = e^{-\alpha z}e^{-j\beta z}. \tag{11}$$

The real part of γ (and consequently $e^{-\alpha z}$) specifies the magnitude of the exponential function $e^{-\gamma z}$; α is therefore known as the *attenuation constant*, measured in nepers per meter. The factor $e^{-j\beta z}$ prescribes the phase of the function and consequently β is known as the *phase constant*, measured in radians per meter.

Equation (9) may be expressed

$$\mathbf{G} = G^+ e^{-\alpha z} e^{j(\omega t - \beta z)} + G^- e^{\alpha z} e^{j(\omega t + \beta z)}. \tag{12}$$

The above solution involves functions of two variables, $f(z,t) \equiv e^{j(\omega t \pm \beta z)}$. Such a function can be easily investigated by keeping one of the variables fixed and studying its variation with respect to the other variable. When $t=0$,

$$f(z, t=0) = e^{\pm j\beta z} \equiv f(z), \tag{13}$$

and when $t = t_1$, some arbitrary value,

$$f(z, t_1) = e^{j(\omega t_1 \pm \beta z)} = e^{\pm j\beta(z \pm \omega t_1/\beta)} \equiv e^{\pm j\beta z_1} \equiv f(z_1), \tag{14}$$

where $z_1 \equiv z \pm \omega t_1/\beta$. Since z and z_1 differ by a constant quantity ($\pm\omega t_1/\beta$), $f(z)$ and $f(z_1)$ must have the same shape. $f(z_1)$ is displaced from $f(z)$ by an amount ($\pm\omega t_1/\beta$) along the z axis. To be more specific, let

$$f^+(z)\equiv Ee^{j(\omega t-\beta z)}|_{t=0}=e^{-j\beta z}, \tag{15}$$

$$f^+(z_1)\equiv e^{j(\omega t-\beta z)}|_{t=t_1}=e^{-j\beta(z-\omega t_1/\beta)}=e^{-j\beta z_1}. \tag{16}$$

The relative location of $f^+(z)$ and $f^+(z_1)$ can be established at some reference point, say, the origin of each variable ($z=0$ and $z_1=0$).

$$z_1=0=z-\omega t_1/\beta \qquad \text{or} \qquad z_1=0 \quad \text{at} \quad z=\omega t_1/\beta. \tag{17}$$

Thus $f^+(z_1)$ is displaced by $\omega t_1/\beta$ in the positive z direction for positive values of t_1. Because $f^+(z,t)\equiv e^{j(\omega t-\beta z)}$ moves in the positive z direction as time increases, it is known as a *positive traveling wave function*.

Similarly, $f^-(z,t)\equiv e^{j(\omega t+\beta z)}$ is known as the *negative traveling wave function*, since the function moves in the negative z direction as time increases.

The speed at which the wave function travels,

$$e^{j\omega t\pm\gamma z}=e^{\pm\alpha z}e^{j(\omega t\pm\beta z)},$$

can be determined by observing the movement of a constant phase point. Suppose at some reference point $t=t_0$, $z=z_0$ the wave function has some phase of value M, so that

$$\omega t_0\pm\beta z_0=M. \tag{18}$$

At a later time $t_1=t_0+\Delta t$, this constant phase point will have traveled to a new position $z_1=z_0+\Delta z$. The phase is then given by

$$\begin{aligned} M&=\omega t_1\pm\beta z_1=\omega(t_0+\Delta t)\pm\beta(z_0\pm\Delta z)=(\omega t_0\pm\beta z_0)+\omega\Delta t\pm\beta\Delta z\\ &=M+\omega\Delta t\pm\beta\Delta z \end{aligned}$$

or

$$\omega\Delta t\pm\beta\Delta z=0$$

$$\Delta z/\Delta t=\mp\omega/\beta. \tag{19}$$

The left-hand side of Eq. (19) has the dimension of velocity and can be interpreted as the average velocity of the constant phase point. In the above derivation, t_1 and z_1 are entirely arbitrary and they may be chosen as small (or large) as possible. If one takes the limit $\Delta t\to 0$, then

$$v_p=\lim_{\Delta t\to 0}\Delta z/\Delta t=dz/dt=\mp\omega/\beta. \tag{20}$$

One calls v_p the *phase velocity* of the traveling wave. The upper sign is for the traveling wave in the negative z direction, while the lower sign is for the positive traveling wave.

1.5. Method of Separation of Variables

The scalar wave equation for a scalar function $g(r)$ in rectangular coordinates is

$$\frac{\partial^2 g}{\partial x^2}+\frac{\partial^2 g}{\partial y^2}+\frac{\partial^2 g}{\partial z^2}-\gamma^2 g=0 \tag{1}$$

or

$$\frac{1}{g}\left(\frac{\partial^2 g}{\partial x^2}+\frac{\partial^2 g}{\partial y^2}+\frac{\partial^2 g}{\partial z^2}\right)-\gamma^2=0. \tag{2}$$

The method of separation of variables assumes a trial solution to be the product of three functions and each one is a function of a single variable only.

$$g(x,y,z)\equiv X(x)Y(y)Z(z). \tag{3}$$

Substitution of the trial solution into Eq. (2) yields

$$\frac{1}{X}\frac{d^2X}{dx^2}+\frac{1}{Y}\frac{d^2Y}{dy^2}+\frac{1}{Z}\frac{d^2Z}{dz^2}-\gamma^2=0. \tag{4}$$

Each of the first three terms can be a function of only one of the independent variables and the fourth term is a constant. Equation (4) requires the sum of all these terms to be constant, independent of all variables. This can be so if and only if each term is a constant itself. This can be verified by differentiating Eq. (4) with respect to one of the variables, say, x:

$$\frac{d}{dx}\left(\frac{1}{X}\frac{d^2X}{dx^2}\right)+\frac{d}{dx}\left(\frac{1}{Y}\frac{d^2Y}{dy^2}\right)+\frac{d}{dx}\left(\frac{1}{Z}\frac{d^2Z}{dz^2}\right)-\frac{d\gamma^2}{dx}=0$$

or

$$\frac{d}{dx}\left(\frac{1}{X}\frac{d^2X}{dx^2}\right)=0. \tag{5}$$

The quantity within the parentheses is thus a constant with respect to x. But by definition $X\equiv X(x)$, and therefore $(1/X)d^2X/dx^2$ is a constant, say, $\gamma_x{}^2$:

$$\frac{1}{X}\frac{d^2X}{dx^2}=\gamma_x{}^2 \quad \text{or} \quad \frac{d^2X}{dx^2}-\gamma_x{}^2X=0. \tag{6}$$

This process can be repeated for the other two variables to obtain

$$\frac{1}{Y}\frac{d^2Y}{dy^2}=\gamma_y{}^2 \qquad \text{or} \qquad \frac{d^2Y}{dy^2}-\gamma_y{}^2Y=0, \tag{7}$$

$$\frac{1}{Z}\frac{d^2Z}{dz^2}=\gamma_z{}^2 \qquad \text{or} \qquad \frac{d^2Z}{dz^2}-\gamma_z{}^2Z=0. \tag{8}$$

Equations (6)–(8) are one-dimensional wave equations as treated in the previous section. Their solutions are found to be

$$X(x)=X^+e^{-\gamma_x x}+X^-e^{\gamma_x x}, \tag{9}$$

$$Y(y)=Y^+e^{-\gamma_y y}+Y^-e^{\gamma_y y}, \tag{10}$$

$$Z(z)=Z^+e^{-\gamma_z z}+Z^-e^{\gamma_z z}, \tag{11}$$

where $X^{\pm}$, $Y^{\pm}$, and $Z^{\pm}$ are constants of integration. The complete solution of Eq. (1) is

$$\begin{aligned} g(x,y,z)&=X(x)Y(y)Z(z)\\ &=C_1e^{\gamma_x x+\gamma_y y+\gamma_z z}+C_2e^{\gamma_x x-\gamma_y y+\gamma_z z}+C_3e^{\gamma_x x+\gamma_y y-\gamma_z z}\\ &\quad+C_4e^{\gamma_x x-\gamma_y y-\gamma_z z}+C_5e^{-\gamma_x x+\gamma_y y+\gamma_z z}+C_6e^{-\gamma_x x-\gamma_y y+\gamma_z z}\\ &\quad+C_7e^{-\gamma_x x-\gamma_y y-\gamma_z z}+C_8e^{-\gamma_x x+\gamma_y y-\gamma_z z}, \end{aligned} \tag{12}$$

where C_i, $i=1,2,\ldots$, is a product of $X^{\pm}$, $Y^{\pm}$, and $Z^{\pm}$. Equation (12) is the most general form of solution and not all the terms exist in an actual problem. A better insight to this complex expression can be obtained by introducing the following vector notation. Let $\mathbf{r}$ be the position vector of a field point,

$$\mathbf{r}=\hat{\mathbf{x}}x+\hat{\mathbf{y}}y+\hat{\mathbf{z}}z. \tag{13}$$

The coordinates (x, y, z) of the field point can take on either positive or negative values depending on the location of the field point.

With the propagation vector $\boldsymbol{\gamma}$ is defined as

$$\boldsymbol{\gamma}\equiv\pm\hat{\mathbf{x}}\gamma_x\pm\hat{\mathbf{y}}\gamma_y\pm\hat{\mathbf{z}}\gamma_z, \tag{14}$$

each term in Eq. (12) can be expressed

$$Ce^{\boldsymbol{\gamma}\cdot\mathbf{r}}.$$

For the problem $\boldsymbol{\gamma}=\hat{\mathbf{x}}\gamma_x+\hat{\mathbf{y}}\gamma_y+\hat{\mathbf{z}}\gamma_z$

$$e^{\boldsymbol{\gamma}\cdot\mathbf{r}}=e^{\gamma_x x+\gamma_y y+\gamma_z z},$$

which is the first term in Eq. (12). In the case where $\boldsymbol{\gamma}=\hat{\mathbf{x}}\gamma_x-\hat{\mathbf{y}}\gamma_y+\hat{\mathbf{z}}\gamma_z$, one can show that $e^{\boldsymbol{\gamma}\cdot\mathbf{r}}$ corresponds to the second term in Eq. (12).

The function

$$\begin{aligned} e^{j\omega t\pm\boldsymbol{\gamma}\cdot\mathbf{r}}&=e^{j\omega t\pm\gamma_x x\pm\gamma_y y\pm\gamma_z z}\\ &=e^{\pm\alpha_x x\pm\alpha_y y\pm\alpha_z z}e^{j(\omega t\pm\beta_x x\pm\beta_y y\pm\beta_z z)}\\ &\equiv m(\boldsymbol{\alpha},\mathbf{r})\tilde{f}(\mathbf{r},t), \end{aligned} \tag{15}$$

where

$$m(\boldsymbol{\alpha},\mathbf{r}) \equiv e^{\pm\alpha_x x \pm \alpha_y y \pm \alpha_z z}, \tag{16}$$

$$\tilde{f}(\mathbf{r},t) \equiv e^{j(\omega t \pm \beta_x x \pm \beta_y y \pm \beta_z z)} \equiv e^{j\omega(t \pm x/v_{px} \pm y/v_{py} \pm z/v_{pz})}. \tag{17}$$

The physical meaning of Eq. (17) can be obtained by analogy to the one-dimensional wave function. Equation (17) represents a traveling wave in all three directions (x, y, and z) with phase velocities v_{px}, v_{py}, and v_{pz}, respectively.

1.6. Nonhomogeneous Media

In nonhomogeneous media, as shown in Section 3, the fields satisfy the following equations:

$$\nabla^2\mathbf{E} + \nabla\left(\frac{1}{\varepsilon}\mathbf{E}\cdot\nabla\varepsilon\right) + \frac{1}{\mu}\nabla\mu \times (\nabla\times\mathbf{E}) - \gamma^2\mathbf{E} = 0, \tag{1}$$

$$\nabla^2\mathbf{H} + \nabla\left(\frac{1}{\mu}\mathbf{H}\cdot\nabla\mu\right) + \frac{1}{\varepsilon}\nabla\varepsilon \times (\nabla\times\mathbf{H}) - \gamma^2\mathbf{H} = 0. \tag{2}$$

These equations reduce to simpler wave equations when constitutive parameters are independent of spatial variables,

$$\nabla^2\mathbf{E} - \gamma^2\mathbf{E} = 0, \tag{3}$$

$$\nabla^2\mathbf{H} - \gamma^2\mathbf{H} = 0. \tag{4}$$

Equations (1) and (2) are much more difficult to solve than Eqs. (3) and (4). It is therefore useful to determine the necessary conditions such that Eqs. (1) and (2) can be approximated by Eqs. (3) and (4).

Many of the practical problems involve media with little loss. The idealized lossless case will serve as a good guideline for the actual problem. For a lossless medium, Eq. (1) becomes

$$\nabla^2\mathbf{E} + \nabla\left(\frac{1}{\epsilon}\nabla\epsilon\cdot\mathbf{E}\right) + \frac{1}{\mu}\nabla\mu \times (\nabla\times\mathbf{E}) + \beta^2\mathbf{E} = 0, \tag{5}$$

and Eq. (3) becomes

$$\nabla^2\mathbf{E} + \beta^2\mathbf{E} = 0. \tag{6}$$

When Eq. (6) is a good approximation for the problem, the fields are given by

$$\mathbf{E} = \mathbf{E}^+ e^{-j\boldsymbol{\beta}\cdot\mathbf{r}} + \mathbf{E}^- e^{j\boldsymbol{\beta}\cdot\mathbf{r}}, \qquad \boldsymbol{\beta} = \hat{\mathbf{x}}\beta_x + \hat{\mathbf{y}}\beta_y + \hat{\mathbf{z}}\beta_z, \quad \mathbf{r} = \hat{\mathbf{x}}x + \hat{\mathbf{y}}y + \hat{\mathbf{z}}z. \tag{7}$$

The relative magnitudes of the last three terms determines whether Eq. (6) is a good approximation of Eq. (5). For the purpose of investigating the order of magnitude of these terms, only the positive traveling wave of **E** will be considered, so that

$$\mathbf{E}=\mathbf{E}^{+}e^{-j\boldsymbol{\beta}\cdot\mathbf{r}}. \tag{8}$$

The magnitude of the last term in Eq. (5) is

$$|\beta^2\mathbf{E}|=|\omega^2\mu\epsilon\mathbf{E}|=\left|(2\pi f/v)^2\mathbf{E}\right|=\left|(2\pi/\lambda)^2\mathbf{E}\right|. \tag{9}$$

The *gradient* of a scalar function is a vector quantity with magnitude equal to the maximum directional derivative of the function and with direction identical to that of the maximum directional derivative. The gradient of a well-behaved function f is thus

$$\nabla f=\left(\frac{df}{dl}\hat{\mathbf{l}}\right)_{\max}, \tag{10}$$

where df/dl is the directional derivative in the direction of $\hat{\mathbf{l}}\,dl$, and the subscript max emphasizes the fact that the quantity within the parentheses corresponds to that at maximum directional derivative. The magnitude of the gradient of a scalar is therefore

$$|\nabla f|=\left|\left(\frac{df}{dl}\right)_{\max}\right|. \tag{11}$$

The magnitude of the second term in Eq. (5) can now be estimated:

$$\begin{aligned}\left|\nabla\left(\frac{1}{\epsilon}\nabla\epsilon\cdot\mathbf{E}\right)\right|&=\left|\left[\frac{\partial}{\partial l}\left(\frac{1}{\epsilon}\nabla\epsilon\cdot\mathbf{E}\right)\right]_{\max}\right|=\left|\left[\left(\frac{\partial\mathbf{E}}{\partial l}\right)\cdot\frac{1}{\epsilon}\nabla\epsilon+\mathbf{E}\cdot\frac{\partial}{\partial l}\left(\frac{\nabla\epsilon}{\epsilon}\right)\right]_{\max}\right|\\&=\left|\left[j\beta\mathbf{E}\cdot\frac{1}{\epsilon}\nabla\epsilon+\mathbf{E}\cdot\frac{\partial}{\partial l}\left(\frac{\nabla\epsilon}{\epsilon}\right)\right]_{\max}\right|,\end{aligned} \tag{12}$$

where $\partial/\partial l$ is assumed to be the maximum directional derivative and Eq. (8) was used to obtain the final form. The second term in Eq. (12) may be expanded to give

$$\begin{aligned}\left|\frac{\partial}{\partial l}\left(\frac{\nabla\epsilon}{\epsilon}\right)\right|&=\left|\frac{1}{\epsilon}\frac{\partial}{\partial l}\nabla\epsilon+\nabla\epsilon\frac{\partial}{\partial l}\left(\frac{1}{\epsilon}\right)\right|=\left|\frac{1}{\epsilon}\frac{\partial}{\partial l}\frac{\partial\epsilon}{\partial l}+\nabla\epsilon\left(\frac{-1}{\epsilon^2}\right)\frac{\partial\epsilon}{\partial l}\right|\\&=\left|\frac{1}{\epsilon}\frac{\partial^2\epsilon}{\partial l^2}-\left(\frac{\nabla\epsilon}{\epsilon}\right)^2\right|.\end{aligned} \tag{13}$$

Thus, Eq. (12) becomes

$$\left|\nabla\left(\frac{1}{\epsilon}\mathbf{E}\cdot\nabla\epsilon\right)\right|=\left|\frac{\mathbf{E}}{\epsilon}\cdot\left(j\frac{\partial\pi}{\lambda}\nabla\epsilon\right)+\frac{\partial^2\epsilon}{\partial l^2}-\frac{(\nabla\epsilon)^2}{\epsilon}\right|. \tag{14}$$

The first term on the right of Eq. (14) will be predominant if

$$\frac{\partial^2\epsilon}{\partial l^2}\ll\frac{2\pi}{\lambda}\nabla\epsilon \tag{15}$$

and

$$\frac{\nabla\epsilon}{\epsilon} \ll \frac{2\pi}{\lambda}. \tag{16}$$

The left-hand side of Eq. (16) can be approximated as

$$\left|\frac{1}{\epsilon}\nabla\epsilon\right| \simeq \left|\frac{1}{\epsilon}\frac{\Delta\epsilon}{\Delta l}\right|, \tag{17}$$

where $\Delta\epsilon$ is the increment of ϵ along the incremental length Δl. Thus Eq. (16) becomes

$$\left|\frac{1}{\epsilon}\frac{\Delta\epsilon}{\Delta l}\right| \ll \frac{2\pi}{\lambda} \qquad \text{or} \qquad \left|\frac{1}{\epsilon}\frac{\Delta\epsilon}{\Delta l}\lambda\right| \ll 2\pi. \tag{18}$$

This implies that the fractional variation of the permittivity per unit wavelength is much smaller than 2π.

When $\Delta l = \lambda$, Eq. (18) becomes

$$\frac{\Delta\epsilon}{\epsilon} \ll 2\pi \qquad \text{or} \qquad \left|\frac{1}{2\pi}\frac{\Delta\epsilon}{\epsilon}\right| \ll 1. \tag{19}$$

It can now be seen that the second term in Eq. (5) may be neglected if Eq. (19) is satisfied.

$$\begin{aligned}\left|\frac{\partial^2\epsilon}{\partial l^2}\right| &= \left|\frac{\partial}{\partial l}\left(\frac{\partial\epsilon}{\partial l}\right)\right| = \left|\frac{\partial}{\partial l}\left(\frac{\nabla\epsilon}{\epsilon}\epsilon\right)\right| \\ &\ll \left|\frac{\partial}{\partial l}\left(\frac{2\pi\epsilon}{\lambda}\right)\right| = \left|\frac{2\pi}{\lambda}\nabla\epsilon\right|.\end{aligned}$$

The magnitude of the third term in Eq. (5) may be estimated as follows:

$$\begin{aligned}\left|\frac{1}{\mu}\nabla\mu\times(\nabla\times\mathbf{E})\right| &= \left|\frac{1}{\mu}\nabla\mu\times\left[\nabla\times(\mathbf{E}^{+}e^{-j\boldsymbol{\beta}\cdot\mathbf{r}})\right]\right| \\ &= \left|\frac{1}{\mu}\nabla\mu\times\left[(\nabla\times\mathbf{E}^{+})e^{-j\boldsymbol{\beta}\cdot\mathbf{r}} + \nabla(e^{-j\boldsymbol{\beta}\cdot\mathbf{r}})\times\mathbf{E}^{+}\right]\right| \\ &= \left|\frac{1}{\mu}\nabla\mu\times(-j\boldsymbol{\beta}\times\mathbf{E})\right| \\ &\simeq \left|\frac{1}{\mu}\nabla\mu\,\beta E\right|.\end{aligned} \tag{20}$$

The third term will be negligible if the following condition is satisfied:

$$\left|\frac{1}{\mu}\nabla\mu\,\beta E\right| \ll \beta^2 E, \qquad \left|\frac{1}{\mu}\nabla\mu\right| \ll \beta = \frac{2\pi}{\lambda},$$

$$\left|\frac{1}{\mu}\frac{\Delta\mu}{\Delta l}\right| \ll \frac{2\pi}{\lambda} \qquad \text{or} \qquad \left|\frac{1}{\mu}\frac{\Delta\mu}{\Delta l}\lambda\right| \ll 2\pi.$$

For $\Delta l = \lambda$,

$$\Delta\mu/2\pi\mu \ll 1. \tag{21}$$

That is, the third term in Eq. (5) is negligible if the variation $\Delta\mu/\mu$ per wavelength is much smaller than 2π.

Thus, when both Eqs. (19) and (21) are satisfied, Eqs. (1) and (2) can be approximated by Eqs. (3) and (4), respectively. This means that the relative incremental variation of the constitutive parameters (ϵ and μ) over a distance of one wavelength must be very much smaller than 2π (say, a few percent). These conditions are usually satisfied by the practical inhomogeneous but continuous media in optical waveguide applications. However, it is unlikely that these conditions will be satisfied at the interface between different media.

1.7. The Poynting Vector

The electromagnetic energy may be expressed in terms of field vectors:

$$P(t)=\int_V \tilde{\mathbf{J}}(\mathbf{r},t)\cdot\tilde{\mathbf{E}}(\mathbf{r},t)\,dv=\int_V\left(\nabla\times\tilde{\mathbf{H}}-\frac{\partial\tilde{\mathbf{D}}}{\partial t}\right)\cdot\tilde{\mathbf{E}}\,dv, \tag{1}$$

where $P(t)$ is the instantaneous power within the volume V. The first term on the right-hand side is one term in the following expansion:

$$\nabla\cdot(\tilde{\mathbf{E}}\times\tilde{\mathbf{H}})=\tilde{\mathbf{H}}\cdot\nabla\times\tilde{\mathbf{E}}-\tilde{\mathbf{E}}\cdot\nabla\times\tilde{\mathbf{H}}. \tag{2}$$

The second term in Eq. (1) is

$$\tilde{\mathbf{E}}\cdot\frac{\partial\tilde{\mathbf{D}}}{\partial t}=\tilde{\mathbf{E}}\cdot\frac{\partial(\epsilon\tilde{\mathbf{E}})}{\partial t}=\frac{\partial}{\partial t}\left(\tfrac{1}{2}\epsilon\tilde{\mathbf{E}}\cdot\tilde{\mathbf{E}}\right)=\frac{\partial\tilde{\bar{w}}_{\mathrm{e}}}{\partial t}, \tag{3}$$

where $\tilde{\bar{w}}_{\mathrm{e}}\equiv\frac{1}{2}\tilde{\mathbf{E}}\cdot\tilde{\mathbf{D}}$ is the electric energy density stored in the medium. If S is the surface of the volume V, Eq. (1) can now be written

$$\begin{aligned}P(t)&=\int_V(\tilde{\mathbf{H}}\cdot\nabla\times\tilde{\mathbf{E}}-\nabla\cdot\tilde{\mathbf{E}}\times\tilde{\mathbf{H}})\,dv-\int_V\frac{\partial\tilde{\bar{w}}_{\mathrm{e}}}{\partial t}\,dv\\&=\int_V\tilde{\mathbf{H}}\cdot\left(-\frac{\partial\tilde{\mathbf{B}}}{\partial t}\right)dv-\oint_S\tilde{\mathbf{E}}\times\tilde{\mathbf{H}}\cdot\hat{n}\,da-\int_V\frac{\partial\tilde{\bar{w}}_{\mathrm{e}}}{\partial t}\,dv\\&=-\int_V\left(\frac{\partial\tilde{\bar{w}}_{\mathrm{m}}}{\partial t}+\frac{\partial\tilde{\bar{w}}_{\mathrm{e}}}{\partial t}\right)dv-\oint_S\tilde{\mathbf{E}}\times\tilde{\mathbf{H}}\cdot\hat{n}\,da,\end{aligned} \tag{4}$$

where $\tilde{\bar{w}}_{\mathrm{m}}\equiv\frac{1}{2}\mathbf{H}\cdot\mathbf{B}$ is the magnetic energy density stored in the medium. The divergence theorem is used to convert the second term into a closed surface integral. The current $\tilde{\mathbf{J}}$ is generally composed of the source $\tilde{\mathfrak{J}}$ and the conduction current density $\tilde{\mathbf{J}}_{\mathrm{c}}$:

$$\tilde{\mathbf{J}}=\tilde{\mathfrak{J}}+\tilde{\mathbf{J}}_{\mathrm{c}}=\tilde{\mathfrak{J}}+\sigma\tilde{\mathbf{E}}. \tag{5}$$

Equation (4) may be rearranged to give

$$-\int_V \tilde{\mathcal{J}}\cdot\tilde{\mathbf{E}}\,dv=\int_V \sigma\tilde{\mathbf{E}}\cdot\tilde{\mathbf{E}}\,dv$$
$$+\int_V\left(\frac{\partial \tilde{\bar{w}}_{\mathrm{e}}}{\partial t}+\frac{\partial \tilde{\bar{w}}_{\mathrm{m}}}{\partial t}\right)dv+\oint_S \tilde{\mathbf{E}}\times\tilde{\mathbf{H}}\cdot\hat{\mathbf{n}}\,da. \tag{6}$$

The term on the left represents the energy supplied by the source within the volume. This term is a positive quantity since $\tilde{\mathcal{J}}$ and $\tilde{\mathbf{E}}$ are in opposite directions within the source region. The first term on the right represents the ohmic losses in the medium. The second integral on the right represents the increase in total stored energy. The last integral yields the total outflow of its integrand which has the dimension of energy per unit area. The integrand of this last integral is known as the *Poynting vector*,

$$\tilde{\mathbf{S}}=\tilde{\mathbf{E}}\times\tilde{\mathbf{H}}. \tag{7}$$

In summary, Eq. (6) states that energy created by the source is balanced by the ohmic losses dissipated, the increase in stored energies, and the outflow of electromagnetic energy through the surface enclosing the region.

Simple harmonic variation is expressed in the form of a complex exponential, $e^{j\omega t}$, which enables Maxwell's equations to be time independent. This simplification is possible for all linear relationships. Applying nonlinear operations as is required in evaluating the Poynting vector, which is the product of two field vectors, requires special caution.

For a steady state sinusoidal $\mathbf{E}$ field,

$$\tilde{\mathbf{E}}=\mathbf{E}(\mathbf{r})\sin\omega t. \tag{8}$$

The corresponding $\mathbf{H}$ field is

$$\tilde{\mathbf{H}}=\mathbf{H}(\mathbf{r})\sin(\omega t+\theta). \tag{9}$$

The instantaneous Poynting vector [Eq. (7)] is

$$\tilde{\mathbf{S}}=\mathbf{E}(\mathbf{r})\times\mathbf{H}(\mathbf{r})\sin\omega t\sin(\omega t+\theta)$$
$$=\tfrac{1}{2}\mathbf{E}\times\mathbf{H}[\cos\theta-\cos(2\omega t+\theta)], \tag{10}$$

where the trigonometric identity

$$\sin A\sin B=\tfrac{1}{2}[\cos(A-B)-\cos(A+B)]$$

was used. The instantaneous Poynting vector is made up of two parts; one part is constant with respect to time and the other varies as the double frequency as that of the field quantities.

The time averaged value of the Poynting vector is the integral of Eq. (10) over a certain interval, say, $T=1/f$, one period.

$$\tilde{\mathbf{S}}_{\mathrm{av}}=\frac{1}{T}\int_0^T\tilde{\mathbf{S}}\,dt=\frac{1}{T}\int_0^T\tfrac{1}{2}\mathbf{E}\times\mathbf{H}[\cos\theta-\cos(2\omega t+\theta)]\,dt$$
$$=\tfrac{1}{2}\mathbf{E}\times\mathbf{H}\cos\theta. \tag{11}$$

The double frequency term $\cos(2\omega t+\theta)$ does not contribute to the average power since the average of a sinusoid is always zero.

Equations (8) and (9) may also be written

$$\mathbf{E}=\operatorname{Im}(\mathbf{E}e^{j\omega t})\equiv\operatorname{Im}\mathbf{M}, \tag{12}$$

$$\mathbf{H}=\operatorname{Im}(\mathbf{H}e^{j(\omega t+\theta)})\equiv\operatorname{Im}\mathbf{N}. \tag{13}$$

Then

$$\mathbf{S}=\operatorname{Im}\mathbf{M}\times\operatorname{Im}\mathbf{N}, \tag{14}$$

but

$$\operatorname{Im}\mathbf{M}=\frac{1}{2j}(\mathbf{M}-\mathbf{M}^*), \qquad \operatorname{Im}\mathbf{N}=\frac{1}{2j}(\mathbf{N}-\mathbf{N}^*).$$

Therefore,

$$\begin{aligned}\mathbf{S}&=\frac{1}{2j}(\mathbf{M}-\mathbf{M}^*)\times\frac{1}{2j}(\mathbf{N}-\mathbf{N}^*)\\&=-\tfrac{1}{4}(\mathbf{M}\times\mathbf{N}+\mathbf{M}^*\times\mathbf{N}^*-\mathbf{M}\times\mathbf{N}^*-\mathbf{M}^*\times\mathbf{N})\\&=-\tfrac{1}{4}(\mathbf{E}\times\mathbf{H}e^{j(2\omega t+\theta)}+\mathbf{E}\times\mathbf{H}e^{-j(2\omega t+\theta)}-\mathbf{E}\times\mathbf{H}e^{-j\theta}-\mathbf{E}\times\mathbf{H}e^{j\theta})\\&=-\tfrac{1}{2}[\mathbf{E}\times\mathbf{H}\cos(2\omega t+\theta)-\mathbf{E}\times\mathbf{H}\cos\theta]\\&=\tfrac{1}{2}\mathbf{E}\times\mathbf{H}[\cos\theta-\cos(2\omega t+\theta)]\\&=\tfrac{1}{2}\operatorname{Re}(\tilde{\mathbf{E}}\times\tilde{\mathbf{H}}^*\mp\tilde{\mathbf{E}}\times\tilde{\mathbf{H}}).\end{aligned} \tag{15}$$

The lower sign of the second term is for the case of cosinusoidal variation, $\tilde{\mathbf{E}}=\mathbf{E}\cos\omega t$ and $\tilde{\mathbf{H}}=\mathbf{H}\cos(\omega t+\theta)$.

The average power is again given by

$$\begin{aligned}\mathbf{S}_{\text{av}}&=\tfrac{1}{2}\mathbf{E}\times\mathbf{H}\cos\theta\\&=\tfrac{1}{2}\operatorname{Re}(\mathbf{E}e^{j\omega t}\times\mathbf{H}e^{-j(\omega t+\theta)})\\&=\tfrac{1}{2}\operatorname{Re}(\tilde{\mathbf{E}}\times\tilde{\mathbf{H}}^*).\end{aligned} \tag{16}$$

1.8. Guided Waves in Homogeneous Media

The solution of Maxwell's equations for a uniform guiding system is obtained by solving the vector wave equations, Eqs. (1.3.12) and (1.3.13).

$$\nabla^2\mathbf{E}-\gamma^2\mathbf{E}=0, \tag{1}$$

$$\gamma^2\equiv-\omega^2\mu\varepsilon.$$

$$\nabla^2\mathbf{H}-\gamma^2\mathbf{H}=0, \tag{2}$$

In a waveguide, the signal is transmitted from the source to the receiver along the axis of the guide. Let the z axis be the direction of transmission; then the field should have the factor

$$e^{j\omega t-\gamma_g z},$$

where $\gamma_g = \alpha_g + j\beta_g$ is the propagation constant of the guide. Let the trial solution be

$$\begin{aligned}\tilde{\mathbf{E}}(\mathbf{r}, t) &= \mathbf{E}(\mathbf{r})e^{j\omega t} = \mathbf{E}(x, y)e^{j\omega t - \gamma_g z},\\ \tilde{\mathbf{H}}(\mathbf{r}, t) &= \mathbf{H}(x, y)e^{j\omega t - \gamma_g z}.\end{aligned} \tag{3}$$

A better insight to the problem may be obtained by expanding Maxwell's equations in rectangular coordinates:

$$\nabla \times \mathbf{E} = -j\omega\mu\mathbf{H},$$

$$\frac{\partial E_z}{\partial y} + \gamma_g E_y = -j\omega\mu H_x, \tag{4a}$$

$$-\gamma_g E_x - \frac{\partial E_z}{\partial x} = -j\omega\mu H_y, \tag{4b}$$

$$\frac{\partial E_y}{\partial x} - \frac{\partial E_x}{\partial y} = -j\omega\mu H_z; \tag{4c}$$

$$\nabla \times \mathbf{H} = j\omega\varepsilon\mathbf{E},$$

$$\frac{\partial H_z}{\partial y} + \gamma_g H_y = j\omega\varepsilon E_x, \tag{4d}$$

$$-\gamma_g H_x - \frac{\partial H_z}{\partial x} = j\omega\varepsilon E_y, \tag{4e}$$

$$\frac{\partial H_y}{\partial x} - \frac{\partial H_x}{\partial y} = j\omega\varepsilon E_z. \tag{4f}$$

The field components transverse to the axial direction can be expressed in terms of the axial components. It is convenient to decompose the general case into two special cases: a) $H_z = 0$ and b) $E_z = 0$.

Case a: $H_z = 0$. Subject to this condition, Eq. (4) is modified as follows.

$$\frac{\partial E_z}{\partial y} + \gamma_g E_y = -j\omega\mu H_x, \tag{5a}$$

$$-\gamma_g E_x - \frac{\partial E_z}{\partial x} = -j\omega\mu H_y, \tag{5b}$$

$$\frac{\partial E_y}{\partial x} - \frac{\partial E_x}{\partial y} = 0; \tag{5c}$$

$$\gamma_g H_y = j\omega\varepsilon E_x, \tag{5d}$$

$$-\gamma_g H_x = j\omega\varepsilon E_y, \tag{5e}$$

$$\frac{\partial H_y}{\partial x} - \frac{\partial H_x}{\partial y} = j\omega\varepsilon E_z. \tag{5f}$$

Rearrange Eq. (5a):

$$\gamma_g E_y + j\omega\mu H_x = -\frac{\partial E_z}{\partial y}. \tag{6}$$

Solve for H_x from Eq. (5e):

$$H_x = \frac{-j\omega\varepsilon}{\gamma_g} E_y. \tag{7}$$

Substitute Eq. (7) into Eq. (6):

$$\frac{1}{\gamma_g}(\gamma_g{}^2 - \gamma^2)E_y = -\frac{\partial E_z}{\partial y}$$

or

$$E_y = \frac{-\gamma_g}{k^2}\frac{\partial E_z}{\partial y}, \tag{8}$$

where

$$\gamma^2 \equiv -\omega^2\mu\varepsilon, \tag{9}$$

$$k^2 \equiv \gamma_g{}^2 - \gamma^2. \tag{10}$$

The elimination of E_y between Eqs. (7) and (8) yields

$$H_x = \frac{j\omega\varepsilon}{k^2}\frac{\partial E_z}{\partial y}. \tag{11}$$

Equation (5b) can be rearranged as

$$-\gamma_g E_x + j\omega\mu H_y = \frac{\partial E_z}{\partial x}. \tag{12}$$

Eliminating H_y between Eqs. (5b) and (5d) yields

$$E_x = \frac{-\gamma_g}{k^2}\frac{\partial E_z}{\partial x}. \tag{13}$$

Eliminating E_x between Eqs. (5d) and (12) yields

$$H_y = \frac{-j\omega\varepsilon}{k^2}\frac{\partial E_z}{\partial x}. \tag{14}$$

The field components transverse to the z direction may be summarized as follows: For $H_z = 0$,

$$\begin{aligned} E_x &= -\frac{\gamma_g}{k^2}\frac{\partial E_z}{\partial x}, \qquad H_x = \frac{j\omega\varepsilon}{k^2}\frac{\partial E_z}{\partial y}, \\ E_y &= -\frac{\gamma_g}{k^2}\frac{\partial E_z}{\partial y}, \qquad H_y = \frac{-j\omega\varepsilon}{k^2}\frac{\partial E_z}{\partial x}, \\ k^2 &\equiv \gamma_g{}^2 - \gamma^2, \qquad \gamma^2 \equiv -\omega^2\mu\varepsilon. \end{aligned} \tag{15}$$

Case b: $E_z = 0$. When E_z does not exist, Eq. (4) becomes

$$\gamma_g E_y = -j\omega\mu H_x, \tag{16a}$$

$$-\gamma_g E_x = -j\omega\mu H_y, \tag{16b}$$

$$\frac{\partial E_y}{\partial x} - \frac{\partial E_x}{\partial y} = -j\omega\mu H_z; \tag{16c}$$

$$\frac{\partial H_z}{\partial y} + \gamma_g H_y = j\omega\varepsilon E_x, \tag{16d}$$

$$-\gamma_g H_x - \frac{\partial H_z}{\partial x} = j\omega\varepsilon E_y, \tag{16e}$$

$$\frac{\partial H_y}{\partial x} - \frac{\partial H_x}{\partial y} = 0. \tag{16f}$$

The elimination of H_y between Eqs. (16b) and (16d) yields

$$E_x = \frac{-j\omega\mu}{k^2}\frac{\partial H_z}{\partial y}. \tag{17}$$

H_y is found from Eqs. (16b) and (17):

$$H_y = \frac{-\gamma_g}{k^2}\frac{\partial H_z}{\partial y}. \tag{18}$$

E_y and H_x are found from Eqs. (16a) and (16e):

$$E_y = \frac{j\omega\mu}{k^2}\frac{\partial H_z}{\partial x}, \tag{19}$$

$$H_x = \frac{-\gamma_g}{k^2}\frac{\partial H_z}{\partial x}. \tag{20}$$

The field components transverse to the axial direction may be summarized as follows: For $E_z = 0$,

$$\begin{aligned} E_x &= -\frac{j\omega\mu}{k^2}\frac{\partial H_z}{\partial y}, \quad & H_x &= \frac{-\gamma_g}{k^2}\frac{\partial H_z}{\partial x}, \\ E_y &= \frac{j\omega\mu}{k^2}\frac{\partial H_z}{\partial x}, \quad & H_y &= \frac{-\gamma_g}{k^2}\frac{\partial H_z}{\partial y}, \\ k^2 &\equiv \gamma_g{}^2 - \gamma^2, \quad & \gamma^2 &\equiv -\omega^2\mu\varepsilon. \end{aligned} \tag{21}$$

In the general case where both E_z and H_z exist, the transverse field components are given by the sum of Eqs. (15) and (21).

$$E_x(x,y) = \frac{1}{k^2}\left(-\gamma_g\frac{\partial E_z}{\partial x} - j\omega\mu\frac{\partial H_z}{\partial y}\right),$$

$$E_y(x,y) = \frac{1}{k^2}\left(-\gamma_g\frac{\partial E_z}{\partial y} + j\omega\mu\frac{\partial H_z}{\partial x}\right),$$

$$H_x(x,y)=\frac{1}{k^2}\left(j\omega\varepsilon\frac{\partial E_z}{\partial y}-\gamma_g\frac{\partial H_z}{\partial x}\right), \tag{22}$$

$$H_y(x,y)=\frac{1}{k^2}\left(-j\omega\varepsilon\frac{\partial E_z}{\partial x}-\gamma_g\frac{\partial H_z}{\partial y}\right),$$

$$k^2\equiv\gamma_g{}^2-\gamma^2 \qquad \text{and} \qquad \gamma^2\equiv-\omega^2\mu\varepsilon.$$

The transverse field components are thus shown to be related directly to the axial field components, in this case E_z and H_z. It is therefore necessary to solve Eqs. (1) and (2) only for the axial field components. The corresponding components in the transverse directions are then given by Eq. (22), and the total field is completely specified. This will be worked out in detail for a particular problem in the next chapter.

Example 1

A region is bounded by two parallel, perfectly conducting, and infinitely large planes (see Figure 1). The conducting surfaces are separated by a distance x_0 in the x direction, and the space between planes is filled with lossless dielectric. Suppose the fields are invariant with respect to y. Determine

a. the TM fields with respect to z;
b. the cutoff frequency and cutoff wavelength;
c. the instantaneous and average Poynting vector and the average power transmitted per unit width of the guide;
d. the cutoff frequency, β_z, and β_x for the TM_2 mode ($n=2$) if $f=10^{12}$ Hz, $x_0=4$ cm, $\mu=\mu_0$, and $\epsilon=\epsilon_0$.

Figure 1. Example 1.

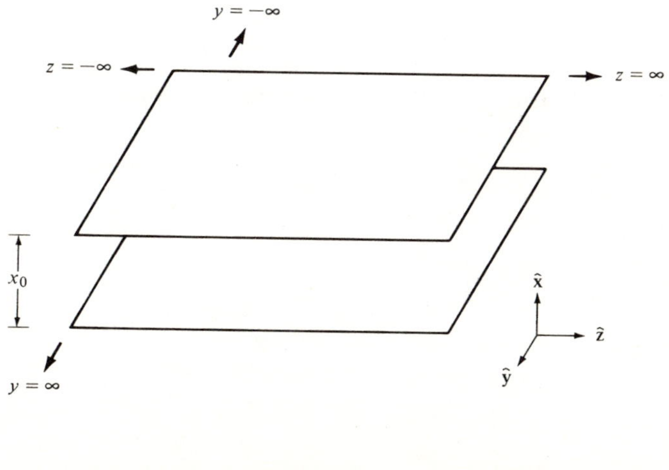

SOLUTION

a. The field satisfies the wave equation

$$\nabla^2\mathbf{E}-\gamma^2\mathbf{E}=0 \qquad (\gamma^2 \equiv -\omega^2\mu\varepsilon). \tag{1}$$

The TM wave solution can be obtained by solving Eq. (1) for the axial component E_z, or

$$\nabla^2 E_z - \gamma^2 E_z = 0, \tag{2}$$

$$\frac{\partial^2 E_z}{\partial x^2}+\frac{\partial^2 E_z}{\partial z^2}-\gamma^2 E_z=0, \qquad E_z \equiv E_z(x,z). \tag{3}$$

Let the trial solution be

$$E_z(x,z)=X(x)Z(z). \tag{4}$$

Substituting the trial solution into Eq. (3) yields

$$\frac{1}{X}\frac{d^2X}{dx^2}+\frac{1}{Z}\frac{d^2Z}{dz^2}-\gamma^2=0. \tag{5}$$

In accordance with Section 1.5, each term in Eq. (5) must be constant. Thus,

$$\frac{1}{X}\frac{d^2X}{dx^2}=\gamma_x{}^2, \qquad X(x)=A_1e^{-\gamma_x x}+A_2e^{\gamma_x x}, \tag{6}$$

$$\frac{1}{Z}\frac{d^2Z}{dz^2}=\gamma_z{}^2, \qquad Z(z)=B_1e^{-\gamma_z z}+B_2e^{\gamma_z z}. \tag{7}$$

Without loss of generality, let the field be excited such that it propagates in the positive z direction. Since the parallel plates extend to $z=\infty$, the reflected wave will not be excited along the z axis. For this reason the second term in Eq. (7) cannot be a solution. (It is a traveling wave in the negative z direction.) Consequently,

$$Z(z)=Be^{-\gamma_z z} \tag{8}$$

and

$$E_z(x,z)=X(x)Z(z)=(C_1e^{-\gamma_x x}+C_2e^{\gamma_x x})e^{-\gamma_z z}. \tag{9}$$

The constants C_1 and C_2 will be determined by applying the boundary conditions at $x=0$ and $x=x_0$. At $x=0$

$$E_z(x=0,z)=0=(C_1+C_2)e^{-\gamma_z z} \qquad \text{or} \qquad C_2=-C_1\equiv -C, \tag{10}$$

so

$$E_z(x,z)=C(e^{-\gamma_x x}-e^{\gamma_x x})e^{-\gamma_z z}. \tag{11}$$

At $x=x_0$

$$E_z(x=x_0,z)=0=C(e^{-\gamma_x x_0}-e^{\gamma_x x_0})e^{-\gamma_z z},$$

$$e^{-\gamma_x x_0}=e^{\gamma_x x_0}. \tag{12}$$

Equation (12) can never be satisfied when the argument of the exponential function is real (except for the trivial case when the argument is zero) and can be fulfilled only if the argument is purely imaginary. Let

$$\gamma_x \equiv j\beta_x \qquad (\beta_x \text{ real}). \tag{13}$$

Then

$$\begin{aligned} E_z(x=x_0, z) = 0 &= C(e^{-j\beta_x x_0} - e^{j\beta_x x_0})e^{-\gamma_z z} \\ &= -2jC\sin\beta_x x_0 e^{-\gamma_z z} \\ &\equiv E_0 \sin\beta_x x_0 e^{-\gamma_z z}. \end{aligned}$$

Therefore,

$$\sin\beta_x x_0 = 0, \qquad \text{or} \qquad \beta_x x_0 = n\pi;$$

and

$$\beta_{nx} = n\pi/x_0, \quad n = 1,2,\ldots. \tag{14}$$

The additional subscript is needed to identify the various modes. The general solution is

$$E_z(x,z) = E_0 \sin\beta_{nx} x e^{-\gamma_z z}. \tag{15}$$

The constant E_0 is usually related to the incident power from the source.

b. Substituting Eq. (15) into Eq. (3) gives the expression for the propagation constant of the guide, $\gamma_g \equiv \gamma_z$.

$$(-\beta_{nx}{}^2 + \gamma_z{}^2 - \gamma^2)E_z = 0,$$

$$\gamma_z = (\beta_{nx}{}^2 + \gamma^2)^{1/2} = \left[\left(\frac{n\pi}{x_0}\right)^2 - \omega^2\mu\varepsilon\right]^{1/2}. \tag{16}$$

For evanescent waves, when $(n\pi/x_0)^2 > \omega^2\mu\varepsilon$ (γ_z is real and $\equiv \alpha_z$),

$$e^{-\gamma_z z} = e^{-\alpha_z z} \text{ (exponential decay).}$$

For cutoff, when $(n\pi/x_0)^2 = \omega^2\mu\varepsilon$ ($\gamma_z = 0$),

$$e^{-\gamma_z z} = 1 \qquad \text{(constant; no propagation).}$$

For propagation, when $(n\pi/x_0)^2 < \omega^2\mu\varepsilon$ (γ_z is imaginary and $\equiv j\beta_z$),

$$e^{-\gamma_z z} = e^{-j\beta_z z} \qquad \text{(traveling waves).}$$

For a parallel plate guide with specified parameters (x_0, μ, ε), the frequency variation will determine the existence of, say, the mth mode. The cutoff frequency of that mode is given by

$$\omega_{mc} = \frac{m\pi/x_0}{\sqrt{\mu\varepsilon}} = 2\pi f_{mc}, \qquad m = 1,2,\ldots; \tag{17}$$

$$\lambda_{mc} = 2\pi/\beta_{mx} = 2x_0/m, \qquad m = 1,2,\ldots. \tag{18}$$

The mth mode exists when the operating frequency is above the cutoff frequency f_{mc}.

The transverse components of the fields can be obtained from Eq. (1.8.15), with $k^2 = \gamma_z{}^2 - \gamma^2 \equiv \beta_{nx}{}^2$:

$$E_x = -\frac{\gamma_z}{\beta_{nx}{}^2}\frac{\partial E_z}{\partial x} = -\frac{\gamma_z}{\beta_{nx}} E_0 \cos\beta_{nx} x e^{-\gamma_z z}, \tag{19}$$

$$E_y = -\frac{\gamma_z}{\beta_{nx}{}^2}\frac{\partial E_z}{\partial y} = 0, \tag{20}$$

$$H_x = \frac{j\omega\varepsilon}{\beta_{nx}{}^2}\frac{\partial E_z}{\partial y} = 0, \tag{21}$$

$$H_y = \frac{-j\omega\varepsilon}{\beta_{nx}{}^2}\frac{\partial E_z}{\partial x} = \frac{-j\omega\varepsilon}{\beta_{nx}} E_0 \cos\beta_{nx} x e^{-\gamma_z z}. \tag{22}$$

c. The instantaneous Poynting vector is given by Eq. (1.7.15):

$$\begin{aligned}\mathbf{S} &= \tfrac{1}{2}\,\mathrm{Re}(\tilde{\mathbf{E}} \times \tilde{\mathbf{H}}^* \mp \tilde{\mathbf{E}} \times \tilde{\mathbf{H}}) \\ &= \tfrac{1}{2}\,\mathrm{Re}\left[(\hat{\mathbf{x}}\tilde{E}_x + \hat{\mathbf{z}}E_z) \times \hat{\mathbf{y}}\tilde{H}_y{}^* \mp (\hat{\mathbf{x}}\tilde{E}_x + \hat{\mathbf{z}}\tilde{E}_z) \times \hat{\mathbf{y}}\tilde{H}_y\right] \\ &= \tfrac{1}{2}\,\mathrm{Re}\left[\hat{\mathbf{z}}(\tilde{E}_x\tilde{H}_y{}^* \mp \tilde{E}_x\tilde{H}_y) - \hat{\mathbf{x}}(\tilde{E}_z\tilde{H}_y{}^* \mp \tilde{E}_z\tilde{H}_y)\right].\end{aligned} \tag{23}$$

The complete expression for the propagating fields ($\gamma_z = j\beta_z$) is

$$\tilde{E}_x = -\frac{j\beta_z}{\beta_{nx}} E_0 \cos\beta_{nx} x\, e^{j(\omega t - \beta_z z)}, \tag{24}$$

$$\tilde{E}_z = E_0 \sin\beta_{nx} x\, e^{j(\omega t - \beta_z z)}, \tag{25}$$

$$\tilde{H}_y = -\frac{j\omega\varepsilon}{\beta_{nx}} E_0 \cos\beta_{nx} x\, e^{j(\omega t - \beta_z z)}. \tag{26}$$

The use of Eqs. (24)–(26) in Eq. (23) produces

$$\begin{aligned}\tilde{\mathbf{S}} = \tfrac{1}{2}\,\mathrm{Re}\Bigg[\hat{\mathbf{z}}\left(\frac{\omega\beta_z\varepsilon^*}{\beta_{nx}{}^2} E_0{}^2 \cos^2\beta_{nx} x \mp \frac{-\omega\beta_z\varepsilon}{\beta_{nx}{}^2} E_0 \cos^2\beta_{nx} x\, e^{2j(\omega t - \beta_z z)}\right) \\ - \hat{\mathbf{x}}\Bigg(\frac{j\omega\varepsilon^*}{\beta_{nx}} E_0{}^2 \sin\beta_{nx} x \cos\beta_{nx} x \\ \mp \frac{-j\omega\varepsilon}{\beta_{nx}} E_0{}^2 \sin\beta_{nx} x \cos\beta_{nx} x\, e^{2j(\omega t - \beta_z z)}\Bigg)\Bigg]\end{aligned}$$

In the lossless dielectric, $\varepsilon=\epsilon$, one has

$$\mathbf{S}=\tfrac{1}{2}\operatorname{Re}\left[\hat{\mathbf{z}}\frac{\omega\beta_z\epsilon}{\beta_{nx}{}^2}E_0{}^2\cos^2\beta_{nx}x\left(1\pm e^{2j(\omega t-\beta_z z)}\right)\right.$$
$$\left.-\hat{\mathbf{x}}\frac{j\omega\epsilon}{\beta_{nx}}E_0{}^2\sin\beta_{nx}x\cos\beta_{nx}x\left(1\pm e^{2j(\omega t-\beta_z z)}\right)\right]$$
$$=\hat{\mathbf{z}}\frac{\omega\beta_z\epsilon}{2\beta_{nx}{}^2}E_0{}^2\cos^2\beta_{nx}x\left[1\pm\cos 2(\omega t-\beta_z z)\right]$$
$$\pm\hat{\mathbf{x}}\frac{\omega\epsilon}{2\beta_{nx}}E_0{}^2\sin\beta_{nx}x\cos\beta_{nx}x\sin 2(\omega t-\beta_z z). \tag{27}$$

The upper sign is for the field that has the form

$$\tilde{\mathbf{E}}=\operatorname{Im}\left[\mathbf{E}(\mathbf{r})e^{j\omega t}\right]$$

and the lower sign for

$$\tilde{\mathbf{E}}=\operatorname{Re}\left[\mathbf{E}(\mathbf{r})e^{j\omega t}\right].$$

The time-averaged Poynting vector is ($\varepsilon=\epsilon$)

$$\mathbf{S}_{\text{av}}=\tfrac{1}{2}\operatorname{Re}(\tilde{\mathbf{E}}\times\tilde{\mathbf{H}}^*)$$
$$=\tfrac{1}{2}\operatorname{Re}\left(\hat{\mathbf{z}}\frac{\omega\beta_z\epsilon}{\beta_{nx}{}^2}E_0{}^2\cos^2\beta_{nx}x-\hat{\mathbf{x}}\frac{j\omega\epsilon}{\beta_{nx}}E_0{}^2\sin\beta_{nx}x\cos\beta_{nx}x\right)$$
$$=\hat{\mathbf{z}}\frac{\omega\beta_z\epsilon}{2\beta_{nx}{}^2}E_0{}^2\cos^2\beta_{nx}x. \tag{28}$$

The average power transmitted per unit width of the guide is

$$P=\int_S\mathbf{S}_{\text{av}}\cdot\hat{\mathbf{n}}\,da$$
$$=\int_{x=0}^{x_0}\int_{y=y_0}^{y_0+1}\mathbf{S}_{\text{av}}\cdot\hat{\mathbf{z}}\,dx\,dy$$
$$=\frac{\omega\beta_z\epsilon}{2\beta_{nx}{}^2}E_0{}^2\int_0^{x_0}\cos^2\beta_{nx}x\,dx$$
$$=\frac{\omega\beta_z\epsilon}{2\beta_{nx}{}^3}E_0{}^2\left(\frac{\beta_{nx}x_0}{2}+\frac{1}{4}\sin 2\beta_{nx}x_0\right)$$
$$=\frac{\omega\beta_z\epsilon}{2\beta_{nx}{}^3}E_0{}^2\frac{n\pi}{2}\qquad(\beta_{nx}x_0=n\pi,\ n=1,2,\ldots);$$

$$P_n=\frac{\omega\epsilon\beta_z E_0{}^2\pi}{4\beta_{nx}{}^3}n\qquad(n=1,2,\ldots). \tag{29}$$

For lossless guide, the incident power per unit width P_i from the source must be entirely transmitted along the guide. Thus, $|E_0|$ can be expressed in terms of this power as

$$|E_0| = \left(\frac{4\beta_{nx}{}^3}{n\pi\omega\epsilon\beta_z} P_i \right)^{1/2}. \tag{30}$$

d. For the propagating modes, $(n\pi/x_0)^2 < \omega^2\mu_0\epsilon_0$.
At cutoff, $\omega_{nc}{}^2\mu_0\epsilon_0 = (n\pi/x_0)^2$, and for $n=2$,

$$\omega_{nc} = \omega_{2c} = \frac{2\pi/x_0}{\sqrt{\mu_0\epsilon_0}} = 2\pi f_{2c},$$

$$f_{2c} = \frac{1}{0.04} \times 3 \times 10^8 = 7.5 \times 10^9 \text{ Hz}.$$

The propagation constant $\gamma_{nz} = j\beta_{nz}$ is

$$\beta_{nz} = \left[\omega^2\mu_0\epsilon_0 - (n\pi/x_0)^2 \right]^{1/2}$$

$$= \left(\omega^2\mu_0\epsilon_0 - \omega_{nc}{}^2\mu_0\epsilon_0 \right)^{1/2}$$

$$= \omega\sqrt{\mu_0\epsilon_0} \left[1 - (f_{nc}/f)^2 \right]^{1/2};$$

$$\beta_{2z} = \frac{2\pi 10^{12}}{3 \times 10^8} \left[1 - \left(\frac{7.5 \times 10^9}{10^{12}} \right)^2 \right]^{1/2} \simeq \frac{2\pi}{3} \times 10^4 \text{ rad/m}.$$

The guided wavelength $\lambda_g = \lambda_z$ is

$$\lambda_{nz} = 2\pi/\beta_{nz} = 2\pi/\left(\tfrac{2}{3}\pi \times 10^4\right) = 3 \times 10^{-4} \text{ m}.$$

The phase constant in the x direction is

$$\beta_{2x} = \left(\omega^2\mu_0\epsilon_0 - \beta_{2z}{}^2 \right)^{1/2}$$

$$= \left\{ \omega^2\mu_0\epsilon_0 - \omega^2\mu_0\epsilon_0\left[1 - (f_{2c}/f)^2 \right] \right\}^{1/2}$$

$$= \omega\sqrt{\mu_0\epsilon_0}\, f_{2c}/f = \tfrac{2}{3}\pi 10^4\left(\tfrac{3}{4} \times 10^{10}\right)/10^{12} = 50\,\pi \text{ rad/m}.$$

The guided wavelength along the x direction is

$$\lambda_{2x} = 2\pi/\beta_{2x} = 2\pi/50\pi = 0.4 \text{ m} = x_0 \qquad \text{(check!)}.$$

References

Further material reviewed in this chapter may found in some of the textbooks listed below.

1. P. Lorrain and D. Corson, *Electromagnetic Fields and Waves* (2nd ed.), Freeman, San Francisco, 1970.

2. R. Plonsey and R. E. Collin, *Principles and Applications of Electromagnetic Fields*, McGraw-Hill, New York, 1961.
3. S. Ramo, J. R. Whinnery, and T. Van Duzer, *Fields and Waves in Communication Electronics*, Wiley, New York, 1967.
4. J. R. Reitz and F. J. Milford, *Foundations of Electromagnetic Theory* (2nd ed.), Addison-Wesley, Reading, Mass., 1967.
5. J. A. Stratton, *Electromagnetic Theory*, McGraw-Hill, New York, 1941.

Other books on electromagnetic theory are also suitable.

2

Dielectric Waveguides in Rectangular Coordinates

One of the simplest types of waveguide is a large sheet of dielectric of uniform thickness. In this chapter, the problem is idealized by assuming the dielectric sheet to be infinite in size. Such an idealized waveguide may not be physically realizable, but its properties provide useful insight into practical optical waveguides, particularly the thin-sheet waveguides widely used in integrated optical circuits.

Guided propagation along an infinitely large dielectric sheet will be investigated by subdividing the general problem into two simpler cases:

a. *Transverse electric waves*: The electric field is polarized in the plane transverse to the direction of propagation $\hat{\gamma}$ and only the magnetic field has a component parallel to $\hat{\gamma}$.
b. *Transverse magnetic waves*: The magnetic field is transverse to $\hat{\gamma}$ and only the electric vector has a component parallel to $\hat{\gamma}$.

Each will be analyzed separately. The general solution is the superposition of these two special cases.

The chapter concludes with the introduction and analysis of radiation modes. The orthogonality properties of both guided and radiation modes will also be examined.

It should be noted that the investigation of dielectric waveguides was carried out as early as the beginning of this century [3, 6, 12]. Aspects of this subject are also to be found in some of the modern textbooks [2, 4, 5, 7, 8, 11].

2.1. Dielectric Sheet Waveguide

The two-dimensional problem of an infinite homogeneous dielectric sheet is illustrated in Figure 2. The dielectric has a thickness of 2δ and extends to infinity in the y and z directions. For an effective waveguide, there must

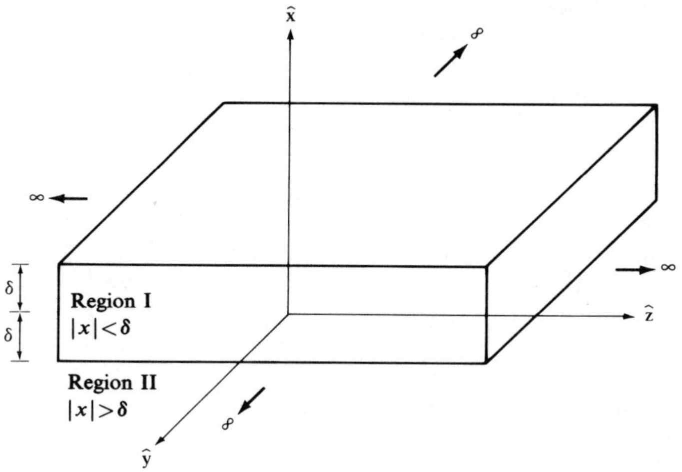

Figure 2. Dielectric sheet waveguide.

be a traveling wave solution along its axis. Let this be the z axis. The solution then has the form

$$\mathbf{E}(\mathbf{r}) = \mathbf{E}(x, y)e^{-\gamma_g z}, \qquad \mathbf{H}(\mathbf{r}) = \mathbf{H}(x, y)e^{-\gamma_g z}, \tag{1}$$

where $\gamma_g = \alpha_g + j\beta_g$ is the propagation constant of the guide.

The simpler case where fields are independent of the y coordinates will be considered. With $\mathbf{E}(x, y) \equiv \mathbf{E}(x)$ and $\mathbf{H}(x, y) \equiv \mathbf{H}(x)$, Maxwell's equations become

$$\nabla \times \mathbf{E} = -j\omega\mu\mathbf{H}, \qquad \nabla \times \mathbf{H} = j\omega\varepsilon\mathbf{E},$$

or

$$\begin{aligned} \gamma_g E_y &= -j\omega\mu H_x, & \gamma_g H_y &= j\omega\varepsilon E_x, \\ -\gamma_g E_x - \frac{dE_z}{dx} &= -j\omega\mu H_y, & -\gamma_g H_x - \frac{dH_z}{dx} &= j\omega\varepsilon E_y, \\ \frac{dE_y}{dx} &= -j\omega\mu H_z, & \frac{dH_y}{dx} &= j\omega\varepsilon E_z. \end{aligned} \tag{2}$$

The above set of equations may be separated into two uncoupled groups:

$$-\gamma_g H_x - \frac{dH_z}{dx} = j\omega\varepsilon E_y, \tag{3a}$$

$$\frac{dE_y}{dx} = -j\omega\mu H_z, \tag{3b}$$

$$\gamma_g E_y = -j\omega\mu H_x, \tag{3c}$$

$$-\gamma_g E_x - \frac{dE_z}{dx} = -j\omega\mu H_y, \tag{4a}$$

$$\frac{dH_y}{dx} = j\omega\varepsilon E_z, \tag{4b}$$

$$\gamma_g H_y = j\omega\varepsilon E_x. \tag{4c}$$

The first group, Eq. (3), does not involve E_z and defines what are known as *transverse electric* (*TE*) *waves* in the z direction. It is convenient to express the **H** field in terms of the only component of the **E** field, E_y.

$$\text{TE:} \qquad H_x = \frac{-\gamma_g}{j\omega\mu} E_y, \qquad H_z = \frac{-1}{j\omega\mu}\frac{dE_y}{dx}. \tag{5}$$

Substitution of Eq. (5) into Eq. (3a) yields

$$\frac{d^2E_y}{dx^2} + k^2E_y = 0, \qquad k^2 \equiv \gamma_g{}^2 - \gamma^2, \quad \gamma^2 = -\omega^2\mu\varepsilon. \tag{6}$$

The second group, Eq. (4), does not involve H_z and defines *transverse magnetic* (*TM*) *waves* in the z direction. The **E** field in this case is expressed in terms of the only component of the **H** field, H_y.

$$\text{TM:} \qquad E_x = \frac{\gamma_g}{j\omega\varepsilon} H_y, \qquad E_z = \frac{1}{j\omega\varepsilon}\frac{dH_y}{dx}, \tag{7}$$

where H_y satisfies the equation

$$\frac{d^2H_y}{dx^2} + k^2H_y = 0. \tag{8}$$

Each mode of solution will be investigated separately.

2.2. Transverse Electric Waves

The transverse electric mode is defined to be the solution whose **E** field does not contain a component in the direction of propagation; that is, $E_z = 0$. The problem is to find the solution of Eq. (2.1.6), which is repeated here:

$$\frac{d^2E_y}{dx^2} + k^2E_y = 0. \tag{1}$$

This is the one-dimensional wave equation treated in Section 1.4. The solution is

$$E_y(x) = E^+e^{-jkx} + E^-e^{jkx} \qquad \text{(exponential form)}$$

$$= E^e\cos kx + E^o\sin kx \qquad \text{(trigonometric form)}. \tag{2}$$

Two forms of solution are identical since $e^{\pm j\theta} = \cos\theta \pm j\sin\theta$, the constant coefficients E^+, E^-, E^e, and E^o are to be determined by boundary conditions. The superscript + (or −) on the coefficient indicates its connection with the positively (or negatively) traveling wave, while e (or o) implies its relation with the even (or odd) function of position in the transverse x direction.

The present problem involves regions filled with different media; therefore Eq. (1) must be considered for each region separately. The boundary conditions will then be applied to the solutions for each region to obtain the final solution of the problem.

Region I: $-\delta<x<\delta$. This is the region within the dielectric sheet. The **E** field in this region satisfies the wave equation

$$\frac{d^2E_{1y}}{dx^2}+k_1{}^2E_{1y}=0, \tag{3}$$

$$k_1{}^2\equiv\gamma_g{}^2-\gamma_1{}^2 \quad \text{and} \quad \gamma_1{}^2\equiv-\omega^2\mu_1\varepsilon_1. \tag{4}$$

In general, k_1 is, a complex quantity,

$$\begin{aligned}
k_1&=\sqrt{\gamma_g{}^2-\gamma_1{}^2} \qquad (\gamma_g\equiv\alpha_g+j\beta_g)\\
&=\left[(\alpha_g+j\beta_g)^2+\omega^2\mu_1\varepsilon_1\right]^{1/2} \qquad (\varepsilon_1\equiv\epsilon_1'+j\epsilon_1'')\\
&=\left[(\alpha_g{}^2-\beta_g{}^2+j2\alpha_g\beta_g)+\omega^2\mu_1\epsilon_1'+j\omega^2\mu_1\epsilon_1''\right]^{1/2}\\
&=\left\{\left[\omega^2\mu_1\epsilon_1'-\beta_g{}^2\left[1-\frac{\alpha_g{}^2}{\beta_g{}^2}\right]\right]\right.\\
&\qquad\left.\times\left[1+j\frac{\omega^2\mu_1\epsilon_1''+2\alpha_g\beta_g}{\omega^2\mu_1\epsilon_1'-\beta_g{}^2(1-\alpha_g{}^2/\beta_g{}^2)}\right]\right\}^{1/2}\\
&\equiv\beta_1+j\alpha_1.
\end{aligned} \tag{5}$$

For low-loss media, that is, if $(\alpha_g/\beta_g)^2\ll1$ and $(\epsilon_1''/\epsilon_1')^2\ll1$, approximate expressions for β_1 and α_1 may be obtained by binomial expansion keeping the first two terms only.

$$(1+x)^n\simeq1+nx,$$

$$\beta_1\simeq\sqrt{\omega^2\mu_1\epsilon_1'-\beta_g{}^2}\,, \quad \alpha_1\simeq\frac{1}{2\beta_1}(\omega^2\mu_1\epsilon_1''+2\alpha_g\beta_g). \tag{6}$$

For the ideal lossless case, $\varepsilon_1=\epsilon_1$ and $\gamma_g=j\beta_g$,

$$k_1=\beta_1=\sqrt{\omega^2\mu_1\epsilon_1-\beta_g{}^2}\,. \tag{7}$$

Thus k_1 is real if $\omega^2\mu_1\epsilon_1 > \beta_g^{\,2}$, when the resulting solution of Eq. (2) will be a traveling wave. On the other hand if $\omega^2\mu_1\epsilon_1 < \beta_g^{\,2}$, then k_1 is purely imaginary and yields an exponentially decaying solution.

Since this region is bounded by the $x = \pm\delta$ planes, a standing-wave solution (in the x direction) is expected. The solution in trigonometric form exhibits this property and will be used to represent E_{1y}.

$$E_{1y}(x) = E^{e}\cos k_1 x + E^{o}\sin k_1 x. \tag{8}$$

Note that the first term is an even function of x and the second term is an odd function; it is therefore convenient to subdivide the solution into even and odd modes.

TE even modes (or symmetric modes).

$$E_{1y}{}^{e}(x) = E^{e}\cos k_1 x, \qquad -\delta < x < \delta. \tag{9}$$

The corresponding **H** field is given by Eq. (2.1.5).

$$\begin{aligned} H_{1x}{}^{e}(x) &= \frac{-\gamma_g}{j\omega\mu_1} E_{1y}{}^{e} = \frac{-\gamma_g}{j\omega\mu_1} E^{e}\cos k_1 x, \\ H_{1z}{}^{e}(x) &= \frac{-1}{j\omega\mu_1}\frac{dE_{1y}{}^{e}}{dx} = \frac{k_1}{j\omega\mu_1} E^{e}\sin k_1 x. \end{aligned} \tag{10}$$

TE odd modes (or antisymmetric modes).

$$\begin{aligned} E_{1y}{}^{o}(x) &= E^{o}\sin k_1 x, \qquad -\delta < x < \delta, \\ H_{1x}{}^{o}(x) &= \frac{-\gamma_g}{j\omega\mu_1} E_{1y}{}^{o} = \frac{-\gamma_g}{j\omega\mu_1} E^{o}\sin k_1 x, \\ H_{1z}{}^{o} &= \frac{-1}{j\omega\mu_1}\frac{dE_{1y}{}^{o}}{dx} = \frac{-k_1}{j\omega\mu_1} E^{o}\cos k_1 x. \end{aligned} \tag{11}$$

The total field in the general case is the sum of both even and odd modes.

$$\begin{aligned} E_{1y}(x) &= E_{1y}{}^{e}(x) + E_{1y}{}^{o}(x), \\ H_{1x}(x) &= H_{1x}{}^{e}(x) + H_{1x}{}^{o}(x), \\ H_{1z}(x) &= H_{1z}{}^{e}(x) + H_{1z}{}^{o}(x). \end{aligned} \tag{12}$$

Region II: $\delta < x < -\delta$. This is the region exterior to the dielectric sheet. The wave equation to be solved is

$$\frac{d^2E_{2y}}{dx^2} + k_2{}^2E_{2y} = 0, \tag{13}$$

$$k_2{}^2 \equiv \gamma_g{}^2 - \gamma_2{}^2 \quad \text{and} \quad \gamma_2{}^2 \equiv -\omega^2\mu_2\varepsilon_2. \tag{14}$$

Since region II extends to infinity in the x direction, an exponential-type solution is usually chosen.

$$\begin{aligned} E_{2y}(x) &= E^{+}e^{-jk_2(x-\delta)} + E^{-}e^{jk_2(x-\delta)}, \qquad x>\delta, \\ &= E^{+}e^{jk_2(x+\delta)} + E^{-}e^{-jk_2(x+\delta)}, \qquad x<-\delta. \end{aligned} \tag{15}$$

When $k_2{}^2>0$, Eq. (15) represents propagating modes in the x direction. The electromagnetic field will thus be radiated from the dielectric slab, which is undesirable for an efficient guiding system.

When $k_2{}^2<0$, then k_2 is purely imaginary and the field decays exponentially. Such a solution is known as the evanescent mode.

$$\begin{aligned} k_2{}^2 &\equiv \gamma_g{}^2 - \gamma_2{}^2 = -(\gamma_2{}^2-\gamma_g{}^2) = -\left[-\omega^2\mu_2\varepsilon_2 - (\alpha_g + j\beta_g)^2\right] \\ &= -\left[\beta_g{}^2(1-\alpha_g{}^2/\beta_g{}^2 - j2\alpha_g/\beta_g) - \omega^2\mu_2\varepsilon_2\right] \equiv -\alpha_2{}^2, \\ k_2 &= \pm j\alpha_2 = \pm j\left[\beta_g{}^2(1-\alpha_g{}^2/\beta_g{}^2 - j2\alpha_g/\beta_g) - \omega^2\mu_2\varepsilon_2\right]^{1/2}. \end{aligned} \tag{16}$$

With α_2 defined above, the field $E_{2y}(x)$ becomes

$$\begin{aligned} E_{2y}(x) &= E^{+}e^{-\alpha_2(x-\delta)} + E^{-}e^{\alpha_2(x-\delta)}, \qquad x>\delta, \\ &= E^{+}e^{\alpha_2(x+\delta)} + E^{-}e^{-\alpha_2(x+\delta)}, \qquad x<-\delta. \end{aligned} \tag{17}$$

The second term in Eq. (17) approaches infinity with the x coordinate and is discarded.

$$\begin{aligned} E_{2y}(x) &= E^{+}e^{-\alpha_2(x-\delta)}, \qquad x>\delta, \\ &= E^{+}e^{\alpha_2(x+\delta)}, \qquad x<-\delta. \end{aligned} \tag{18}$$

The coefficient E^{+} is related to E^{e} and E^{o} by the continuity condition at the interface between regions I and II. The tangential components of the **E** fields must be continuous across the boundary, that is, $E_{1y}(\pm\delta) = E_{2y}(\pm\delta)$.

$$\text{TE even modes:} \qquad E^{+} = E^{e}\cos k_1\delta, \tag{19}$$

$$\begin{aligned} \text{TE odd modes:} \qquad E^{+} &= E^{o}\sin k_1\delta, \qquad x>\delta, \\ &= -E^{o}\sin k_1\delta, \qquad x<-\delta. \end{aligned} \tag{20}$$

The evanescent fields in region II are summarized as follows.

TE even modes: $\delta<x<-\delta$.

$$\begin{aligned} E_{2y}{}^{e}(x) &= E^{e}\cos k_1\delta\, e^{-\alpha_2(x-\delta)}, \qquad x>\delta, \\ &= E^{e}\cos k_1\delta\, e^{\alpha_2(x+\delta)}, \qquad x<-\delta; \end{aligned} \tag{21}$$

$$H_{2x}{}^{e}(x)=\frac{-\gamma_g}{j\omega\mu_2}E_{2y}{}^{e}$$

$$=\frac{-\gamma_g}{j\omega\mu_2}E^{e}\cos k_1\delta\, e^{-\alpha_2(x-\delta)}, \qquad x>\delta, \tag{22}$$

$$=\frac{-\gamma_g}{j\omega\mu_2}E^{e}\cos k_1\delta\, e^{\alpha_2(x+\delta)}, \qquad x<-\delta;$$

$$H_{2z}{}^{e}(x)=\frac{-1}{j\omega\mu_2}\frac{dE_{2y}{}^{e}}{dx}$$

$$=\frac{\alpha_2}{j\omega\mu_2}E^{e}\cos k_1\delta\, e^{-\alpha_2(x-\delta)}, \qquad x<\delta, \tag{23}$$

$$=\frac{-\alpha_2}{j\omega\mu_2}E^{e}\cos k_1\delta\, e^{\alpha_2(x+\delta)}, \qquad x<-\delta.$$

TE odd modes: $\delta<x<-\delta$.

$$E_{2y}{}^{o}(x)=E^{o}\sin k_1\delta\, e^{-\alpha_2(x-\delta)}, \qquad x>\delta,$$

$$=-E^{o}\sin k_1\delta\, e^{\alpha_2(x+\delta)}, \qquad x<-\delta; \tag{24}$$

$$H_{2x}{}^{o}(x)=\frac{-\gamma_g}{j\omega\mu_2}E_{2y}{}^{o}$$

$$=\frac{-\gamma_g}{j\omega\mu_2}E^{o}\sin k_1\delta\, e^{-\alpha_2(x-\delta)}, \qquad x>\delta, \tag{25}$$

$$=\frac{\gamma_g}{j\omega\mu_2}E^{o}\sin k_1\delta\, e^{\alpha_2(x+\delta)}, \qquad x<-\delta;$$

$$H_{2z}{}^{o}(x)=\frac{-1}{j\omega\mu_2}\frac{dE_{2y}{}^{o}}{dx}$$

$$=\frac{\alpha_2}{j\omega\mu_2}E^{o}\sin k_1\delta\, e^{-\alpha_2(x-\delta)}, \qquad x>\delta, \tag{26}$$

$$=\frac{\alpha_2}{j\omega\mu_2}E^{o}\sin k_1\delta\, e^{\alpha_2(x+\delta)}, \qquad x<-\delta.$$

The condition for the existence of evanescent mode can be obtained from Eq. (16). This equation involves complex quantities, which makes it complicated to interpret. Greater insight may be obtained by considering the idealized lossless case where $\gamma_g=j\beta_g$ and $\varepsilon_2=\epsilon_2$; then

$$k_{20}\equiv k_2|_{\text{lossless}}=\pm j\sqrt{\beta_g{}^2-\omega^2\mu_2\epsilon_2}\equiv\pm j\alpha_{20}, \tag{27}$$

and when $\alpha_{20}{}^2>0$, that is, when $\beta_g{}^2>\omega^2\mu_2\epsilon_2$, the fields exterior to the dielectric sheet is attenuated exponentially in the x direction. On the other hand, if $\alpha_{20}{}^2<0$, or $\beta_g{}^2<\omega^2\mu_2\epsilon_2$, radiation modes exist.

In the case when the media are slightly lossy, then α_2 is a complex quantity with a predominant real part and a small imaginary part. This will yield damped propagating modes.

2.2a. TE Eigenmodes

The solutions for each region of the dielectric sheet guide were determined in the previous section. These were expressed in terms of a traveling wave in the z direction. The propagation constant of the guide γ_g can be determined by the continuity condition of the tangential component of the **H** field at the interface, $H_{1z}(\pm\delta)=H_{2z}(\pm\delta)$.

$$\text{TE even modes:} \qquad \tan k_1\delta=\alpha_2\mu_1/k_1\mu_2; \tag{1}$$

$$\text{TE odd modes:} \qquad \tan k_1\delta=-k_1\mu_2/\alpha_2\mu_1. \tag{2}$$

For common dielectrics, that is, for $\mu_1=\mu_2=\mu_0$,

$$\text{TE even modes:} \qquad \tan k_1\delta=\alpha_2/k_1; \tag{3}$$

$$\text{TE odd modes:} \qquad \tan k_1\delta=-k_1/\alpha_2. \tag{4}$$

A solution of the above equations can be obtained graphically, by plotting each side of the equations. The intersections of these curves yield the solutions. The left-hand side of these equations can easily be evaluated for a given thickness δ. The right-hand side can be expressed as a function of k_1 by using Eqs. (2.2.4) and (2.2.16).

$$\gamma_g{}^2=k_1{}^2+\gamma_1{}^2,$$

$$\alpha_2{}^2=\gamma_2{}^2-\gamma_g{}^2=\gamma_2{}^2-(k_1{}^2+\gamma_1{}^2)=\omega^2\mu_0(\varepsilon_1-\varepsilon_2)-k_1{}^2,$$

$$\alpha_2/k_1=(1/k_1)\left[\omega^2\mu_0(\varepsilon_1-\varepsilon_2)-k_1{}^2\right]^{1/2}. \tag{5}$$

Greater insight is obtained by considering the lossless case, $\varepsilon_1=\epsilon_1=\epsilon_0\epsilon_{1r}$ and $\varepsilon_2=\epsilon_2=\epsilon_0\epsilon_{2r}$.

$$\alpha_2/k_1=(1/k_1)\left[\omega^2\mu_0\epsilon_0(\epsilon_{1r}-\epsilon_{2r})-k_1{}^2\right]^{1/2}. \tag{6}$$

This indicates that the ratio α_2/k_1 is zero when the frequency is equal to

$$\omega=k_1/\left[\mu_0\epsilon_0(\epsilon_{1r}-\epsilon_{2r})\right]^{1/2}, \qquad \epsilon_{1r}>\epsilon_{2r}. \tag{7}$$

The right-hand side of Eq. (6) becomes imaginary when the frequency falls below the value given by Eq. (7). This is unacceptable since α_2 is defined to be a real quantity by Eq. (2.2.16). Therefore, the evanescent mode does not exist for Eqs. (3) and (4) at frequencies below that specified by Eq. (7).

With Eq. (6), the eigenvalue equations (3) and (4) become

$$\text{TE even modes:} \quad \tan k_1\delta = \frac{1}{k_1}[\omega^2\mu_0\epsilon_0(\epsilon_{1r}-\epsilon_{2r})-k_1{}^2]^{1/2}; \tag{8}$$

$$\text{TE odd modes:} \quad \tan k_1\delta = -k_1[\omega^2\mu_0\epsilon_0(\epsilon_{1r}-\epsilon_{2r})-k_1{}^2]^{-1/2}. \tag{9}$$

If the thickness δ is given, then k_1 can be treated as the variable. The α_2/k_1 curve vanishes at

$$k_1 = \omega^2\mu_0\epsilon_0(\epsilon_{1r}-\epsilon_{2r})^{1/2} \equiv k_{1\,\max}, \tag{10}$$

where $k_{1\,\max}$ is the upper limiting value of k_1 at the specified frequency, and any value of k_1 greater than $k_{1\,\max}$ will cause α_2 to become imaginary. The ratio α_2/k_1 increases as k_1 decreases, becoming infinite at $k_1=0$.

The k_1/α_2 curve behaves inversely as the α_2/k_1 curve; it is zero at $k_1=0$ and becomes infinite at $k_1=k_{1\,\max}$.

The graphical representation of each side of Eq. (8) is shown in Figure 3.

The number of intersections between the α_2/k_1 and $\tan k_1\delta$ curves represents the number of guided modes (or number of solutions). More modes (for a given thickness δ) are possible for a higher value of $k_{1\,\max}$,

Figure 3. Graphical method for determining the propagation constant.

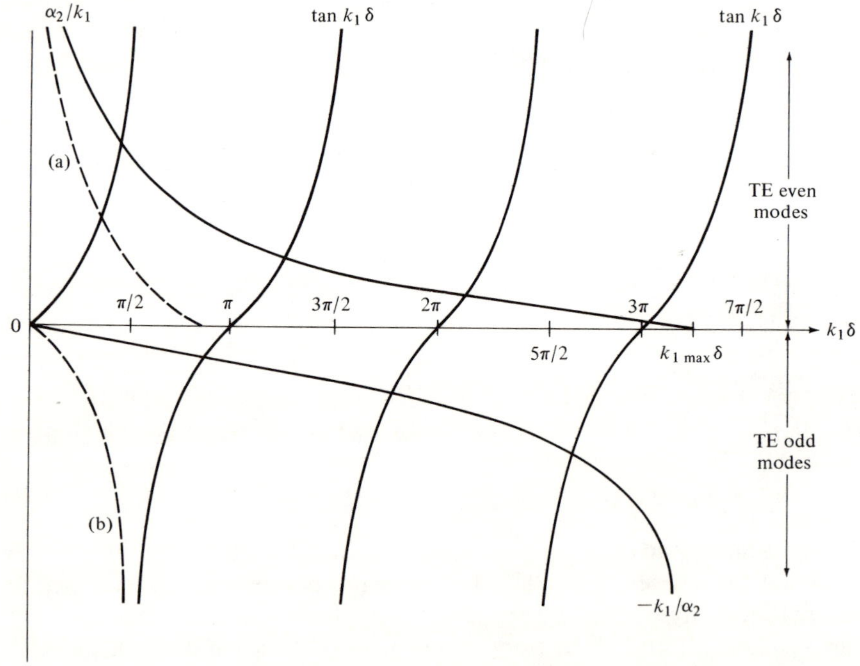

which can be accomplished by increasing the frequency ω or by increasing the difference of the permittivities of the media, $\epsilon_1 - \epsilon_2$.

On the other hand, when $k_{1\,\max}\delta$ is made less than π, only a single mode can propagate. The lowest TE even mode can propagate at any frequency and is never cut off, as is obvious from the dotted curve (a) in Figure 3.

The eigenvalue equation for the TE odd modes, Eq. (9), is also plotted in Figure 3. For the TE odd modes, the value of $k_{1\,\max}$ again determines the number of propagating modes. When $k_{1\,\max}\delta < \pi/2$, there cannot be any propagating TE odd mode, since it is impossible to have an intersection between the $-k_1/\alpha_2$ and $\tan k_1\delta$ curves [see the dotted curve (b)]. This is known as the cutoff condition for the TE odd modes.

2.2b. Power of TE Guided Modes

The power transmitted along the axis of the guide is obtained by integrating the time average value of the Poynting vector over the cross section of the guide. Because the slab guide extends to infinity in the y direction, an infinite amount of power is transmitted along the guide. To avoid this infinite result, the power transmitted per unit width of the guide will be evaluated.

$$
\begin{aligned}
P &= \tfrac{1}{2}\int_S \mathrm{Re}(\tilde{\mathbf{E}}\times\tilde{\mathbf{H}}^*)\cdot\hat{\mathbf{n}}\,da \\
&= \tfrac{1}{2}\int_{x=-\infty}^{\infty}\int_{y=0}^{1}\mathrm{Re}\left[\hat{\mathbf{y}}\tilde{E}_y\times(\hat{\mathbf{x}}\tilde{H}_x^* + \hat{\mathbf{z}}\tilde{H}_z^*)\right]\cdot\hat{\mathbf{z}}\,dx\,dy \\
&= -\tfrac{1}{2}\int_{-\infty}^{\infty}\mathrm{Re}(\tilde{E}_y\tilde{H}_x^*)\,dx \\
&= -\tfrac{1}{2}\left[\int_{-\delta}^{\delta}\mathrm{Re}(\tilde{E}_{1y}\tilde{H}_{1x}^*)\,dx + \left(\int_{-\infty}^{-\delta} + \int_{\delta}^{\infty}\right)\mathrm{Re}(\tilde{E}_{2y}\tilde{H}_{2x}^*)\,dx\right] \\
&\equiv P_1 + P_2.
\end{aligned}
\qquad (1)
$$

The power transmitted for the TE even modes and TE odd modes will be evaluated separately.

TE even modes. The fields for the TE even modes are given by Eqs. (2.2.9), (2.2.10), and (2.2.21)–(2.2.23). These are repeated here with the factor $e^{j\omega t - \gamma_g z}$ incorporated.

$$\tilde{E}_{1y}^{\,e} = E^e \cos k_1 x\, e^{j\omega t - \gamma_g z}, \qquad (2)$$

$$\tilde{H}_{1x}^{\,e} = \frac{-\gamma_g}{j\omega\mu_1} E^e \cos k_1 x\, e^{j\omega t - \gamma_g z}, \qquad (3)$$

$$\tilde{E}_{2y}^{\,e} = E^e \cos k_1\delta\, e^{-\alpha_2(x-\delta)} e^{j\omega t - \gamma_g z}, \qquad x > \delta, \qquad (4)$$

$$\tilde{H}_{2x}{}^{e}=\frac{-\gamma_g}{j\omega\mu_2}E^e\cos k_1\delta\, e^{-\alpha_2(x-\delta)}e^{j\omega t-\gamma_g z},\qquad x>\delta, \tag{5}$$

$$\begin{aligned}P_1{}^{e}&=-\tfrac{1}{2}\int_{-\delta}^{\delta}\operatorname{Re}\left(\tilde{E}_{1y}{}^{e}\tilde{H}_{1x}{}^{e*}\right)dx\\&=-\int_0^{\delta}\operatorname{Re}\Bigg[\left(E^e\cos k_1x\, e^{j\omega t-\gamma_g z}\right)\\&\qquad\times\left(\frac{-\gamma_g}{j\omega\mu_1}E^e\cos k_1x\, e^{j\omega t-\gamma_g z}\right)^*\Bigg]dx.\end{aligned} \tag{6}$$

Let $k_1\equiv\beta_1+j\alpha_1$. Then

$$\begin{aligned}\cos k_1x&=\cos(\beta_1+j\alpha_1)x\\&=\cos\beta_1x\cosh\alpha_1x-j\sin\beta_1x\sinh\alpha_1x,\\\cos k_1{}^*x&=\cos(\beta_1-j\alpha_1)x\\&=\cos\beta_1x\cosh\alpha_1x+j\sin\beta_1x\sinh\alpha_1x,\\\cos k_1x\cos k_1{}^*x&=\cos^2\beta_1x\cosh^2\alpha_1x+\sin^2\beta_1x\sinh^2\alpha_1x\\&=\cos^2\beta_1x\cosh^2\alpha_1x+\left(1-\cos^2\beta_1x\right)\sinh^2\alpha_1x\\&=\sinh^2\alpha_1x+\cos^2\beta_1x.\end{aligned} \tag{7}$$

Substituting Eq. (7) into Eq. (6) yields

$$\begin{aligned}P_1{}^{e}&=-\left(\frac{-\beta_g}{\omega\mu_1}|E^e|^2e^{-2\alpha_g z}\right)\int_0^{\delta}\left(\sinh^2\alpha_1x+\cos^2\beta_1x\right)dx\\&=\frac{\beta_g}{\omega\mu_1}|E^e|^2e^{-2\alpha_g z}\Bigg[\frac{1}{\alpha_1}\left(\frac{\sinh 2\alpha_1x}{4}-\frac{\alpha_1x}{2}\right)\\&\qquad+\frac{1}{\beta_1}\left(\frac{\beta_1x}{2}+\frac{\sin 2\beta_1x}{4}\right)\Bigg]_0^{\delta}\\&=\frac{\beta_g}{\omega\mu_1}|E^e|^2e^{-2\alpha_g z}\left(\frac{\delta}{2}\,\frac{\sinh 2\alpha_1\delta}{2\alpha_1\delta}+\frac{\sin 2\beta_1\delta}{4\beta_1}\right),\end{aligned} \tag{8}$$

where $\gamma_g\equiv\alpha_g+j\beta_g$.

$$\begin{aligned}P_2{}^{e}&=-\tfrac{1}{2}\left(\int_{-\infty}^{-\delta}+\int_{\delta}^{\infty}\right)\operatorname{Re}\left(\tilde{E}_{2y}{}^{e}\tilde{H}_{2x}{}^{e*}\right)dx\\&=-\int_{\delta}^{\infty}\operatorname{Re}\Bigg[\left(E^e\cos k_1\delta\, e^{-\alpha_2(x-\delta)}e^{j\omega t-\gamma_g z}\right)\\&\qquad\times\left(\frac{-\gamma_g{}^*}{-j\omega\mu_2}E^{e*}\cos k_1{}^*\delta\, e^{-\alpha_2(x-\delta)}e^{-j\omega t-\gamma_g{}^*z}\right)\Bigg]dx\\&=\frac{\beta_g}{\omega\mu_2}|E^e|^2e^{-2\alpha_g z}\left(\sinh^2\alpha_1\delta+\cos^2\beta_1\delta\right)\int_{\delta}^{\infty}e^{-2\alpha_2(x-\delta)}dx\\&=\frac{\beta_g}{\omega\mu_2}|E^e|^2\left(\sinh^2\alpha_1\delta+\cos^2\beta_1\delta\right)e^{-2\alpha_g z}\frac{1}{2\alpha_2}.\end{aligned} \tag{9}$$

The total power transmitted is given by the sum of Eqs. (8) and (9),

$$\begin{aligned} P^e &= P_1{}^e + P_2{}^e \\ &= \frac{\beta_g |E^e|^2}{\omega} e^{-2\alpha_g z}\left[\left(\frac{\delta}{2}\,\frac{\sinh 2\alpha_1\delta}{2\alpha_1\delta} + \frac{\sin 2\beta_1\delta}{4\beta_1}\right)\frac{1}{\mu_1}\right. \\ &\qquad \left. + \frac{1}{2\alpha_2\mu_2}(\sinh^2\alpha_1\delta + \cos^2\beta_1\delta)\right]. \end{aligned} \tag{10}$$

For a lossless system, $\alpha_g = 0$, $\alpha_1 = 0$, $k_1 = \beta_1$, and $\alpha_2 = \sqrt{\beta_g{}^2 - \omega^2\mu_2\epsilon_2}$, the power transmitted is simplified to

$$\begin{aligned} P^e &= \frac{\beta_g}{\omega}|E^e|^2\left[\frac{1}{\mu_1}\left(\frac{\delta}{2} + \frac{2\sin\beta_1\delta\cos\beta_1\delta}{4\beta_1}\right) + \frac{\cos^2\beta_1\delta}{2\alpha_2\mu_2}\right] \\ &= \frac{\beta_g}{\omega}|E^e|^2\left[\frac{1}{\mu_1}\left(\frac{\delta}{2} + \frac{\sin^2\beta_1\delta}{2\beta_1}\,\frac{1}{\tan\beta_1\delta}\right) + \frac{\cos^2\beta_1\delta}{2\alpha_2\mu_2}\right] \\ &= \frac{\beta_g}{2\omega}|E^e|^2\left[\frac{1}{\mu_1}\left(\delta + \frac{\sin^2\beta_1\delta}{\alpha_2}\,\frac{\mu_2}{\mu_1}\right) + \frac{\cos^2\beta_1\delta}{\alpha_2\mu_2}\right]. \end{aligned} \tag{11}$$

Equation (2.2a.3) is used to obtain the last expression.

For the conventional dielectric ($\mu_1 = \mu_2 = \mu_0$), Eq. (11) may be further simplified.

$$P^e = \frac{\beta_g}{2\omega\mu_0}|E^e|^2(\delta + 1/\alpha_2). \tag{12}$$

The magnitude $|E^e|$ may be expressed in terms of the power transmitted.

$$|E^e| = \left[\frac{2\omega\mu_0 P^e}{\beta_g(\delta + 1/\alpha_2)}\right]^{1/2} \qquad \text{(lossless guide).} \tag{13}$$

TE odd modes. The fields for the TE odd modes are given by Eqs. (2.2.11) and (2.2.24)–(2.2.26) and are repeated here.

$$\tilde{E}_{1y}{}^o = E^o \sin k_1 x\, e^{j\omega t - \gamma_g z}, \tag{14}$$

$$\tilde{H}_{1x}{}^o = \frac{-\gamma_g}{j\omega\mu_1} E^o \sin k_1 x\, e^{j\omega t - \gamma_g z}, \tag{15}$$

$$\tilde{E}_{2y}{}^o = E^o \sin k_1\delta\, e^{-\alpha_2(x-\delta)} e^{j\omega t - \gamma_g z}, \qquad x > \delta, \tag{16}$$

$$\tilde{H}_{2x}{}^o = \frac{-\gamma_g}{j\omega\mu_2} E^o \sin k_1\delta\, e^{-\alpha_2(x-\delta)} e^{j\omega t - \gamma_g z}, \qquad x > \delta. \tag{17}$$

The power of TE odd modes are obtained similarly.

$$P_1{}^{\circ} = -\int_0^{\delta} \mathrm{Re}\left[\tilde{E}_{1y}{}^{\circ}\tilde{H}_{1x}{}^{\circ *}\right] dx$$

$$= \frac{\beta_g}{\omega\mu_1}|E^{\circ}|^2 e^{-2\alpha_g z}\int_0^{\delta}\sin k_1 x \sin k_1{}^* x\, dx$$

$$= \frac{\beta_g}{\omega\mu_1}|E^{\circ}|^2 e^{-2\alpha_g z}\int_0^{\delta}\left(\sinh^2\alpha_1 x + \sin^2\beta_1 x\right) dx \tag{18}$$

$$= \frac{\beta_g}{\omega\mu_1}|E^{\circ}|^2 e^{-2\alpha_g z}\left(\frac{\delta}{2}\,\frac{\sinh 2\alpha_1\delta}{2\alpha_1\delta} - \frac{\sin 2\beta_1\delta}{4\beta_1}\right). \tag{19}$$

The identities

$$\sin(a \pm jb) = \sin a \cosh b \pm j \cos a \sinh b$$

and

$$\sin(a+jb)\sin(a-jb) = \sin^2 a + \sinh^2 b$$

were used to obtain Eq. (18).

$$P_2{}^{\circ} = \int_{\delta}^{\infty} \mathrm{Re}\left[\tilde{E}_{2y}{}^{\circ}\tilde{H}_{2x}{}^{\circ *}\right] dx$$

$$= \frac{\beta_g}{\omega\mu_2}|E^{\circ}|^2\left(\sinh^2\alpha_1\delta + \sin^2\beta_1\delta\right)\frac{1}{2\alpha_2}e^{-2\alpha_g z}. \tag{20}$$

The total transmitted power of TE odd modes is the sum of Eqs. (19) and (20),

$$P^{\circ} = P_1{}^{\circ} + P_2{}^{\circ}$$

$$= \frac{\beta_g}{\omega}|E^{\circ}|^2 e^{-2\alpha_g z}\left[\frac{1}{\mu_1}\left(\frac{\delta}{2}\,\frac{\sinh 2\alpha_1\delta}{2\alpha_1\delta} - \frac{\sin 2\beta_1\delta}{4\beta_1}\right) + \frac{1}{2\alpha_2\mu_2}\left(\sinh^2\alpha_1\delta + \sin^2\beta_1\delta\right)\right]. \tag{21}$$

Equation (21) may be further simplified for the lossless system using conventional dielectrics ($\mu_1 = \mu_2 = \mu_0$, $\alpha_g = 0$, $\alpha_1 = 0$).

$$P^{\circ} = \frac{\beta_g}{\omega\mu_0}|E^{\circ}|^2\left(\frac{\delta}{2} - \frac{\sin 2\beta_1\delta}{4\beta_1} + \frac{\sin^2\beta_1\delta}{2\alpha_2}\right)$$

$$= \frac{\beta_g}{\omega\mu_0}|E^{\circ}|^2\left(\frac{\delta}{2} - \frac{\cos^2\beta_1\delta\tan\beta_1\delta}{2\beta_1} + \frac{\sin^2\beta_1\delta}{2\alpha_2}\right)$$

$$= \frac{\beta_g}{2\omega\mu_0}|E^{\circ}|^2\left(\delta - \frac{1}{\alpha_2}\right). \tag{22}$$

Equation (2.2a.4) is used to obtain the final form of Eq. (22).

The magnitude $|E^{\circ}|$ may be expressed in terms of the power as

$$E^{\circ}=\left(\frac{2\omega\mu_0 P^{\circ}}{\beta_g(\delta-1/\alpha_2)}\right)^{1/2}. \tag{23}$$

2.3. Transverse Magnetic Waves

The transverse magnetic mode is defined as the solution such that the **H** field has no component in the direction of propagation, that is, the solution for which $H_z=0$. The equation to be solved is Eq. (2.1.8), which is repeated here:

$$\frac{d^2H_y}{dx^2}+k^2H_y=0, \qquad k^2\equiv\gamma_g^2-\gamma^2, \quad \gamma^2\equiv-\omega^2\mu\varepsilon. \tag{1}$$

The solutions will be sought separately for each region (see Figure 2).

Region I: $-\delta<x<\delta$. The **H** field in this region satisfies the wave equation

$$\frac{d^2H_{1y}}{dx^2}+k_1{}^2H_{1y}=0, \tag{2}$$

$$k_1{}^2\equiv\gamma_g{}^2-\gamma_1{}^2 \qquad \text{and} \qquad \gamma_1{}^2\equiv-\omega^2\mu_1\varepsilon_1. \tag{3}$$

The solution of Eq. (2) in trigonometric form is found to consist of the sum of an even and an odd function of x,

$$H_{1y}(x)=H^{e}\cos k_1x+H^{o}\sin k_1x. \tag{4}$$

The even and odd modes of the solution will be treated independently.

TM even modes (symmetric modes): $-\delta<x<\delta$.

$$H_{1y}{}^{e}(x)=H^{e}\cos k_1x, \qquad -\delta<x<\delta. \tag{5}$$

The corresponding **E** field is given by Eq. (2.1.7).

$$E_{1x}{}^{e}(x)=\frac{\gamma_g}{j\omega\varepsilon_1}H_{1y}{}^{e}=\frac{\gamma_g}{j\omega\varepsilon_1}H^{e}\cos k_1x,$$

$$E_{1z}{}^{e}(x)=\frac{1}{j\omega\varepsilon_1}\frac{dH_{1y}{}^{e}}{dx}=\frac{-k_1}{j\omega\varepsilon_1}H^{e}\sin k_1x. \tag{6}$$

TM odd modes (antisymmetric modes).

$$H_{1y}{}^{\mathrm{o}}(x)=H^{\mathrm{o}}\sin k_1x, \qquad -\delta<x<\delta,$$

$$E_{1x}{}^{\mathrm{o}}(x)=\frac{\gamma_g}{j\omega\varepsilon_1}H_{1y}{}^{\mathrm{o}}=\frac{\gamma_g}{j\omega\varepsilon_1}H^{\mathrm{o}}\sin k_1x, \tag{7}$$

$$E_{1z}{}^{\mathrm{o}}(x)=\frac{1}{j\omega\varepsilon_1}\frac{dH_{1y}{}^{\mathrm{o}}}{dx}=\frac{k_1}{j\omega\varepsilon_1}H^{\mathrm{o}}\cos k_1x.$$

Region II: $\delta<x<-\delta$. The wave equation for the region exterior to the dielectric sheet is

$$\frac{d^2H_{2y}}{dx^2}+k_2{}^2H_{2y}=0, \tag{8}$$

$$k_2{}^2\equiv\gamma_g{}^2-\gamma_2{}^2 \qquad \text{and} \qquad \gamma_2{}^2\equiv-\omega^2\mu_2\varepsilon_2. \tag{9}$$

The solution in exponential form is

$$\begin{aligned} H_{2y}(x)&=H^+e^{-jk_2(x-\delta)}+H^-e^{jk_2(x-\delta)}, && x>\delta,\\ &=H^+e^{jk_2(x+\delta)}+H^-e^{-jk_2(x+\delta)}, && x<-\delta. \end{aligned} \tag{10}$$

The parameter k_2 bears the same physical interpretation as in the TE modes (Section 2.2). For evanescent modes the solutions assume the following forms:

$$\begin{aligned} H_{2y}(x)&=H^+e^{-\alpha_2(x-\delta)}, && x>\delta,\\ &=H^+e^{\alpha_2(x+\delta)}, && x<-\delta, \end{aligned} \tag{11}$$

where α_2 is given by [see Eq. (2.2.16)]

$$\alpha_2\equiv\left[\beta_g{}^2(1-\alpha_g{}^2/\beta_g{}^2-j\alpha_g/\beta_g)-\omega^2\mu_2\varepsilon_2\right]^{1/2}. \tag{12}$$

The coefficient H^+ is evaluated by applying the boundary conditions at the interface, $H_{1y}(\pm\delta)=H_{2y}(\pm\delta)$.

$$\text{TM even modes:} \quad H^+=H^{\mathrm{e}}\cos k_1\delta; \tag{13}$$

$$\begin{aligned} \text{TM odd modes:} \quad H^+&=H^{\mathrm{o}}\sin k_1\delta, && x>\delta,\\ &=-H^{\mathrm{o}}\sin k_1\delta, && x<-\delta. \end{aligned} \tag{14}$$

The evanescent field in region II are summarized as follows.

TM even modes: $\delta<x<-\delta$.

$$\begin{aligned} H_{2y}{}^{\mathrm{e}}(x)&=H^{\mathrm{e}}\cos k_1\delta\, e^{-\alpha_2(x-\delta)}, && x>\delta,\\ &=H^{\mathrm{e}}\cos k_1\delta\, e^{\alpha_2(x+\delta)}, && x<-\delta; \end{aligned} \tag{15}$$

$$E_{2x}{}^{e}(x)=\frac{\gamma_g}{j\omega\varepsilon_2}H_{2y}{}^{e}$$

$$=\frac{\gamma_g}{j\omega\varepsilon_2}H^{e}\cos k_1\delta\, e^{-\alpha_2(x-\delta)}, \qquad x>\delta, \tag{16}$$

$$=\frac{\gamma_g}{j\omega\varepsilon_2}H^{e}\cos k_1\delta\, e^{\alpha_2(x+\delta)}, \qquad x<-\delta;$$

$$E_{2z}{}^{e}(x)=\frac{1}{j\omega\varepsilon_2}\frac{dH_{2y}{}^{e}}{dx}$$

$$=\frac{-\alpha_2}{j\omega\varepsilon_2}H^{e}\cos k_1\delta\, e^{-\alpha_2(x-\delta)}, \qquad x>\delta, \tag{17}$$

$$=\frac{\alpha_2}{j\omega\varepsilon_2}H^{e}\cos k_1\delta\, e^{\alpha_2(x+\delta)}, \qquad x>-\delta.$$

TM odd modes: $\delta<x<-\delta$.

$$H_{2y}{}^{o}(x)=H^{o}\sin k_1\delta\, e^{-\alpha_2(x-\delta)}, \qquad x>\delta,$$

$$=-H^{o}\sin k_1\delta\, e^{\alpha_2(x+\delta)}, \qquad x<-\delta; \tag{18}$$

$$E_{2x}{}^{o}(x)=\frac{\gamma_g}{j\omega\varepsilon_2}H^{o}\sin k_1\delta\, e^{-\alpha_2(x-\delta)}, \qquad x>\delta,$$

$$=\frac{-\gamma_g}{j\omega\varepsilon_2}H^{o}\sin k_1\delta\, e^{\alpha_2(x+\delta)}, \qquad x<-\delta; \tag{19}$$

$$E_{2z}{}^{o}(x)=\frac{-\alpha_2}{j\omega\varepsilon_2}H^{o}\sin k_1\delta\, e^{-\alpha_2(x-\delta)}, \qquad x>\delta,$$

$$=\frac{-\alpha_2}{j\omega\varepsilon_2}H^{o}\sin k_1\delta\, e^{\alpha_2(x+\delta)}, \qquad x<-\delta. \tag{20}$$

2.3a. TM Eigenmodes

The guide-propagation constant γ_g for the TM modes is determined by applying the boundary condition on the tangential components of the **E** fields at the interface, $E_{1z}(\pm\delta)=E_{2z}(\pm\delta)$.

$$\text{TM even modes:} \qquad \tan k_1\delta=\frac{\varepsilon_1}{\varepsilon_2}\frac{\alpha_2}{k_1}; \tag{1}$$

$$\text{TM odd modes:} \qquad \tan k_1\delta=-\frac{\varepsilon_2}{\varepsilon_1}\frac{k_1}{\alpha_2}. \tag{2}$$

This set of equations is similar to that for the TE modes, Eqs. (2.2a.1) and (2.2a.2), with μ_1 and μ_2 replaced by ε_1 and ε_2, respectively. For the lossless case, the fraction $\varepsilon_1/\varepsilon_2=\epsilon_1/\epsilon_2$ can be treated as a scalar factor. The analysis for the TE modes (Section 2.2a) is also valid for the TM modes. Equation (1) shows that TM even modes can exist at arbitrarily small frequencies.

2.3b. Power of TM Guided Modes

The power transmitted along the axis of the guide can be calculated as follows:

$$
\begin{aligned}
P &= \tfrac{1}{2}\int_S \mathrm{Re}(\tilde{\mathbf{E}}\times\tilde{\mathbf{H}}^*)\cdot\hat{\mathbf{n}}\,da \\
&= \tfrac{1}{2}\int_{x=-\infty}^{\infty}\int_{y=0}^{1}\mathrm{Re}\left[(\hat{\mathbf{x}}\tilde{E}_x+\hat{z}\tilde{E}_z)\times\hat{\mathbf{y}}\tilde{H}_y{}^*\right]\cdot\hat{\mathbf{z}}\,dx\,dy \\
&= \tfrac{1}{2}\int_{-\infty}^{\infty}\mathrm{Re}(\tilde{E}_x\tilde{H}_y{}^*)\,dx \\
&= \tfrac{1}{2}\int_{-\infty}^{\infty}\mathrm{Re}\left(\frac{\gamma_g}{j\omega\varepsilon}\tilde{H}_y\tilde{H}_y{}^*\right)dx \\
&= \mathrm{Re}\left(\frac{\gamma_g}{j\omega\varepsilon}\int_0^{\infty}|\tilde{H}_y|^2\,dx\right) \\
&= \mathrm{Re}\left(\frac{\gamma_g}{j\omega\varepsilon_1}\int_0^{\delta}|\tilde{H}_{1y}|^2\,dx\right)+\mathrm{Re}\left(\frac{\gamma_g}{j\omega\varepsilon_2}\int_{\delta}^{\infty}|\tilde{H}_{2y}|^2\,dx\right) \\
&\equiv P_1+P_2.
\end{aligned}
\tag{1}
$$

The even and odd modes will be treated separately.

TM even modes. The fields of TM even modes are given by Eqs. (2.3.5) and (2.3.15), and are repeated here.

$$\tilde{H}_{1y}{}^{\mathrm{e}} = H^{\mathrm{e}}\cos k_1 x\, e^{j\omega t-\gamma_g z}, \qquad -\delta<x<\delta, \tag{2}$$

$$\tilde{H}_{2y}{}^{\mathrm{e}} = H^{\mathrm{e}}\cos k_1\delta\, e^{-\alpha_2(x-\delta)}e^{j\omega t-\gamma_g z}, \qquad x>\delta; \tag{3}$$

$$
\begin{aligned}
P_1{}^{\mathrm{e}} &= \mathrm{Re}\left(\frac{\gamma_g}{j\omega\varepsilon_1}\int_0^{\delta}|H_{1y}{}^{\mathrm{e}}|^2\,dx\right) \\
&= \mathrm{Re}\left(\frac{\gamma_g}{j\omega\varepsilon_1}\int_0^{\delta}|H^{\mathrm{e}}|^2 e^{-2\alpha_g z}\cos k_1 x\cos k_1{}^* x\,dx\right) \\
&= \mathrm{Re}\left[\frac{\gamma_g}{j\omega\varepsilon_1}\int_0^{\delta}|H^{\mathrm{e}}|^2 e^{-2\alpha_g z}(\sinh^2\alpha_1 x+\cos^2\beta_1 x)\,dx\right] \\
&= \mathrm{Re}\left[\frac{\gamma_g}{j\omega\varepsilon_1}|H^{\mathrm{e}}|^2 e^{-2\alpha_g z}\left(\frac{\delta}{2}\frac{\sinh 2\alpha_1\delta}{2\alpha_1\delta}+\frac{\sin 2\beta_1\delta}{4\beta_1}\right)\right],
\end{aligned}
\tag{4}
$$

$$P_2{}^{e} = \mathrm{Re}\left(\frac{\gamma_g}{j\omega\varepsilon_2}\int_\delta^\infty |H_{2y}{}^{e}|^2\,dx\right)$$

$$= \mathrm{Re}\left(\frac{\gamma_g}{j\omega\varepsilon_2}|H^e|^2 e^{-2\alpha_g z}\int_\delta^\infty \cos k_1\delta \cos k_1{}^*\delta\, e^{-2\alpha_2(x-\delta)}\,dx\right)$$

$$= \mathrm{Re}\left[\frac{\gamma_g}{j\omega\varepsilon_2}|H^e|^2 e^{-2\alpha_g z}\frac{1}{2\alpha_2}(\sinh^2\alpha_1\delta + \cos^2\beta_1\delta)\right]. \tag{5}$$

The total transmitted power is the sum of Eqs. (4) and (5),

$$P^e = P_1{}^{e} + P_2{}^{e}$$

$$= |H^e|^2 e^{-2\alpha_g z}\left[\mathrm{Re}\left(\frac{\gamma_g}{j\omega\varepsilon_1}\right)\left(\frac{\delta}{2}\frac{\sinh 2\alpha_1\delta}{2\alpha_1\delta} + \frac{\sin 2\beta_1\delta}{4\beta_1}\right)\right.$$

$$\left. + \mathrm{Re}\left(\frac{\gamma_g}{2\,j\omega\varepsilon_2\alpha_2}\right)(\sinh^2\alpha_1\delta + \cos^2\beta_1\delta)\right]. \tag{6}$$

For a lossless system—$k_1=\beta_1$, $\varepsilon_1=\epsilon_1$, $\varepsilon_2=\epsilon_2$, $\gamma_g = j\beta_g$, and $\alpha_2 = \sqrt{\beta_g{}^2 - \omega^2\mu_2\epsilon_2}$ —Eq. (6) simplifies to

$$P^e = |H^e|^2\frac{\beta_g}{2\omega}\left[\frac{1}{\epsilon_1}\left(\delta + \frac{1}{\beta_1}\sin\beta_1\delta\cos\beta_1\delta\right) + \frac{1}{\epsilon_2\alpha_2}\cos^2\beta_1\delta\right]$$

$$= |H^e|^2\frac{\beta_g}{2\omega\epsilon_1}\left(\delta + \frac{1}{\beta_1}\tan\beta_1\delta\cos^2\beta_1\delta + \frac{\epsilon_1}{\epsilon_2\alpha_2}\cos^2\beta_1\delta\right) \tag{7}$$

$$= |H^e|^2\frac{\beta_g}{2\omega\epsilon_1}\left[\delta + \left(\frac{1}{\beta_1}\frac{\epsilon_1\alpha_2}{\epsilon_2\beta_1} + \frac{\epsilon_1}{\epsilon_2\alpha_2}\right)\cos^2\beta_1\delta\right] \tag{8}$$

$$= \frac{\beta_g}{2\omega\epsilon_1}|H^e|^2\left\{\delta + \frac{(\alpha_2{}^2 + \beta_1{}^2)\epsilon_1}{\alpha_2\beta_1{}^2\epsilon_2}\frac{1}{1 + [(\epsilon_1/\epsilon_2)(\alpha_2/\beta_1)]^2}\right\} \tag{9}$$

$$= \frac{\beta_g}{2\omega\epsilon_1}|H^e|^2\left[\delta + \frac{\alpha_2{}^2 + \beta_1{}^2}{\alpha_2}\frac{\epsilon_1\epsilon_2}{(\epsilon_1\alpha_2)^2 + (\epsilon_2\beta_1)^2}\right]$$

$$= \frac{\beta_g}{2\omega\epsilon_1}|H^e|^2\left[\delta + \frac{\epsilon_1\epsilon_2}{\alpha_2}\frac{\alpha_2{}^2 + \beta_1{}^2}{(\epsilon_1\alpha_2)^2 + (\epsilon_2\beta_1)^2}\right]. \tag{10}$$

The identity $\cos^2\theta = 1/(1+\tan^2\theta)$ and Eq. (2.3a.1) are used to obtain Eqs. (8) and (9).

The magnitude $|H^e|$ can be expressed in terms of the power as

$$|H^e| = \left\{2\omega\epsilon_1 P^e \Big/ \beta_g\left[\delta + \frac{\epsilon_1\epsilon_2}{\alpha_2}\frac{\alpha_2{}^2 + \beta_1{}^2}{(\epsilon_1\alpha_2)^2 + (\epsilon_2\beta_1)^2}\right]\right\}^{1/2}. \tag{11}$$

TM odd modes. The fields of TM odd modes are given by Eqs. (2.3.7) and (2.3.18)–(2.3.20).

$$\tilde{H}_{1y}{}^{\circ} = H^{\circ} \sin k_1 x \, e^{j\omega t - \gamma_g z}, \tag{12}$$

$$\tilde{H}_{2y}{}^{\circ} = H^{\circ} \sin k_1 \delta \, e^{-\alpha_2(x-\delta)} e^{j\omega t - \gamma_g z}, \qquad x > \delta, \tag{13}$$

$$|\tilde{H}_{1y}{}^{\circ}|^2 = \tilde{H}_{1y}{}^{\circ}\tilde{H}_{1y}{}^{\circ *} = |H^{\circ}|^2 e^{-2\alpha_g z} \sin k_1 x \sin k_1{}^* x$$

$$= |H^{\circ}|^2 e^{-2\alpha_g z}(\sinh^2 \alpha_1 x + \sin^2 \beta_1 x),$$

$$\int_0^{\delta} |\tilde{H}_{1y}{}^{\circ}|^2 \, dx = |H^{\circ}|^2 e^{-2\alpha_g z}\left(\frac{\delta}{2}\,\frac{\sinh 2\alpha_1\delta}{2\alpha_1\delta} - \frac{\sin 2\beta_1\delta}{4\beta_1}\right), \tag{14}$$

$$|\tilde{H}_{2y}{}^{\circ}|^2 = |H^{\circ}|^2 \sin k_1\delta \sin k_1{}^*\delta \, e^{-2\alpha_g z} e^{-2\alpha_2(x-\delta)}$$

$$= |H^{\circ}|^2 (\sinh^2 \alpha_1\delta + \sin^2 \beta_1\delta) e^{-2\alpha_2(x-\delta)} e^{-2\alpha_g z},$$

$$\int_{\delta}^{\infty} |\tilde{H}_{2y}{}^{\circ}| \, dx = |H^{\circ}|^2 \frac{1}{2\alpha_2}(\sinh^2 \alpha_1\delta + \sin^2 \beta_1\delta) e^{-2\alpha_g z}. \tag{15}$$

The power transmitted is given by

$$P^{\circ} = \mathrm{Re}\left(\frac{\gamma_g}{j\omega\varepsilon_1}\int_0^{\delta} |H_{1y}{}^{\circ}|^2 \, dx\right) + \mathrm{Re}\left(\frac{\gamma_g}{j\omega\varepsilon_2}\int_{\delta}^{\infty} |H_{2y}{}^{\circ}|^2 \, dx\right)$$

$$= |H^{\circ}|^2 e^{-2\alpha_g z}\left[\mathrm{Re}\left(\frac{\gamma_g}{j\omega\varepsilon_1}\right)\left(\frac{\delta}{2}\,\frac{\sinh 2\alpha_1\delta}{2\alpha_1\delta} - \frac{\sin 2\beta_1\delta}{4\beta_1}\right)\right.$$

$$\left. + \mathrm{Re}\left(\frac{\gamma_g}{j\omega\varepsilon_2}\right)(\sinh^2 \alpha_1\delta + \sin^2 \beta_1\delta)\frac{1}{2\alpha_2}\right]. \tag{16}$$

For a lossless system—$k_1 = \beta_1$, $\gamma_g = j\beta_g$, $\varepsilon_1 = \epsilon_1$, $\varepsilon_2 = \epsilon_2$, and $\alpha_2 = \sqrt{\beta_g{}^2 - \omega^2\mu_2\epsilon_2}$ —Eq. (16) simplifies to

$$P^{\circ} = \frac{\beta_g}{2\omega}|H^{\circ}|^2 \frac{1}{\epsilon_1}\left(\delta - \frac{1}{\beta_1}\sin^2 \beta_1\delta \frac{1}{\tan \beta_1\delta} + \frac{\epsilon_1}{\epsilon_2\alpha_2}\sin^2 \beta_1\delta\right)$$

$$= \frac{\beta_g}{2\omega\epsilon_1}|H^{\circ}|^2$$

$$\times \left\{\delta + \sin^2 \beta_1\delta\left[\frac{-1}{\beta_1}\,\frac{1}{(-\epsilon_2/\epsilon_1)(\beta_1/\alpha_2)} + \frac{\epsilon_1}{\epsilon_2\alpha_2}\right]\right\}$$

$$= \frac{\beta_g}{2\omega\epsilon_1}|H^{\circ}|^2\left[\delta + \sin^2 \beta_1\delta \frac{\epsilon_1}{\epsilon_2}\left(\frac{\alpha_2}{\beta_1{}^2} + \frac{1}{\alpha_2}\right)\right], \tag{17}$$

$$\sin^2 \beta_1\delta = \frac{1}{1 + \cot^2 \beta_1\delta} = \frac{1}{1 + (-\epsilon_1\alpha_2/\epsilon_2\beta_1)^2} = \frac{(\epsilon_2\beta_1)^2}{(\epsilon_1\alpha_2)^2 + (\epsilon_2\beta_1)^2}. \tag{18}$$

Equation (2.3a.2) is used to obtain Eqs. (17) and (18). Substituting Eq. (18) into Eq. (17) yields

$$P^{\mathrm{o}} = \frac{\beta_{\mathrm{g}}}{2\omega\epsilon_1}|H^{\mathrm{o}}|^2\left[\delta + \frac{\epsilon_1}{\epsilon_2}\frac{\alpha_2{}^2+\beta_1{}^2}{\alpha_2\beta_1{}^2}\frac{(\epsilon_2\beta_1)^2}{(\epsilon_1\alpha_2)^2+(\epsilon_2\beta_1)^2}\right]$$

$$= \frac{\beta_{\mathrm{g}}}{2\omega\epsilon_1}|H^{\mathrm{o}}|^2\left[\delta + \frac{\epsilon_1\epsilon_2}{\alpha_2}\frac{\alpha_2{}^2+\beta_1{}^2}{(\epsilon_1\alpha_2)^2+(\epsilon_2\beta_1)^2}\right] \tag{19}$$

and

$$|H^{\mathrm{o}}| = \left\{2\omega\epsilon_1 P^{\mathrm{o}}\Big/\beta_{\mathrm{g}}\left[\delta + \frac{\epsilon_1\epsilon_2}{\alpha_2}\frac{\alpha_2{}^2+\beta_1{}^2}{(\epsilon_1\alpha_2)^2+(\epsilon_2\beta_1)^2}\right]\right\}^{1/2}. \tag{20}$$

2.4. Radiation Modes

It has been pointed out [1, 2, 4, 5, 7, 9, 10] that *evanescent modes* (surface waves) are not by themselves sufficient to account for the existence of a given field near a guiding system. A field in the neighborhood of a launching device within a rectangular waveguide can be represented as the sum of an infinite number of evanescent modes. However, only a finite number of evanescent modes can exist in a given dielectric slab guide, and these are insufficient to represent the field from an arbitrary source. This requires *radiation modes* of solution.

The radiation modes for TE and TM waves will be investigated separately.

2.5. TE Radiation Modes

The radiation field will be represented as

$$\tilde{\mathbf{E}}^{\mathrm{r}} = \mathbf{e}(x)e^{j\omega t-\gamma_{\mathrm{r}}z}, \tag{1}$$

where the superscript r indicates radiation modes, lower-case letters, for example, $\mathbf{e}\equiv\mathbf{e}(x)$, are used to represent the radiation modes for the more frequently used quantities, and γ_{r} is the propagation constant in the z direction for the radiation modes.

The TE field exterior to the dielectric sheet (region II) is specified by [Eq. (2.2.13)]

$$\frac{d^2e_{2y}}{dx^2} - \rho_2{}^2 e_{2y} = 0, \tag{2}$$

$$\rho_2{}^2 \equiv \gamma_2{}^2 - \gamma_{\mathrm{r}}{}^2 \quad \text{and} \quad \gamma_2{}^2 \equiv -\omega^2\mu_2\varepsilon_2. \tag{3}$$

To differentiate the radiation modes from those of evanescent, a slight change of notation is introduced. The solution of Eq. (2) is

$$\begin{aligned} e_{2y}(x) &= E^+e^{-\rho_2(x-\delta)} + E^-e^{\rho_2(x-\delta)}, \qquad x>\delta,\\ &= E^+e^{\rho_2(x+\delta)} + E^-e^{-\rho_2(x+\delta)}, \qquad x<-\delta, \end{aligned} \tag{4}$$

where

$$\begin{aligned} \rho_2 &\equiv \sqrt{\gamma_2{}^2-\gamma_r{}^2} \qquad (\gamma_r \equiv \alpha_r + j\beta_r)\\ &= \left[-\omega^2\mu_2\varepsilon_2 - (\alpha_r + j\beta_r)^2\right]^{1/2} \qquad (\varepsilon_2 \equiv \epsilon_2{}' + j\epsilon_2{}'')\\ &= \left[-\omega^2\mu_2(\epsilon_2{}'+j\epsilon_2{}'') + \beta_r{}^2(1-\alpha_r{}^2/\beta_r{}^2 - j2\alpha_r/\beta_r)\right]^{1/2}\\ &= j\left\{\left[\omega^2\mu_2\epsilon_2{}' - \beta_r{}^2(1-\alpha_r{}^2/\beta_r{}^2)\right] + j(\omega^2\mu_2\epsilon_2{}'' + 2\alpha_r\beta_r)\right\}^{1/2}\\ &= j\left[\omega^2\mu_2\epsilon_2{}' - \beta_r{}^2\left(1-\frac{\alpha_r{}^2}{\beta_r{}^2}\right)\right]^{1/2}\\ &\quad \times\left(1 + j\frac{\omega^2\mu_2\epsilon_2{}'' + 2\alpha_r\beta_r}{\omega^2\mu_2\epsilon_2{}' - \beta_r{}^2(1-\alpha_r{}^2/\beta_r{}^2)}\right)^{1/2}\\ &= \alpha_\rho + j\beta_\rho. \end{aligned} \tag{5}$$

For the case of a low-loss system, $\epsilon_2{}''/\epsilon_2{}' \ll 1$ and $(\alpha_r/\beta_r)^2 \ll 1$, Eq. (5) may be approximated by neglecting the higher-order terms:

$$\begin{aligned} \rho_2 &\simeq j(\omega^2\mu_2\epsilon_2{}' - \beta_r{}^2)^{1/2}\left(1 + j\frac{\omega^2\mu_2\epsilon_2{}'' + 2\alpha_r\beta_r}{\omega^2\mu_2\epsilon_2{}' - \beta_r{}^2}\right)^{1/2}\\ &= j(\omega^2\mu_2\epsilon_2{}' - \beta_r{}^2)^{1/2}\left[1 + \tfrac{1}{2}j\left(\frac{\omega^2\mu_2\epsilon_2{}'' + 2\alpha_r\beta_r}{\omega^2\mu_2\epsilon_2{}' - \beta_r{}^2}\right) + \cdots\right]\\ &\equiv \alpha_\rho + j\beta_\rho,\\ \alpha_\rho &\simeq \frac{1}{2\beta_\rho}(\omega^2\mu_2\epsilon_2{}'' + 2\alpha_r\beta_r),\\ \beta_\rho &\simeq (\omega^2\mu_2\epsilon_2{}' - \beta_r{}^2)^{1/2}. \end{aligned} \tag{6}$$

For the ideal lossless case, $\epsilon_2{}'' = 0$ and $\alpha_r = 0$, one has $\rho_2 = j\beta_\rho = j(\omega^2\mu_2\epsilon_2{}' - \beta_r{}^2)^{1/2}$ and Eq. (4) becomes

$$\begin{aligned} e_{2y}(x) &= E^+e^{-j\beta_\rho(x-\delta)} + E^-e^{j\beta_\rho(x-\delta)}, \qquad x>\delta,\\ &= E^+e^{j\beta_\rho(x+\delta)} + E^-e^{-j\beta_\rho(x+\delta)}, \qquad x<-\delta. \end{aligned} \tag{7}$$

Note that the second term in Eq. (4) increases markedly at large values of x if ρ_2 contains a real part. This is acceptable for the present problem.

In formulating this waveguide problem, it is assumed that the region of interest does not contain any sources. The field within the guiding system is created by sources exterior to the guide, but the guiding system extends to infinity; thus the field within the system must be generated by the sources at infinity. This calls for infinite sources simply because the system under consideration is infinitely large (which is also physically unrealizable).

The corresponding **H** field of the radiation modes is represented by

$$\tilde{\mathbf{H}}^{\mathrm{r}} \equiv \mathbf{h}(x)e^{j\omega t-\gamma_{\mathrm{r}}z}. \tag{8}$$

The accompanying **H** field is given by Eq. (2.1.5).

$$\begin{aligned} h_{2x}(x) &= \frac{-\gamma_{\mathrm{r}}}{j\omega\mu_2}e_{2y} \\ &= \frac{-\gamma_{\mathrm{r}}}{j\omega\mu_2}\left(E^{+}e^{-\rho_2(x-\delta)}+E^{-}e^{\rho_2(x-\delta)}\right), \qquad x>\delta, \\ &= \frac{-\gamma_{\mathrm{r}}}{j\omega\mu_2}\left(E^{+}e^{\rho_2(x+\delta)}+E^{-}e^{-\rho_2(x+\delta)}\right), \qquad x<-\delta; \end{aligned} \tag{9}$$

$$\begin{aligned} h_{2z}(x) &= \frac{-1}{j\omega\mu_2}\frac{de_{2y}}{dx} \\ &= \frac{\rho_2}{j\omega\mu_2}\left(E^{+}e^{-\rho_2(x-\delta)}-E^{-}e^{\rho_2(x-\delta)}\right), \qquad x>\delta, \\ &= \frac{-\rho_2}{j\omega\mu_2}\left(E^{+}e^{\rho_2(x+\delta)}-E^{-}e^{-\rho_2(x+\delta)}\right), \qquad x<-\delta. \end{aligned} \tag{10}$$

The coefficients E^{+} and E^{-} can be determined from the boundary conditions for the fields. This will be carried out for the even and the odd modes separately.

TE even modes. The fields within the dielectric sheet (region I) are [Eqs. (2.29) and (2.2.10)]

$$e_{1y}{}^{\mathrm{e}}(x)=E^{\mathrm{e}}\cos k_1x, \tag{11}$$

$$h_{1x}{}^{\mathrm{e}}(x)=\frac{-\gamma_{\mathrm{r}}}{j\omega\mu_1}E^{\mathrm{e}}\cos k_1x, \tag{12}$$

$$h_{1z}{}^{\mathrm{e}}(x)=\frac{k_1}{j\omega\mu_1}E^{\mathrm{e}}\sin k_1x, \tag{13}$$

$$k_1{}^2=\gamma_{\mathrm{r}}{}^2-\gamma_1{}^2 \qquad \text{and} \qquad \gamma_1{}^2=-\omega^2\mu_1\varepsilon_1. \tag{14}$$

The tangential field components in the two regions must be continuous at the interface $x=\pm\delta$:

$$\begin{aligned} e_{1y}{}^{e}(\delta)&=e_{2y}{}^{e}(\delta), & E^{e}\cos k_1\delta&=E^{+}+E^{-},\\ h_{1z}{}^{e}(\delta)&=h_{2z}{}^{e}(\delta), & \frac{k_1}{j\omega\mu_1}E^{e}\sin k_1\delta&=\frac{\rho_2}{j\omega\mu_2}(E^{+}-E^{-}), \end{aligned} \tag{15}$$

or

$$\frac{k_1\mu_2}{\rho_2\mu_1}E^{e}\sin k_1\delta=E^{+}-E^{-}. \tag{16}$$

The sum and the difference of Eqs. (15) and (16) yield E^{+} and E^{-}, respectively.

$$\begin{aligned} E^{+}&=\tfrac{1}{2}E^{e}\left(\cos k_1\delta+\frac{k_1\mu_2}{\rho_2\mu_1}\sin k_1\delta\right),\\ E^{-}&=\tfrac{1}{2}E^{e}\left(\cos k_1\delta-\frac{k_1\mu_2}{\rho_2\mu_1}\sin k_1\delta\right). \end{aligned} \tag{17}$$

For lossless media, $k_1=\beta_1$ and $\rho_2=j\beta_\rho$, so that

$$\begin{aligned} E^{+}&=\tfrac{1}{2}E^{e}\left(\cos\beta_1\delta+\frac{\beta_1\mu_2}{j\beta_\rho\mu_1}\sin\beta_1\delta\right)\equiv\tfrac{1}{2}E^{e}(f-jg),\\ E^{-}&=\tfrac{1}{2}E^{e}\left(\cos\beta_1\delta-\frac{\beta_1\mu_2}{j\mu_1\beta_\rho}\sin\beta_1\delta\right)\equiv\tfrac{1}{2}E^{e}(f+jg), \end{aligned} \tag{18}$$

where

$$f\equiv\cos\beta_1\delta \qquad \text{and} \qquad g\equiv(\beta_1\mu_2/\beta_\rho\mu_1)\sin\beta_1\delta, \tag{19}$$

and it is obvious that

$$E^{-}=E^{+*}. \tag{20}$$

Following the same procedure, one can show that the coefficients E^{+} and E^{-} for the fields in the region $x<-\delta$ are identical to those given by Eq. (18).

In summary, the radiation fields of TE even modes are given by Eqs. (11)–(14) for region I and Eqs. (4), (9), and (10) for region II. The coefficients E^{+} and E^{-} are given by Eq. (17) for the general case and by Eq. (18) for the lossless case.

In the next section the coefficient E^{e} will be evaluated in terms of the transmitted power.

TE odd modes. The fields within the dielectric sheet (region I) are given by Eq. (2.2.11),

$$e_{1y}{}^{o}(x)=E^{o}\sin k_1 x, \tag{21}$$

$$h_{1x}{}^{o}(x)=\frac{-\gamma_r}{j\omega\mu_1}E^{o}\sin k_1 x, \tag{22}$$

$$h_{1z}{}^{o}(x)=\frac{-k_1}{j\omega\mu_1}E^{o}\cos k_1 x, \tag{23}$$

$$k_1{}^2=\gamma_r{}^2-\gamma_1{}^2 \qquad \text{and} \qquad \gamma_1{}^2=-\omega^2\mu_1\varepsilon_1. \tag{24}$$

The fields in region II are given by Eqs. (4), (9), and (10), which are repeated here with a change of notation.

$$\begin{aligned} e_{2y}{}^{o}(x)&=E_o{}^{+}e^{-\rho_2(x-\delta)}+E_o{}^{-}e^{\rho_2(x-\delta)}, \qquad x>\delta,\\ &=E_o{}^{+}e^{\rho_2(x+\delta)}+E_o{}^{-}e^{-\rho_2(x+\delta)}, \qquad x<-\delta; \end{aligned} \tag{25}$$

$$\begin{aligned} h_{2x}{}^{o}(x)&=\frac{-\gamma_r}{j\omega\mu_2}\left[E_o{}^{+}e^{-\rho_2(x-\delta)}+E_o{}^{-}e^{\rho_2(x-\delta)}\right], \qquad x>\delta,\\ &=\frac{-\gamma_r}{j\omega\mu_2}\left[E_o{}^{+}e^{\rho_2(x+\delta)}+E_o{}^{-}e^{-\rho_2(x+\delta)}\right], \qquad x<-\delta; \end{aligned} \tag{26}$$

$$\begin{aligned} h_{2z}{}^{o}(x)&=\frac{\rho_2}{j\omega\mu_2}\left[E_o{}^{+}e^{-\rho_2(x-\delta)}-E_o{}^{-}e^{+\rho_2(x-\delta)}\right], \qquad x>\delta,\\ &=\frac{-\rho_2}{j\omega\mu_2}\left[E_o{}^{+}e^{\rho_2(x+\delta)}-E_o{}^{-}e^{-\rho_2(x+\delta)}\right], \qquad x<-\delta. \end{aligned} \tag{27}$$

These fields must satisfy the continuity conditions at the interface $x=\pm\delta$:

$$\begin{aligned} e_{1y}{}^{o}(\delta)=e_{2y}{}^{o}(\delta): &\qquad E^{o}\sin k_1\delta=E_o{}^{+}+E_o{}^{-};\\ h_{1z}{}^{o}(\delta)=h_{2z}{}^{o}(\delta): &\qquad \frac{-k_1}{j\omega\mu_1}E^{o}\cos k_1\delta=\frac{\rho_2}{j\omega\mu_2}(E_o{}^{+}-E_o{}^{-}) \end{aligned} \tag{28}$$

or

$$\frac{-k_1\mu_2}{\rho_2\mu_1}E^{o}\cos k_1\delta=E_o{}^{+}-E_o{}^{-}. \tag{29}$$

Therefore,

$$E_o{}^{+}=\tfrac{1}{2}E^{o}\left(\sin k_1\delta-\frac{k_1\mu_2}{\rho_2\mu_1}\cos k_1\delta\right), \tag{30}$$

$$E_o{}^{-}=\tfrac{1}{2}E^{o}\left(\sin k_1\delta+\frac{k_1\mu_2}{\rho_2\mu_1}\cos k_1\delta\right). \tag{31}$$

For the lossless case ($\gamma_r = j\beta_r$, $k_1 = \beta_1$, and $\rho_2 = j\beta_\rho$), Eqs. (30) and (31) become

$$E_o^{\ +} = \tfrac{1}{2}E^o\left(\sin\beta_1\delta + j\frac{\beta_1\mu_2}{\beta_\rho\mu_1}\cos\beta_1\delta\right) = \tfrac{1}{2}E^o(f_o + jg_o),$$

$$E_o^{\ -} = \tfrac{1}{2}E^o\left(\sin\beta_1\delta - j\frac{\beta_1\mu_2}{\beta_\rho\mu_1}\cos\beta_1\delta\right) = \tfrac{1}{2}E^o(f_o - jg_o) = E_o^{\ +*}, \quad (32)$$

$$f_o = \sin\beta_1\delta \quad \text{and} \quad g_o = \frac{\beta_1\mu_2}{\beta_\rho\mu_1}\cos\beta_1\delta. \quad (33)$$

The radiation fields of TE odd modes are given by Eqs. (21)–(23) for region I and by Eqs. (25)–(27) for region II. The coefficients $E_o^{\ +}$ and $E_o^{\ -}$ are given by Eqs. (30) and (31) in general, and by Eq. (32) for the lossless case. In the next section the coefficient E^o will be evaluated in terms of the power carried by the fields.

2.5a. *Power of TE Radiation Modes*

The power of TE radiation modes propagated in the z direction is evaluated by integrating the Poynting vector over the cross section of the guide.

$$\begin{aligned}
P &= \tfrac{1}{2}\int_S \operatorname{Re}(\tilde{\mathbf{E}}\times\tilde{\mathbf{H}}^*)\cdot\hat{\mathbf{n}}\,da \\
&= \tfrac{1}{2}\int_{x=-\infty}^{\infty}\int_{y=0}^{1} \operatorname{Re}\left[\hat{\mathbf{y}}\tilde{E}_y^{\ r}\times(\hat{\mathbf{x}}\tilde{H}_x^{\ r} + \hat{\mathbf{z}}\tilde{H}_z^{\ r})^*\right]\cdot\hat{\mathbf{z}}\,dx\,dy \\
&= -\tfrac{1}{2}\int_{-\infty}^{\infty}\operatorname{Re}\left(\tilde{E}_y^{\ r}\tilde{H}_x^{\ r*}\right)dx \qquad \left(\tilde{H}_x^{\ r} = \frac{-\gamma_r}{j\omega\mu}\tilde{E}_y^{\ r}\right) \\
&= -\tfrac{1}{2}\int_{-\infty}^{\infty}\operatorname{Re}(-\gamma_r^*/-j\omega\mu)\left(\tilde{E}_y^{\ r}\right)^2 dx. \qquad (1)
\end{aligned}$$

The even and odd modes will be treated separately. Owing to the complexity of the problem, only lossless case will be considered.

TE even modes. The fields of TE even radiation modes for the lossless case —$k_1 = \beta_1$, $\rho_2 = j\beta_\rho$, and $\gamma_r = j\beta_r$—are given by Eqs. (2.5.1), (2.5.4), (2.5.11), and (2.5.18):

$$E_{iy}^{\ r} = e_{iy}^{\ e}e^{j(\omega t - \beta_r z)}, \qquad i = 1 \text{ or } 2, \quad (2)$$

$$e_{1y}^{\ e}(x) = E^e\cos\beta_1 x, \qquad |x| < \delta, \quad (3)$$

$$e_{2y}^{\ e}(x) = E^+e^{-j\beta_\rho(x-\delta)} + E^-e^{j\beta_\rho(x-\delta)}, \qquad x > \delta, \quad (4)$$

where

$$\beta_\rho = (\omega^2 \mu_2 \epsilon_2' - \beta_r{}^2)^{1/2}, \tag{5}$$

$$E^+ = E^{-*} = \tfrac{1}{2} E^e \left(\cos\beta_1\delta - j\frac{\beta_1\mu_2}{\beta_\rho\mu_1} \sin\beta_1\delta \right) \equiv \tfrac{1}{2} E^e (f - jg). \tag{6}$$

Now consider the integral in Eq. (1) for the general case where the fields are of different modes,

$$\tilde{E}_{iy}{}^r(\beta_\rho, \beta_r) = e_{iy}{}^e(\beta_r, \beta_\rho) e^{j(\omega t - \beta_r z)},$$

$$\tilde{E}_{iy}{}^r(\beta_\rho', \beta_r') = e_{iy}{}^e(\beta_r', \beta_\rho') e^{j(\omega t - \beta_r' z)};$$

$$P = -\tfrac{1}{2}\int_{-\infty}^{\infty} \mathrm{Re}\left\{ \tilde{E}_y{}^r(\beta_r, \beta_\rho) \left[\frac{-j\beta_r'}{j\omega\mu} \tilde{E}_y{}^{r*}(\beta_\rho', \beta_r') \right] \right\} dx$$

$$= \tfrac{1}{2}\int_{-\delta}^{\delta} \mathrm{Re}\left[\frac{\beta_r'}{\omega\mu_1} \tilde{E}_{1y}{}^r(\beta_r, \beta_\rho) E_{1y}{}^{r*}(\beta_r', \beta_\rho') \right] dx$$

$$+ \tfrac{1}{2}\left(\int_{-\infty}^{-\delta} + \int_{\delta}^{\infty} \right) \mathrm{Re}\left[\tilde{E}_{2y}{}^r(\beta_r, \beta_\rho) \frac{\beta_r'}{\omega\mu_2} \tilde{E}_{2y}{}^{r*}(\beta_r', \beta_\rho') \right] dx, \tag{7}$$

$$P = \frac{1}{2}\frac{\beta_r'}{\omega}\left\{ \int_{-\delta}^{\delta} \frac{1}{\mu_1} \mathrm{Re}\left[e_{1y}{}^e(\beta_r, \beta_\rho) e^{j(\omega t - \beta_r z)} \right.\right.$$

$$\left. \times e_{1y}{}^{e*}(\beta_r') e^{-j(\omega t - \beta_r' z)} \right] dx$$

$$+ \frac{1}{\mu_2}\left(\int_{-\infty}^{-\delta} + \int_{\delta}^{\infty} \right) \mathrm{Re}\left[e_{2y}{}^e(\beta_r, \beta_\rho) e^{j(\omega t - \beta_r z)} \right.$$

$$\left.\left. \times e_{2y}{}^{e*}(\beta_r', \beta_\rho') e^{-j(\omega t - \beta_r' z)} \right] dx \right\}$$

$$= \frac{\beta_r'}{\omega} \cos(\beta_r' - \beta_r) z \left[\frac{1}{\mu_1} \int_0^{\delta} e_{1y}{}^e(\beta_r) e_{1y}{}^{e*}(\beta_r') \, dx \right.$$

$$\left. + \frac{1}{\mu_2} \int_{\delta}^{\infty} e_{2y}{}^e(\beta_r, \beta_\rho) e_{2y}{}^{e*}(\beta_r', \beta_\rho') \, dx \right]$$

$$= \frac{\beta_r'}{\omega} \cos(\beta_r' - \beta_r) z \left(\frac{1}{\mu_1} I_1^e + \frac{1}{\mu_2} I_2^e \right); \tag{8}$$

$$I_1{}^{\mathrm{e}} \equiv \int_0^\delta e_{1y}{}^{\mathrm{e}}(\beta_{\mathrm{r}}) e_{1y}{}^{\mathrm{e}*}(\beta_{\mathrm{r}}')\,dx$$

$$= \int_0^\delta |E^{\mathrm{e}}|^2 \cos\beta_1 x \cos\beta_1' x\,dx$$

$$(\beta_1{}^2 = \omega^2\mu_1\epsilon_1 - \beta_{\mathrm{r}}{}^2,\ \beta_1'^2 = \omega^2\mu_1\epsilon_1 - \beta_{\mathrm{r}}'^2)$$

$$= \tfrac{1}{2}|E^{\mathrm{e}}|^2 \int_0^\delta [\cos(\beta_1+\beta_1')x + \cos(\beta_1-\beta_1')x]\,dx$$

$$= \tfrac{1}{2}|E^{\mathrm{e}}|^2 \left[\frac{\sin(\beta_1+\beta_1')x}{\beta_1+\beta_1'} + \frac{\sin(\beta_1-\beta_1')x}{\beta_1-\beta_1'} \right]_0^\delta. \tag{9}$$

For $\beta_1 \neq \beta_1'$,

$$I_1{}^{\mathrm{e}} = \tfrac{1}{2}|E^{\mathrm{e}}|^2 \left[\frac{\sin(\beta_1+\beta_1')\delta}{\beta_1+\beta_1'} + \frac{\sin(\beta_1-\beta_1')\delta}{\beta_1-\beta_1'} \right]$$

$$= \frac{|E^{\mathrm{e}}|^2}{2(\beta_1{}^2-\beta_1'^2)} [(\beta_1-\beta_1')(\sin\beta_1\delta\cos\beta_1'\delta + \sin\beta_1'\delta\cos\beta_1\delta)$$

$$+ (\beta_1+\beta_1')(\sin\beta_1\delta\cos\beta_1'\delta - \sin\beta_1'\delta\cos\beta_1\delta)]$$

$$= \frac{|E^{\mathrm{e}}|^2}{(\beta_1{}^2-\beta_1'^2)} (\beta_1\sin\beta_1\delta\cos\beta_1'\delta - \beta_1'\sin\beta_1'\delta\cos\beta_1\delta). \tag{10}$$

For $\beta_1 = \beta_1'$,

$$I_1{}^{\mathrm{e}} = \tfrac{1}{2}|E^{\mathrm{e}}|^2 \left\{ \left[\frac{\sin(\beta_1+\beta_1')x}{\beta_1+\beta_1'} + \frac{\sin(\beta_1-\beta_1')x}{\beta_1-\beta_1'} \right]_{\beta_1\to\beta_1'} \right\}\Bigg|_0^\delta$$

$$= \tfrac{1}{2}|E^{\mathrm{e}}|^2 \left[\frac{\sin 2\beta_1 x}{2\beta_1} + \frac{(\beta_1-\beta_1')x}{\beta_1-\beta_1'}\bigg|_{\beta_1\to\beta_1'} \right]_0^\delta$$

$$= \tfrac{1}{2}|E^{\mathrm{e}}|^2 \left(\delta + \frac{\sin 2\beta_1\delta}{2\beta_1} \right). \tag{11}$$

The second integral in Eq. (8) is

$$I_2{}^{\mathrm{e}} \equiv \int_\delta^\infty e_{2y}{}^{\mathrm{e}}(\beta_{\mathrm{r}},\beta_\rho) e_{2y}{}^{\mathrm{e}*}(\beta_{\mathrm{r}}',\beta_\rho')\,dx$$

$$= \int_\delta^\infty (E^+ e^{-j\beta_\rho(x-\delta)} + E^- e^{j\beta_\rho(x-\delta)})$$

$$\times (E'^+ e^{-j\beta_\rho'(x-\delta)} + E'^- e^{j\beta_\rho'(x-\delta)})^*\,dx$$

$$= \tfrac{1}{4}|E^{\mathrm{e}}|^2 \int_0^\infty [(f+jg)(f'-jg')e^{j(\beta_\rho'-\beta_\rho)s}$$

$$+ (f+jg)(f'-jg')e^{j(\beta_\rho-\beta_\rho')s}$$

$$+(f-jg)(f'-jg')e^{-j(\beta_\rho+\beta_\rho')s}$$
$$+(f+jg)(f'+jg')e^{j(\beta_\rho+\beta_\rho')s}\Big]\,ds \qquad (s\equiv x-\delta) \tag{12}$$
$$=\tfrac{1}{4}|E^{\mathrm{e}}|^2\int_0^\infty\{[(ff'+gg')+j(fg'-f'g)]e^{j(\beta_\rho'-\beta_\rho)s}$$
$$+[(ff'+gg')-j(fg'-f'g)]e^{-j(\beta_\rho'-\beta_\rho)s}$$
$$+[(ff'-gg')-j(fg'+f'g)]e^{-j(\beta_\rho+\beta_\rho')s}$$
$$+[(ff'-gg')+j(fg'+f'g)]e^{j(\beta_\rho+\beta_\rho')s}\}\,ds$$
$$=\tfrac{1}{2}|E^{\mathrm{e}}|^2\int_0^\infty\{[(ff'+gg')\cos(\beta_\rho'-\beta_\rho)s-(fg'-f'g)\sin(\beta_\rho'-\beta_\rho)s$$
$$+(ff'-gg')\cos(\beta_\rho+\beta_\rho')s-(fg'+f'g)\sin(\beta_\rho+\beta_\rho')s]\}\,ds$$
$$=\tfrac{1}{2}|E^{\mathrm{e}}|^2\Big(\int_0^\infty\{ff'[\cos(\beta_\rho'-\beta_\rho)s+\cos(\beta_\rho'+\beta_\rho)s]$$
$$+gg'[\cos(\beta_\rho'-\beta_\rho)s-\cos(\beta_\rho'+\beta_\rho')s]\}\,ds+2I_A{}^{\mathrm{e}}\Big)$$
$$=\tfrac{1}{2}|E^{\mathrm{e}}|^2\left[\int_0^\infty(2ff'\cos\beta_\rho's\cos\beta_\rho s+2gg'\sin\beta_\rho's\sin\beta_\rho s)\,ds+2I_A{}^{\mathrm{e}}\right], \tag{13}$$

where f' and g' are defined as in Eq. (6), that is, by

$$f'\equiv\cos\beta'_1\delta, \qquad g'\equiv\frac{\beta_1'\mu_2}{\beta_\rho'\mu_1}\sin\beta'_1\delta, \tag{14}$$

and

$$2I_A{}^{\mathrm{e}}\equiv-\int_0^\infty[(fg'-f'g)\sin(\beta_\rho'-\beta_\rho)s$$
$$+(fg'+f'g)\sin(\beta_\rho'+\beta_\rho)s]\,ds. \tag{15}$$

Relations (A.2.10) and (A.2.11),

$$\int_0^\infty\cos\omega x\cos\Omega x\,dx=\frac{\pi}{2}\delta(\omega-\Omega),$$
$$\int_0^\infty\sin\omega x\sin\Omega x\,dx=\frac{\pi}{2}\delta(\omega-\Omega), \tag{16}$$

will be used to evaluate the first two integrals in Eq. (13). [The behavior of the delta function $\delta(\omega)$ is summarized in Appendix 2.] Thus

$$I_2{}^{\mathrm{e}}=|E^{\mathrm{e}}|^2[(\pi/2)(ff'+gg')\delta(\beta_\rho-\beta_\rho')+I_A{}^{\mathrm{e}}]. \tag{17}$$

Evaluation of the integral $I_A{}^{e}$ yields

$$I_A{}^{e} = -\tfrac{1}{2}(fg'+f'g)\int_0^{\infty}\sin(\beta_\rho'+\beta_\rho)s\,ds$$

$$-\tfrac{1}{2}(fg'-f'g)\int_0^{\infty}\sin(\beta_\rho'-\beta_\rho)s\,ds$$

$$=\frac{fg'+f'g}{2(\beta_\rho'+\beta_\rho)}\left[\cos(\beta_\rho'+\beta_\rho)s\right]_0^{\infty}$$

$$+\frac{fg'-f'g}{2(\beta_\rho'-\beta_\rho)}\left[\cos(\beta_\rho'-\beta_\rho)s\right]_0^{\infty}$$

$$=\frac{fg'+f'g}{2(\beta_\rho'+\beta_\rho)}\cos(\beta_\rho'+\beta_\rho)\infty-\frac{f'g-fg'}{2(\beta_\rho'-\beta_\rho)}\cos(\beta_\rho'-\beta_\rho)\infty$$

$$+\frac{\beta_\rho f'g-\beta_\rho' fg'}{\beta_\rho'^2-\beta_\rho{}^2}. \tag{18}$$

Substitution of Eq. (18) into Eq. (17) yields

$$I_2{}^{e}=\frac{|E^{e}|^2}{2}\Bigg[\pi(ff'+gg')\delta(\beta_\rho-\beta_\rho')+2\frac{\beta_\rho f'g-\beta_\rho' fg'}{\beta_\rho'^2-\beta_\rho{}^2}$$

$$+\frac{fg'+f'g}{\beta_\rho'+\beta_\rho}\cos(\beta_\rho'+\beta_\rho)\infty-\frac{f'g-fg'}{\beta_\rho'-\beta_\rho}\cos(\beta_\rho'-\beta_\rho)\infty\Bigg]. \tag{19}$$

Recall from Appendix 2 that the delta function is defined as

$$\delta(s)=0 \qquad \text{for} \quad s\neq 0,$$

$$\delta(s)=\infty \qquad \text{for} \quad s=0;$$

$$\int_{-\infty}^{\infty}\delta(s)\,ds=1. \tag{20}$$

The delta function thus has a finite value only when it appears as an integrand. This in turn implies that the power expression, Eq. (19), is meaningful only when it appears under an integral. The last two terms in Eq. (19) oscillate at an infinite frequency and contribute nothing to an integral. Therefore,

$$I_2{}^{e}=\frac{|E^{e}|^2}{2}\left[\pi(ff'+gg')\delta(\beta_\rho-\beta_\rho')+2\frac{\beta_\rho f'g-\beta_\rho' fg'}{\beta_\rho'^2-\beta_\rho{}^2}\right]. \tag{21}$$

The second term in Eq. (21) may be expanded to give

$$\beta_\rho f'g-\beta_\rho' fg'=\beta_\rho'\cos\beta_1'\delta\frac{\beta_1\mu_2}{\beta_\rho\mu_1}\sin\beta_1\delta-\beta_\rho'\cos\beta_1\delta\frac{\beta_1'\mu_2}{\beta_\rho'\mu_1}\sin\beta_1'\delta$$

$$=\frac{\mu_2}{\mu_1}(\beta_1\cos\beta_1'\delta\sin\beta_1\delta-\beta_1'\cos\beta_1\delta\sin\beta_1'\delta), \tag{22}$$

$$P^{\mathrm{e}}(\beta_\rho \neq \beta_\rho') = \frac{\beta_{\mathrm{r}}'}{\omega} \cos(\beta_{\mathrm{r}}' - \beta_{\mathrm{r}})z \times \left[\frac{1}{\mu_1} I_1{}^{\mathrm{e}}(\beta_\rho \neq \beta_\rho') + \frac{1}{\mu_2} I_2{}^{\mathrm{e}}(\beta_\rho \neq \beta_\rho') \right]$$

$$= \frac{\beta_{\mathrm{r}}'}{\omega} \cos(\beta_{\mathrm{r}}' - \beta_{\mathrm{r}})z \left[\frac{|E^{\mathrm{e}}|^2}{\mu_1(\beta_1{}^2 - \beta_1{}'^2)} \times (\beta_1 \sin\beta_1\delta \cos\beta_1'\delta - \beta_1' \sin\beta_1'\delta \cos\beta_1\delta) + \frac{|E^{\mathrm{e}}|^2}{\mu_2} \frac{\mu_2}{\mu_1} \frac{1}{\beta_\rho'^2 - \beta_\rho{}^2} \times (\beta_1 \cos\beta_1'\delta \sin\beta_1\delta - \beta_1' \cos\beta_1\delta \sin\beta_1'\delta) \right]$$

$$= 0. \tag{23}$$

Note that

$$\begin{aligned} \beta_1{}^2 - \beta_1{}'^2 &= (\omega^2\mu_1\epsilon_1 - \beta_{\mathrm{r}}{}^2) - (\omega^2\mu_1\epsilon_1 - \beta_{\mathrm{r}}{}'^2) = \beta_{\mathrm{r}}{}'^2 - \beta_{\mathrm{r}}{}^2, \\ \beta_\rho{}'^2 - \beta_\rho{}^2 &= (\omega^2\mu_2\epsilon_2 - \beta_{\mathrm{r}}{}'^2) - (\omega^2\mu_2\epsilon_2 - \beta_{\mathrm{r}}{}^2) = \beta_{\mathrm{r}}{}^2 - \beta_{\mathrm{r}}{}'^2, \end{aligned} \tag{24}$$

$$P^{\mathrm{e}}(\beta_\rho = \beta_\rho') = \frac{\beta_{\mathrm{r}}}{\omega} \left[\frac{1}{\mu_1} \frac{|E^{\mathrm{e}}|^2}{2} \left(\delta + \frac{\sin 2\beta_1\delta}{2\beta_1} \right) + \frac{|E^{\mathrm{e}}|^2}{2\mu_2} \left[\pi(ff' + gg')\delta(\beta_\rho - \beta_\rho') + I_A{}^{\mathrm{e}}(\beta_\rho = \beta_\rho') \right] \right], \tag{25}$$

$$I_A{}^{\mathrm{e}}(\beta_\rho = \beta_\rho') = -\tfrac{1}{2} \int_0^\infty 2fg \sin 2\beta_\rho s \, ds = \frac{fg}{2\beta_\rho}(\cos 2\beta_\rho \infty - 1)$$

$$= \frac{-fg}{2\beta_\rho}, \tag{26}$$

$$P^{\mathrm{e}}(\beta_\rho = \beta_\rho') = \frac{\beta_{\mathrm{r}}|E^{\mathrm{e}}|^2}{2\omega\mu_2} \left[\pi(f^2 + g^2)\delta(\beta_\rho - \beta_\rho') - \frac{fg}{2\beta_\rho} + \frac{\mu_2}{\mu_1}\delta + \frac{\mu_2}{\mu_1} \frac{\sin 2\beta_1\delta}{2\beta_1} \right]$$

$$= \frac{\pi\beta_{\mathrm{r}}|E^{\mathrm{e}}|^2}{2\omega\mu_2} \left[\cos^2\beta_1\delta + \left(\frac{\beta_1\mu_2}{\beta_\rho\mu_1} \sin\beta_1\delta \right)^2 \right] \delta(\beta_\rho - \beta_\rho'). \tag{27}$$

The last three terms are negligible compared to the delta function. Relations (23) and (27) show that TE even radiation modes are orthogonal to each other.

The magnitude $|E^{\mathrm{e}}|$ may be expressed in terms of the power P^{e}:

$$|E^{\mathrm{e}}|=\left\{\frac{2\omega\mu_2 P^{\mathrm{e}}}{\pi\beta_{\mathrm{r}}}\left[\cos^2\beta_1\delta+\left(\frac{\beta_1\mu_2}{\beta_\rho\mu_1}\sin\beta_1\delta\right)^2\right]^{-1}\right\}^{1/2}. \tag{28}$$

TE odd modes. The fields of TE odd radiation modes in a lossless system are given by Eqs. (2.5.1) and (2.5.21)–(2.5.27).

$$\tilde{E}_{iy}{}^{\mathrm{r}}=e_{iy}{}^{\mathrm{o}}e^{j(\omega t-\beta_{\mathrm{r}}z)}, \qquad i=1 \text{ or } 2, \tag{29}$$

$$e_{1y}{}^{\mathrm{o}}(x)=E^{\mathrm{o}}\sin\beta_1 x, \qquad |x|<\delta, \tag{30}$$

$$e_{2y}{}^{\mathrm{o}}(x)=E_{\mathrm{o}}{}^{+}e^{-j\beta_\rho(x-\delta)}+E_{\mathrm{o}}{}^{-}e^{j\beta_\rho(x-\delta)}, \qquad x>\delta, \tag{31}$$

where

$$\beta_\rho=(\omega^2\mu_2\epsilon_2{}'-\beta_{\mathrm{r}}{}^2), \tag{32}$$

$$E_{\mathrm{o}}{}^{+}=E_{\mathrm{o}}{}^{-*}=\tfrac{1}{2}E^{\mathrm{o}}(f_{\mathrm{o}}+jg_{\mathrm{o}}), \qquad f_{\mathrm{o}}\equiv\sin\beta_1\delta, \quad g_{\mathrm{o}}\equiv\frac{\beta_1\mu_2}{\beta_\rho\mu_1}\cos\beta_1\delta. \tag{33}$$

Evaluation of Eq. (8) for the TE odd modes yields

$$\begin{aligned}
I_1{}^{\mathrm{o}}&\equiv\int_0^\delta e_{1y}{}^{\mathrm{o}}(\beta_{\mathrm{r}})e_{1y}{}^{\mathrm{o}*}(\beta_{\mathrm{r}}')\,dx\\
&=\int_0^\delta |E^{\mathrm{o}}|^2\sin\beta_1 x\sin\beta_1' x\,dx\\
&=\frac{|E^{\mathrm{o}}|^2}{2}\int_0^\delta[\cos(\beta_1-\beta_1')x-\cos(\beta_1+\beta_1')x]\,dx\\
&=\frac{|E^{\mathrm{o}}|^2}{2}\left[\frac{\sin(\beta_1-\beta_1')x}{\beta_1-\beta_1'}-\frac{\sin(\beta_1+\beta_1')x}{\beta_1+\beta_1'}\right]_0^\delta;
\end{aligned}$$

$$\begin{aligned}
I_1{}^{\mathrm{o}}(\beta_1\neq\beta_1')&=\frac{|E^{\mathrm{o}}|^2}{2}\left[\frac{\sin(\beta_1-\beta_1')\delta}{\beta_1-\beta_1'}-\frac{\sin(\beta_1+\beta_1')\delta}{\beta_1+\beta_1'}\right]\\
&=\frac{|E^{\mathrm{o}}|^2}{\beta_1{}^2-\beta_1'^2}(\beta'_1\sin\beta_1\delta\cos\beta_1'\delta-\beta_1\sin\beta_1'\delta\cos\beta_1\delta),
\end{aligned} \tag{34}$$

$$I_1{}^{\mathrm{o}}(\beta_1=\beta_1')=\frac{|E^{\mathrm{o}}|^2}{2}\left(\delta-\frac{\sin 2\beta_1\delta}{2\beta_1}\right); \tag{35}$$

$$\begin{aligned}
I_2{}^{\circ} &\equiv \int_{\delta}^{\infty} e_{2y}{}^{\circ}(\beta_r, \beta_\rho) e_{2y}{}^{\circ *}(\beta_r', \beta_\rho')\, dx \qquad (s \equiv x-\delta) \\
&= \int_0^{\infty} (E_o{}^{+} e^{-j\beta_\rho s} + E_o{}^{-} e^{j\beta_\rho s})(E_o'{}^{+} e^{-j\beta_\rho' s} + E_o'{}^{-} e^{j\beta_\rho' s})^*\, ds \\
&= \frac{|E^{\circ}|^2}{4} \int_0^{\infty} \big[(f_o + jg_o)(f_o' - jg_o') e^{-j(\beta_\rho - \beta_\rho')s} \\
&\qquad + (f_o - jg_o)(f_o' + jg_o') e^{j(\beta_\rho - \beta_\rho')s} \\
&\qquad + (f_o + jg_o)(f_o' + jg_o') e^{-j(\beta_\rho + \beta'_\rho)s} \\
&\qquad + (f_o - jg_o)(f_o' - jg_o') e^{j(\beta_\rho + \beta_\rho')s} \big]\, ds \\
&= |E^{\circ}|^2 \Big[\int_0^{\infty} (f_o f_o' \cos\beta_\rho s \cos\beta_\rho' s \\
&\qquad + g_o g_o' \sin\beta_\rho s \sin\beta_\rho' s)\, ds + I_A{}^{\circ} \Big],
\end{aligned} \tag{36}$$

where f_o' and g_o' are defined as Eq. (33), that is,

$$f_o' = \sin\beta_1'\delta, \qquad g_o' = \frac{\beta_1'\mu_2}{\beta_\rho'\mu_1} \cos\beta_1'\delta, \tag{37}$$

and

$$\begin{aligned}
I_A{}^{\circ} &\equiv \tfrac{1}{2} \int_0^{\infty} \big[(f_o g_o' - f_o' g_o) \sin(\beta_\rho' - \beta_\rho)s \\
&\qquad + (f_o g_o' + f_o' g_o) \sin(\beta_\rho' + \beta_\rho)s \big]\, ds \\
&= \frac{1}{2} \bigg(\frac{f_o g_o' + f_o' g}{\beta_\rho' + \beta_\rho} [1 - \cos(\beta_\rho' + \beta_\rho)\infty] \\
&\qquad + \frac{f_o g_o' - f_o' g_o}{\beta_\rho' - \beta_\rho} [1 - \cos(\beta_\rho' - \beta_\rho)\infty] \bigg) \\
&= \frac{\beta_\rho' f_o g_o' - \beta_\rho f_o' g_o}{\beta_\rho'^2 - \beta_\rho{}^2} \\
&= \frac{1}{\beta_\rho'^2 - \beta_\rho{}^2} \frac{\mu_2}{\mu_1} [\beta_1' \sin\beta_1\delta \cos\beta_1'\delta \\
&\qquad - \beta_1 \sin\beta_1'\delta \cos\beta_1\delta] \qquad (\beta_\rho \neq \beta_\rho').
\end{aligned} \tag{38}$$

The terms with $\cos[(\beta_\rho' \pm \beta_\rho)\infty]$ have been dropped, as explained for the TE even radiation modes.

Evaluation of the integral $I_2{}^{\circ}$ by using Eqs. (16) and (38) yields

$$I_2{}^{\circ}=|E^{\circ}|^2\Bigg[(\pi/2)(f_{\mathrm{o}}f_{\mathrm{o}}'+g_{\mathrm{o}}g_{\mathrm{o}}')\delta(\beta_\rho-\beta_\rho')+\frac{\mu_2}{\mu_1}\frac{1}{\beta_\rho'^2-\beta_\rho{}^2}$$

$$\times(\beta_1'\sin\beta_1\delta\cos\beta_1'\delta-\beta_1\sin\beta_1'\delta\cos\beta_1\delta)\Bigg]. \quad (39)$$

The power P° for the TE odd radiation modes may be obtained from Eq. (8) with the use of Eqs. (34) or (35) and Eq. (39).

$$P^{\circ}(\beta_\rho\neq\beta_\rho')=\frac{\beta_{\mathrm{r}}'}{\omega}\cos(\beta_{\mathrm{r}}'-\beta_{\mathrm{r}})z\left(\frac{1}{\mu_1}I_1{}^{\circ}+\frac{1}{\mu_2}I_2{}^{\circ}\right)$$

$$=|E^{\circ}|^2\frac{\beta_{\mathrm{r}}'}{\omega}\cos(\beta_{\mathrm{r}}'-\beta_{\mathrm{r}})z$$

$$\times\Bigg[\frac{1}{\mu_1}\frac{\beta'_1\sin\beta_1\delta\cos\beta_1'\delta-\beta_1\sin\beta_1'\delta\cos\beta_1\delta}{\beta_1{}^2-\beta_1'^2}$$

$$+\frac{1}{\mu_2}\frac{\pi}{2}(f_{\mathrm{o}}f_{\mathrm{o}}'+g_{\mathrm{o}}g_{\mathrm{o}}')\delta(\beta_\rho-\beta_\rho')+\frac{1}{\mu_1}\frac{1}{\beta_\rho'^2-\beta_\rho{}^2}$$

$$\times(\beta_1'\sin\beta_1\delta\cos\beta'_1\delta-\beta_1\sin\beta_1'\delta\cos\beta_1\delta)\Bigg]$$

$$=0. \quad (40)$$

For the case where $\beta_\rho=\beta_\rho'$, the integral $I_A{}^{\circ}$ is

$$I_A{}^{\circ}(\beta_\rho=\beta_\rho')=f_{\mathrm{o}}g_{\mathrm{o}}\int_0^{\infty}\sin 2\beta_\rho s\,ds=\frac{f_{\mathrm{o}}g_{\mathrm{o}}}{2\beta_\rho}(1-\cos 2\beta_\rho\infty)$$

$$=\frac{f_{\mathrm{o}}g_{\mathrm{o}}}{2\beta_\rho} \quad (41)$$

and

$$I_2{}^{\circ}(\beta_\rho=\beta_\rho')=|E^{\circ}|^2\left[\frac{\pi}{2}(f_{\mathrm{o}}{}^2+g_{\mathrm{o}}{}^2)\delta(\beta_\rho-\beta_\rho')+\frac{f_{\mathrm{o}}g_{\mathrm{o}}}{2\beta_\rho}\right]. \quad (42)$$

The power P° is then given by

$$P^{\circ}(\beta_\rho=\beta_\rho')=\frac{|E^{\circ}|^2\beta_{\mathrm{r}}}{\omega}\Bigg\{\frac{1}{2\mu_1}\left(\delta-\frac{\sin 2\beta_1\delta}{2\beta_1}\right)$$

$$+\frac{1}{\mu_2}\left[\frac{\pi}{2}(f_{\mathrm{o}}{}^2+g_{\mathrm{o}}{}^2)\delta(\beta_\rho-\beta_\rho')+\frac{f_{\mathrm{o}}g_{\mathrm{o}}}{2\beta_\rho}\right]\Bigg\}$$

$$=\frac{|E^{\circ}|^2\beta_{\mathrm{r}}\pi}{2\omega\mu_2}\left[\sin^2\beta_1\delta+\left(\frac{\beta_1\mu_2}{\beta_\rho\mu_1}\cos\beta_1\delta\right)^2\right]\delta(\beta_\rho-\beta_\rho'). \quad (43)$$

The magnitude $|E^{\circ}|$ can be expressed in terms of the power P° as

$$|E^{\circ}|=\left\{\frac{2\omega\mu_2 P^{\circ}}{\pi\beta_r}\left[\sin^2\beta_1\delta+\left(\frac{\beta_1\mu_2}{\beta_\rho\mu_1}\cos\beta_1\delta\right)^2\right]^{-1}\right\}^{1/2}. \tag{44}$$

2.6. TM Radiation Modes

The analysis of TM radiation modes closely follows that for TE radiation modes. The radiation fields are represented by

$$\tilde{\mathbf{H}}=\mathbf{h}(x)e^{j\omega t-\gamma_r z}, \qquad \tilde{\mathbf{E}}^{r}=\mathbf{e}(x)e^{j\omega t-\gamma_r z}. \tag{1}$$

It has assumed that the fields are invariant with respect to y. The function $\mathbf{h}(x)$ will be determined from the wave equation, instead of $\mathbf{e}(x)$ as in the case of TE modes.

The TM fields external to the dielectric sheet (Region II) satisfy the wave equation (2.3.8).

$$\frac{d^2h_{2y}}{dx^2}-\rho_2{}^2h_{2y}=0, \tag{2}$$

$$\rho_2{}^2\equiv\gamma_2{}^2-\gamma_r{}^2 \qquad \text{and} \qquad \gamma_2{}^2=-\omega^2\mu_2\varepsilon_2. \tag{3}$$

The solution is

$$\begin{aligned} h_{2y}(x)&=H^{+}e^{-\rho_2(x-\delta)}+H^{-}e^{\rho_2(x-\delta)}, \qquad x>\delta,\\ &=H^{+}e^{\rho_2(x+\delta)}+H^{-}e^{-\rho_2(x+\delta)}, \qquad x<-\delta, \end{aligned} \tag{4}$$

where

$$\rho_2=\sqrt{-\omega^2\mu_2\varepsilon_2-(\alpha_r+j\beta_r)^2}\equiv\alpha_\rho+j\beta_\rho. \tag{5}$$

The expressions for α_ρ and β_ρ are given in Section 2.5. The accompanying E field is given by Eqs. (2.1.7),

$$\begin{aligned} e_{2x}(x)&=\frac{\gamma_r}{j\omega\varepsilon_2}h_{2y}(x)\\ &=\frac{\gamma_r}{j\omega\varepsilon_2}(H^{+}e^{-\rho_2(x-\delta)}+H^{-}e^{\rho_2(x-\delta)}), \qquad x>\delta,\\ &=\frac{\gamma_r}{j\omega\varepsilon_2}(H^{+}e^{\rho_2(x+\delta)}+H^{-}e^{-\rho_2(x+\delta)}), \qquad x<-\delta; \end{aligned} \tag{6}$$

$$\begin{aligned} e_{2z}(x)&=\frac{1}{j\omega\varepsilon_2}\frac{dh_{2y}(x)}{dx}\\ &=\frac{-\rho_2}{j\omega\varepsilon_2}(H^{+}e^{-\rho_2(x-\delta)}-H^{-}e^{\rho_2(x-\delta)}), \qquad x>\delta,\\ &=\frac{\rho_2}{j\omega\varepsilon_2}(H^{+}e^{\rho_2(x+\delta)}-H^{-}e^{-\rho_2(x+\delta)}), \qquad x<-\delta. \end{aligned} \tag{7}$$

The coefficients H^+ and H^- can be determined from the boundary conditions at the interface. The even and odd modes will be treated separately.

TM even modes. The TM even modes within the dielectric sheet (region I) are given by Eqs. (2.3.5) and (2.3.6),

$$h_{1y}{}^{e}(x)=H^{e}\cos k_1 x, \tag{8}$$

$$e_{1x}{}^{e}(x)=\frac{\gamma_r}{j\omega\varepsilon_1}H^{e}\cos k_1 x, \tag{9}$$

$$e_{1z}{}^{e}(x)=\frac{-k_1}{j\omega\varepsilon_1}H^{e}\sin k_1 x, \tag{10}$$

$$k_1{}^2\equiv\gamma_r{}^2-\gamma_1{}^2 \quad \text{and} \quad \gamma_1{}^2=-\omega^2\mu_1\varepsilon_1. \tag{11}$$

The continuity of tangential field components at the interface yields the following relations:

$$h_{1y}{}^{e}(\delta)=h_{2y}{}^{e}(\delta), \qquad H^{e}\cos k_1\delta=H^{+}+H^{-}, \tag{12}$$

$$e_{1z}{}^{e}(\delta)=e_{2z}{}^{e}(\delta), \qquad H^{e}\frac{k_1\varepsilon_2}{\rho_2\varepsilon_1}\sin k_1\delta=H^{+}-H^{-}. \tag{13}$$

Therefore,

$$\begin{aligned} H^{+}&=\tfrac{1}{2}H^{e}\left(\cos k_1\delta+\frac{k_1\varepsilon_2}{\rho_2\varepsilon_1}\sin k_1\delta\right),\\ H^{-}&=\tfrac{1}{2}H^{e}\left(\cos k_1\delta-\frac{k_1\varepsilon_2}{\rho_2\varepsilon_1}\sin k_1\delta\right). \end{aligned} \tag{14}$$

For lossless media, $k_1=\beta_1$ and $\rho_2=j\beta_\rho$; Eq. (14) then becomes

$$\begin{aligned} H^{+}&=\tfrac{1}{2}H^{e}\left(\cos\beta_1\delta-j\frac{\epsilon_2\beta_1}{\epsilon_1\beta_\rho}\sin\beta_1\delta\right)\equiv\tfrac{1}{2}H^{e}(q-jr)\equiv H,\\ H^{-}&\equiv\tfrac{1}{2}H^{e}(q+jr)=H^{+*}\equiv H^{*}, \end{aligned} \tag{15}$$

$$q\equiv\cos\beta_1\delta \quad \text{and} \quad r\equiv\frac{\epsilon_2\beta_1}{\epsilon_1\beta_\rho}\sin\beta_1\delta. \tag{16}$$

The TM fields for the lossless case are

Region I:

$$h_{1y}{}^{e}(x)=H^{e}\cos\beta_1 x, \tag{17}$$

$$e_{1x}{}^{e}(x)=\frac{\beta_r}{\omega\epsilon_1}H^{e}\cos\beta_1 x, \tag{18}$$

$$e_{1z}{}^{e}(x)=\frac{-\beta_1}{j\omega\epsilon_1}H^{e}\sin\beta_1 x. \tag{19}$$

Region II:

$$h_{2y}{}^{e}(x)=He^{-j\beta_\rho(x-\delta)}+H^*e^{j\beta_\rho(x-\delta)}, \qquad x>\delta,$$
$$=He^{j\beta_\rho(x+\delta)}+H^*e^{-j\beta_\rho(x+\delta)}, \qquad x<-\delta; \tag{20}$$

$$e_{2x}{}^{e}(x)=\frac{\beta_r}{\omega\epsilon_2}\left(He^{-j\beta_\rho(x-\delta)}+H^*e^{j\beta_\rho(x-\delta)}\right), \qquad x>\delta,$$
$$=\frac{\beta_r}{\omega\epsilon_2}\left(He^{j\beta_\rho(x+\delta)}+H^*e^{-j\beta_\rho(x+\delta)}\right), \qquad x<-\delta; \tag{21}$$

$$e_{2z}{}^{e}(x)=\frac{-\beta_\rho}{\omega\epsilon_2}\left(He^{-j\beta_\rho(x-\delta)}-H^*e^{j\beta_\rho(x-\delta)}\right), \qquad x>\delta,$$
$$=\frac{\beta_\rho}{\omega\epsilon_2}\left(He^{j\beta_\rho(x+\delta)}-H^*e^{-j\beta_\rho(x+\delta)}\right), \qquad x<-\delta. \tag{22}$$

TM odd modes. The field of TM odd modes within the dielectric sheet are given by Eq. (2.3.7):

$$h_{1y}{}^{o}(x)=H^{o}\sin k_1x, \tag{23}$$

$$e_{1x}{}^{o}(x)=\frac{\gamma_r}{j\omega\varepsilon_1}H^{o}\sin k_1x, \tag{24}$$

$$e_{1z}{}^{o}(x)=\frac{k_1}{j\omega\varepsilon_1}H^{o}\cos k_1x. \tag{25}$$

The boundary conditions on the tangential field components yield the following relations:

$$h_{1y}{}^{o}(\delta)=h_{2y}{}^{o}(\delta), \qquad H^{o}\sin k_1\delta=H^{+}+H^{-}, \tag{26}$$

$$e_{1z}{}^{o}(\delta)=e_{2z}{}^{o}(\delta), \qquad H^{o}\frac{-\varepsilon_2k_1}{\varepsilon_1\rho_2}\cos k_1\delta=H^{+}-H^{-}. \tag{27}$$

Therefore,

$$H^{+}=\tfrac{1}{2}H^{o}\left(\sin k_1\delta-\frac{\varepsilon_2k_1}{\varepsilon_1\rho_2}\cos k_1\delta\right),$$
$$H^{-}=\tfrac{1}{2}H^{o}\left(\sin k_1\delta+\frac{\varepsilon_2k_1}{\varepsilon_1\rho_2}\cos k_1\delta\right). \tag{28}$$

For lossless media, $k_1=\beta_1$ and $\rho_2=j\beta_\rho$, define

$$\hat{H}\equiv H^{+}=\tfrac{1}{2}H^{o}(u+jv) \qquad \text{and} \qquad \hat{H}^*=H^{-}, \tag{29}$$

$$u\equiv\sin\beta_1\delta \qquad \text{and} \qquad v\equiv\frac{\epsilon_2\beta_1}{\epsilon_1\beta_\rho}\cos\beta_1\delta. \tag{30}$$

The fields for the lossless case are as follows.

Region I:

$$h_{1y}{}^{\mathrm{o}}(x)=H^{\mathrm{o}}\sin\beta_1 x, \tag{31}$$

$$e_{1x}{}^{\mathrm{o}}=\frac{\beta_{\mathrm{r}}}{\omega\epsilon_1}H^{\mathrm{o}}\sin\beta_1 x, \tag{32}$$

$$e_{1z}{}^{\mathrm{o}}=\frac{\beta_1}{j\omega\epsilon_1}H^{\mathrm{o}}\cos\beta_1 x. \tag{33}$$

Region II:

$$\begin{aligned} h_{2y}{}^{\mathrm{o}}(x)&=\hat{H}e^{-j\beta_\rho(x-\delta)}+\hat{H}^*e^{j\beta_\rho(x-\delta)}, && x>\delta,\\ &=\hat{H}e^{j\beta_\rho(x+\delta)}+\hat{H}^*e^{-j\beta_\rho(x+\delta)}, && x<-\delta; \end{aligned} \tag{34}$$

$$\begin{aligned} e_{2x}{}^{\mathrm{o}}(x)&=\frac{\beta_{\mathrm{r}}}{\omega\epsilon_2}\left[\hat{H}e^{-j\beta_\rho(x-\delta)}+\hat{H}^*e^{j\beta_\rho(x-\delta)}\right], && x>\delta,\\ &=\frac{\beta_{\mathrm{r}}}{\omega\epsilon_2}\left[\hat{H}e^{j\beta_\rho(x+\delta)}+\hat{H}^*e^{-j\beta_\rho(x+\delta)}\right], && x<-\delta; \end{aligned} \tag{35}$$

$$\begin{aligned} e_{2z}{}^{\mathrm{o}}(x)&=\frac{-\beta_\rho}{\omega\epsilon_2}\left[\hat{H}e^{-j\beta_\rho(x-\delta)}-\hat{H}^*e^{j\beta_\rho(x-\delta)}\right], && x>\delta,\\ &=\frac{\beta_\rho}{\omega\epsilon_2}\left[\hat{H}e^{j\beta_\rho(x+\delta)}-\hat{H}^*e^{-j\beta_\rho(x+\delta)}\right], && x<-\delta. \end{aligned} \tag{36}$$

The coefficients $\hat{H}$ and $\hat{H}^*$ are as in Eq. (29). The constants H^{o} and H^{e} can be expressed in terms of the power P^{o} and P^{e} carried by the odd and even modes, respectively. This will be done in the next section.

2.6a. Power of TM Radiation Modes

The power of TM radiation modes propagating along the z direction is given by

$$\begin{aligned} P&=\tfrac{1}{2}\int_S \mathrm{Re}(\tilde{\mathbf{E}}^{\mathrm{r}}\times\tilde{\mathbf{H}}^{\mathrm{r}*})\cdot\hat{\mathbf{n}}\,da\\ &=\tfrac{1}{2}\int_S \mathrm{Re}\left[(\hat{\mathbf{x}}\tilde{E}_x{}^{\mathrm{r}}+\hat{\mathbf{z}}\tilde{E}_z{}^{\mathrm{r}})\times\hat{\mathbf{y}}\tilde{H}_y{}^{\mathrm{r}*}\right]\cdot\hat{\mathbf{z}}\,da\\ &=\tfrac{1}{2}\int_S \mathrm{Re}\left(\tilde{E}_x{}^{\mathrm{r}}\tilde{H}_y{}^{\mathrm{r}*}\right)da. \end{aligned} \tag{1}$$

For the lossless case, the even and the odd modes will again be treated separately.

TM even modes. The fields of TM even modes are given by Eqs. (2.6.1) and (2.6.17)–(2.6.22):

$$\tilde{E}_{1x}{}^{\mathrm{r}} = \frac{\beta_{\mathrm{r}}}{\omega\epsilon_1} H^{\mathrm{e}} \cos\beta_1 x\, e^{j(\omega t-\beta_{\mathrm{r}} z)}, \tag{2}$$

$$\tilde{H}_{1y}{}^{\mathrm{r}} = H^{\mathrm{e}} \cos\beta_1 x\, e^{j(\omega t-\beta_{\mathrm{r}} z)}, \tag{3}$$

$$\tilde{E}_{2x}{}^{\mathrm{r}} = \frac{\beta}{\omega\epsilon_2}\left(He^{-j\beta_\rho(x-\delta)} + H^* e^{j\beta_\rho(x-\delta)}\right)e^{j(\omega t-\beta_{\mathrm{r}} z)}, \qquad x>\delta, \tag{4}$$

$$\tilde{H}_{2y}{}^{\mathrm{r}} = \left(He^{-j\beta_\rho(x-\delta)} + H^* e^{j\beta_\rho(x-\delta)}\right)e^{j(\omega t-\beta_{\mathrm{r}} z)}, \qquad x>\delta. \tag{5}$$

The expression for the power now becomes

$$\begin{aligned} P^{\mathrm{e}} &= \tfrac{1}{2}\int_{x=-\delta}^{\delta}\int_{y=0}^{1} \mathrm{Re}\left(\tilde{E}_{1x}{}^{\mathrm{r}}\tilde{H}_{1y}{}^{\mathrm{r}*}\right) dx\,dy \\ &\quad + \tfrac{1}{2}\int_{y=0}^{1}\left(\int_{-\infty}^{-\delta} + \int_{\delta}^{\infty}\right)\mathrm{Re}\left(\tilde{E}_{2x}{}^{\mathrm{r}}\tilde{H}_{2y}{}^{\mathrm{r}*}\right) dx\,dy \\ &\equiv \frac{1}{2}\frac{\beta_{\mathrm{r}}}{\omega\epsilon_1}P_1 + \frac{1}{2}\frac{\beta_{\mathrm{r}}}{\omega\epsilon_2}\mathrm{Re}\,P_2; \end{aligned} \tag{6}$$

$$\begin{aligned} P_1 &\equiv \int_{y=0}^{1}\int_{x=-\delta}^{\delta} \mathrm{Re}\Big[\left(H^{\mathrm{e}}\cos\beta_1 x\, e^{j(\omega t-\beta_{\mathrm{r}} z)}\right) \\ &\qquad \times \left(H^{\mathrm{e}*}\cos\beta_1' x\, e^{-j(\omega t-\beta_{\mathrm{r}}' z)}\right)\Big]\, dx\,dy \\ &= 2|H^{\mathrm{e}}|^2\cos(\beta_{\mathrm{r}}'-\beta_{\mathrm{r}})z\int_0^{\delta}\cos\beta_1 x\cos\beta_1' x\,dx, \end{aligned}$$

$$\begin{aligned} P_1(\beta_1\neq\beta_1') &= 2|H^{\mathrm{e}}|^2\cos(\beta_{\mathrm{r}}'-\beta_{\mathrm{r}})z\frac{1}{\beta_1{}^2-\beta_1'^2} \\ &\quad \times\left(\beta_1\sin\beta_1\delta\cos\beta_1'\delta - \beta_1'\sin\beta_1'\delta\cos\beta_1\delta\right). \end{aligned} \tag{7}$$

For the case where $\beta_1 = \beta_1'$,

$$\begin{aligned} P_1(\beta_1=\beta_1') &= 2|H^{\mathrm{e}}|^2\int_0^{\delta}\cos^2\beta_1 x\,dx \\ &= 2|H^{\mathrm{e}}|^2\left(\frac{\delta}{2} + \frac{\sin 2\beta_1\delta}{4\beta_1}\right), \end{aligned} \tag{8}$$

$$\begin{aligned} P_2 &\equiv 2\int_{\delta}^{\infty}\Big[\left(He^{-j\beta_\rho(x-\delta)} + H^* e^{j\beta_\rho(x-\delta)}\right)e^{j(\omega t-\beta_{\mathrm{r}} z)} \\ &\qquad \times\left(H'^* e^{j\beta_\rho'(x-\delta)} + H' e^{-j\beta_\rho'(x-\delta)}\right)e^{-j(\omega t-\beta_{\mathrm{r}}' z)}\Big]\,dx \\ &= 2e^{j(\beta_{\mathrm{r}}'-\beta_{\mathrm{r}})z}\int_0^{\infty}\Big[\left(He^{-j\beta_\rho s} + H^* e^{j\beta_\rho s}\right) \\ &\qquad \times\left(H'^* e^{j\beta_\rho' s} + H' e^{-j\beta_\rho' s}\right)\Big]\,ds, \end{aligned} \tag{9}$$

where $s \equiv x-\delta$ and H is defined by Eq. (2.6.15).

Expanding Eq. (9) yields

$$
\begin{aligned}
P_2(\beta_\rho \neq \beta_\rho') &= 2e^{j(\beta_\rho' - \beta_\rho)z} \int_0^\infty \left[HH'^* e^{-j(\beta_\rho - \beta_\rho')s} + H^* H' e^{j(\beta_\rho - \beta_\rho')s} \right. \\
&\qquad \left. + HH' e^{-j(\beta_\rho + \beta_\rho')s} + H^* H'^* e^{j(\beta_\rho + \beta_\rho')s} \right] ds \\
&= \tfrac{1}{2} |H^{\mathrm{e}}|^2 e^{j(\beta_\rho' - \beta_\rho)z} \int_0^\infty \left[(q - jr)(q' + jr') e^{-j(\beta_\rho - \beta_\rho')s} \right. \\
&\qquad + (q + jr)(q' - jr') e^{j(\beta_\rho - \beta_\rho')s} \\
&\qquad + (q - jr)(q' - jr') e^{-j(\beta_\rho + \beta_\rho')s} \\
&\qquad \left. + (q + jr)(q' + jr') e^{j(\beta_\rho + \beta_\rho')s} \right] ds \\
&= 2|H^{\mathrm{e}}|^2 e^{j(\beta_\rho' - \beta_\rho)z} \left[\int_0^\infty (qq' \cos\beta_\rho x \cos\beta_\rho' s \right. \\
&\qquad \left. + rr' \sin\beta_\rho s \sin\beta_\rho' s)\, ds + I_B{}^{\mathrm{e}} \right] \\
&= 2|H^{\mathrm{e}}|^2 e^{j(\beta_\rho' - \beta_\rho)z} \left[(qq' + rr')(\pi/2)\delta(\beta_\rho - \beta_\rho') + I_B{}^{\mathrm{e}} \right],
\end{aligned} \tag{10}
$$

where

$$
\begin{aligned}
I_B{}^{\mathrm{e}} \equiv \tfrac{1}{2} \int_0^\infty & \left[(qr' - q'r) \sin(\beta_\rho - \beta_\rho')s \right. \\
& \left. - (qr' + q'r) \sin(\beta_\rho + \beta_\rho')s \right] ds,
\end{aligned} \tag{11}
$$

$$
\begin{aligned}
I_B{}^{\mathrm{e}}(\beta_\rho \neq \beta_\rho') &= \tfrac{1}{2} \left[-(qr' - q'r) \frac{\cos(\beta_\rho - \beta_\rho')s}{\beta_\rho - \beta_\rho'} \right. \\
&\qquad \left. + (qr' + q'r) \frac{\cos(\beta_\rho + \beta_\rho')s}{\beta_\rho + \beta_\rho'} \right]_0^\infty \\
&= C_1 \cos(\beta_\rho - \beta_\rho')\infty + C_2 \cos(\beta_\rho + \beta_\rho')\infty \\
&\qquad + \left(\frac{-\beta_\rho q' r + \beta_\rho' q r'}{\beta_\rho{}^2 - \beta_\rho'^2} \right),
\end{aligned} \tag{12}
$$

where

$$
C_1 \equiv \frac{q'r - qr'}{2(\beta_\rho - \beta_\rho')} \qquad \text{and} \qquad C_2 \equiv \frac{qr' + q'r}{2(\beta_\rho + \beta_\rho')}. \tag{13}
$$

When $\beta_\rho = \beta_\rho'$,

$$
I_B{}^{\mathrm{e}}(\beta_\rho = \beta_\rho') = -qr \int_0^\infty \sin 2\beta_\rho s\, ds = \frac{qr}{2\beta_\rho}(\cos 2\beta_\rho \infty - 1). \tag{14}
$$

The complete expression of P_2 is

$$P_2(\beta_\rho \neq \beta_\rho') = 2|H^e|^2 e^{j(\beta_\rho' - \beta_\rho)z} \times \left[(qq' + rr')\frac{\pi}{2}\delta(\beta_\rho - \beta_\rho') + \left(\frac{-\beta_\rho q'r + \beta_\rho' qr'}{\beta_\rho^{\,2} - \beta_\rho'^2} \right) + C_1 \cos(\beta_\rho - \beta_\rho')\infty + C_2 \cos(\beta_\rho + \beta_\rho')\infty \right]$$

$$= 2|H^e|^2 e^{j(\beta_\rho' - \beta_\rho)z} \frac{\beta_\rho q'r - \beta_\rho' qr'}{\beta_\rho'^2 - \beta_\rho^{\,2}}, \tag{15}$$

$$P_2(\beta_\rho = \beta_\rho') = 2|H^e|^2 \left[(qq' + rr')\frac{\pi}{2}\delta(\beta_\rho - \beta_\rho') + \frac{qr}{2\beta_\rho} \right]. \tag{16}$$

The trigonometric terms with infinite frequency are dropped for the same reasons as given in Section 2.5a. Substitution of P_1 and P_2 into Eq. (6) yields

$$P^e(\beta \neq \beta') = \frac{1}{2} \frac{\beta_r}{\omega\epsilon_1} 2|H^e|^2 \cos(\beta_r' - \beta_r)z \frac{1}{\beta_1^{\,2} - \beta_1'^2} \times (\beta_1 \sin\beta_1\delta \cos\beta_1'\delta - \beta_1' \sin\beta_1'\delta \cos\beta_1\delta) + \frac{1}{2} \frac{\beta_r}{\omega\epsilon_2} 2|H^e|^2 \cos(\beta_r' - \beta_r)z \times \frac{1}{\beta_\rho^{\,2} - \beta_\rho'^2}(-\beta_\rho q'r + \beta_\rho q'r) = 0, \tag{17}$$

$$P^e(\beta = \beta') = |H^e|^2 \frac{\beta_r}{\omega\epsilon_2}(q^2 + r^2)\frac{\pi}{2}\delta(\beta_\rho - \beta_\rho'). \tag{18}$$

$|H^e|$ can be expressed in terms of P^e as follows:

$$|H^e| = \left\{ 2\omega\epsilon_2 P^e / \beta_r \pi \left[\cos^2\beta_1\delta + \left(\frac{\epsilon_2\beta_1}{\epsilon_1\beta_\rho} \sin\beta_1\delta \right)^2 \right] \right\}^{1/2}. \tag{19}$$

TM odd modes. The fields of TM odd modes are given by Eqs. (2.6.1) and (2.6.31)–(2.6.36),

$$\tilde{E}_{1x}{}^{r} = \frac{\beta_r}{\omega\epsilon_1} H^o \sin\beta_1 x\, e^{j(\omega t - \beta_r z)}, \tag{20}$$

$$\tilde{H}_{1y}{}^{r} = H^o \sin\beta_1 x\, e^{j(\omega t - \beta_r z)}, \tag{21}$$

$$\tilde{E}_{2x}{}^{\mathrm{r}}=\frac{\beta_{\mathrm{r}}}{\omega\epsilon_2}(\hat{H}e^{-j\beta_\rho(x-\delta)}+\hat{H}^*e^{j\beta_\rho(x-\delta)})e^{j(\omega t-\beta_{\mathrm{r}}z)},\qquad x>\delta,\tag{22}$$

$$\tilde{H}_{2y}{}^{\mathrm{r}}=(\hat{H}e^{-j\beta_\rho(x-\delta)}+\hat{H}^*e^{j\beta_\rho(x-\delta)})e^{j(\omega t-\beta_{\mathrm{r}}z)},\qquad x>\delta.\tag{23}$$

The power of TM odd modes is given by

$$\begin{aligned}P^{\mathrm{o}}&=\int_0^\delta \mathrm{Re}\left(\tilde{E}_{1x}{}^{\mathrm{r}}\tilde{H}_{1y}{}^{\mathrm{r}*}\right)dx+\int_\delta^\infty \mathrm{Re}\left(\tilde{E}_{2x}{}^{\mathrm{r}}\tilde{H}_{2y}{}^{\mathrm{r}*}\right)dx\\&=\frac{\beta_{\mathrm{r}}}{\omega\epsilon_1}P_1+\frac{\beta_{\mathrm{r}}}{\omega\epsilon_2}\mathrm{Re}\,P_2;\end{aligned}\tag{24}$$

$$\begin{aligned}P_1&\equiv\int_0^\delta \mathrm{Re}\left[\left(H^{\mathrm{o}}\sin\beta_1 x\,e^{j(\omega t-\beta_{\mathrm{r}}z)}\right)\right.\\&\qquad\left.\times\left(H^{\mathrm{o}}\sin\beta_1' x\,e^{-j(\omega t-\beta_{\mathrm{r}}'z)}\right)\right]dx\\&=|H^{\mathrm{o}}|^2\cos(\beta_{\mathrm{r}}'-\beta_{\mathrm{r}})z\int_0^\delta \sin\beta_1 x\sin\beta_1' x\,dx,\end{aligned}\tag{25}$$

$$\begin{aligned}P_1(\beta_1\neq\beta_1')&=|H^{\mathrm{o}}|^2\cos(\beta_{\mathrm{r}}'-\beta_{\mathrm{r}})z\\&\quad\times\frac{1}{2}\left[\frac{\sin(\beta_1-\beta_1')\delta}{\beta_1-\beta_1'}-\frac{\sin(\beta_1+\beta_1')\delta}{\beta_1+\beta_1'}\right]\\&=|H^{\mathrm{o}}|^2\cos(\beta_{\mathrm{r}}'-\beta_{\mathrm{r}})z\frac{1}{\beta_1{}^2-\beta_1'^2}\\&\quad\times(\beta_1'\sin\beta_1\delta\cos\beta_1'\delta-\beta_1\sin\beta_1'\delta\cos\beta_1\delta).\end{aligned}\tag{26}$$

For $\beta_1=\beta_1'$,

$$\begin{aligned}P_1(\beta_1=\beta_1')&=|H^{\mathrm{o}}|^2\int_0^\delta \sin^2\beta_1 x\,dx\\&=|H^{\mathrm{o}}|^2\left[\frac{\delta}{2}-\frac{\sin 2\beta_1\delta}{4\beta_1}\right],\end{aligned}\tag{27}$$

$$\begin{aligned}P_2&\equiv\int_\delta^\infty(\hat{H}e^{-j\beta_\rho(x-\delta)}+H^*e^{j\beta_\rho(x-\delta)})e^{j(\omega t-\beta_{\mathrm{r}}z)}\\&\quad\times(\hat{H}'^*e^{j\beta_\rho'(x-\delta)}+\hat{H}'e^{-j\beta_\rho'(x-\delta)})e^{-j(\omega t-\beta_{\mathrm{r}}'z)}\,dx\\&=e^{-j(\beta_{\mathrm{r}}-\beta_{\mathrm{r}}')z}\int_0^\infty(\hat{H}\hat{H}'^*e^{-j(\beta_\rho-\beta_\rho')s}+\hat{H}^*\hat{H}'e^{j(\beta_\rho-\beta_\rho')s}\\&\quad+\hat{H}\hat{H}'e^{-j(\beta_\rho+\beta_\rho')s}+H^*H'^*e^{j(\beta_\rho+\beta_\rho')s})\,ds\\&=|H^{\mathrm{o}}|^2e^{-j(\beta_{\mathrm{r}}-\beta_{\mathrm{r}}')z}\left[(uu'+vv')\frac{\pi}{2}\delta(\beta_\rho-\beta_\rho')+I_B{}^{\mathrm{o}}\right],\end{aligned}\tag{28}$$

$$I_B{}^{\circ} \equiv \tfrac{1}{2}(vu'-uv')\int_0^{\infty}\sin(\beta_\rho-\beta_\rho')s\,ds$$

$$+\tfrac{1}{2}(vu'+uv')\int_0^{\infty}\sin(\beta_\rho+\beta_\rho')s\,ds, \tag{29}$$

$$I_B{}^{\circ}(\beta_\rho\neq\beta_\rho')=\frac{vu'-uv'}{2(\beta_\rho-\beta_\rho')}\left[1-\cos(\beta_\rho-\beta_\rho')\infty\right]$$

$$+\frac{vu'+uv'}{2(\beta_\rho+\beta_\rho')}\left[1-\cos(\beta_\rho+\beta_\rho')\infty\right]$$

$$=\frac{\beta_\rho vu'-\beta_\rho' uv'}{\beta_\rho{}^2-\beta_\rho'^2}-C_3\cos(\beta_\rho-\beta_\rho')\infty$$

$$-C_4\cos(\beta_\rho+\beta_\rho')\infty, \tag{30}$$

$$C_3\equiv\frac{vu'-uv'}{2(\beta_\rho-\beta_\rho')}\quad\text{and}\quad C_4\equiv\frac{vu'+uv'}{2(\beta_\rho+\beta_\rho')}, \tag{31}$$

$$I_B{}^{\circ}(\beta_\rho=\beta_\rho')=\tfrac{1}{2}(vu'+uv')\int_0^{\infty}\sin 2\beta_\rho s\,ds=(uv/2\beta_\rho)(1-\cos 2\beta_\rho\infty). \tag{32}$$

The expression for P_2 is

$$P_2(\beta_\rho\neq\beta_\rho')=|H^{\circ}|^2e^{-j(\beta_r-\beta_r')z}$$

$$\times\left[(uu'+vv')\frac{\pi}{2}\delta(\beta_\rho-\beta_\rho')+\frac{\beta_\rho vu'-\beta_\rho' uv'}{\beta_\rho{}^2-\beta_\rho'^2}\right], \tag{33}$$

$$P_2(\beta_\rho=\beta_\rho')=|H^{\circ}|^2\left[(u^2+v^2)\frac{\pi}{2}\delta(\beta_\rho-\beta_\rho')+\frac{uv}{2\beta_\rho}\right]. \tag{34}$$

The power of the TM odd modes is now given by

$$P^{\circ}(\beta_\rho\neq\beta_\rho')=\frac{\beta_r}{\omega\epsilon_1}|H^{\circ}|^2\cos(\beta_r'-\beta_r)z\frac{1}{\beta_1{}^2-\beta_1'^2}$$

$$\times\left[\beta_1'\sin\beta_1\delta\cos\beta_1'\delta-\beta_1\sin\beta_1'\delta\cos\beta_1\delta\right]$$

$$+\frac{\beta_r}{\omega\epsilon_2}|H^{\circ}|^2\cos(\beta_r'-\beta_r)z\frac{1}{\beta_\rho{}^2-\beta_\rho'^2}(\beta_\rho vu'-\beta_\rho' uv')=0, \tag{35}$$

$$P^\circ(\beta_\rho=\beta_\rho')=\frac{\beta_r}{\omega\epsilon_1}|H^\circ|^2\left(\frac{\delta}{2}-\frac{\sin 2\beta_1\delta}{4\beta_1}\right)+\frac{\beta_r}{\omega\epsilon_2}|H^\circ|^2$$
$$\times\left[(u^2+v^2)\frac{\pi}{2}\delta(\beta_\rho-\beta_\rho')+\frac{uv}{2\beta_\rho}\right]$$
$$=\frac{\beta_r}{\omega\epsilon_2}|H^\circ|^2$$
$$\times\left\{\left[\sin^2\beta_1\delta+\left(\frac{\epsilon_2\beta_1}{\epsilon_1\beta_\rho}\cos\beta_1\delta\right)^2\right]\frac{\pi}{2}\delta(\beta_\rho-\beta_\rho')\right\}, \quad (36)$$

and

$$|H^\circ|=\left\{2\omega\epsilon_2 P^\circ/\pi\beta_r\left[\sin^2\beta_1\delta+\left(\frac{\epsilon_2\beta_1}{\epsilon_1\beta_\rho}\cos\beta_1\delta\right)^2\right]\right\}^{1/2}. \quad (37)$$

The functions u and v are defined by Eq. (2.6.30).

2.7. Orthogonality Relations

The guided and radiation modes of a dielectric waveguide are orthogonal to each other. This property will now be explored.

Let $\mathbf{E}_m,\mathbf{H}_m$, and $\mathbf{E}_k,\mathbf{H}_k$ be the fields of the mth and kth modes, respectively. Then Maxwell's equations for these fields are

$$\nabla\times\mathbf{E}_m=-j\omega\mu\mathbf{H}_m,\qquad \nabla\times\mathbf{H}_m=j\omega\varepsilon\mathbf{E}_m,$$
$$\nabla\times\mathbf{E}_k=-j\omega\mu\mathbf{H}_k,\qquad \nabla\times\mathbf{H}_k=j\omega\varepsilon\mathbf{E}_k. \quad (1)$$

Form the products

$$\mathbf{H}_k^*\cdot\nabla\times\mathbf{E}_m=-j\omega\mu\mathbf{H}_m\cdot\mathbf{H}_k^*,\qquad \mathbf{E}_k^*\cdot\nabla\times\mathbf{H}_m=j\omega\varepsilon\mathbf{E}_m\cdot\mathbf{E}_k^*,$$
$$\mathbf{H}_m\cdot\nabla\times\mathbf{E}_k^*=j\omega\mu\mathbf{H}_m\cdot\mathbf{H}_k^*,\qquad \mathbf{E}_m\cdot\nabla\times\mathbf{H}_k^*=-j\omega\varepsilon\mathbf{E}_m\cdot\mathbf{E}_k^*. \quad (2)$$

Then

$$\nabla(\mathbf{E}_m\times\mathbf{H}_k^*)=\mathbf{H}_k^*\cdot\nabla\times\mathbf{E}_m-\mathbf{E}_m\cdot\nabla\times\mathbf{H}_k^*$$
$$=j\omega(\varepsilon\mathbf{E}_m\cdot\mathbf{E}_k^*-\mu\mathbf{H}_m\cdot\mathbf{H}_k^*),$$
$$\nabla(\mathbf{E}_k^*\times\mathbf{H}_m)=\mathbf{H}_m\cdot\nabla\times\mathbf{E}_k^*-\mathbf{E}_k^*\cdot\nabla\times\mathbf{H}_m$$
$$=-j\omega(\varepsilon\mathbf{E}_m\cdot\mathbf{E}_k^*-\mu\mathbf{H}_m\cdot\mathbf{H}_k^*), \quad (3)$$

and

$$\nabla\cdot(\mathbf{E}_m\times\mathbf{H}_k^*)+\nabla\cdot(\mathbf{E}_k^*\times\mathbf{H}_m)=0. \quad (4)$$

It is convenient to introduce the following notation.

$$\nabla\cdot\mathbf{G}=\left[\left(\hat{\mathbf{x}}\frac{\partial}{\partial x}+\hat{\mathbf{y}}\frac{\partial}{\partial y}\right)+\hat{\mathbf{z}}\frac{\partial}{\partial z}\right]\cdot\left[(\hat{\mathbf{x}}G_x+\hat{\mathbf{y}}G_y)+\hat{\mathbf{y}}G_z\right]$$

$$\equiv\left(\nabla_{\mathrm{t}}+\hat{\mathbf{z}}\frac{\partial}{\partial z}\right)\cdot(\mathbf{G}_{\mathrm{t}}+\hat{\mathbf{z}}G_z)$$

$$=\nabla_{\mathrm{t}}\cdot\mathbf{G}_{\mathrm{t}}+\frac{\partial G_z}{\partial z}. \tag{5}$$

Then Eq. (4) becomes

$$\nabla\cdot[(\mathbf{E}_m\times\mathbf{H}_k{}^*)+(\mathbf{E}_k{}^*\times\mathbf{H}_m)]=\nabla_{\mathrm{t}}\cdot[(\mathbf{E}_m\times\mathbf{H}_k{}^*)_{\mathrm{t}}+(\mathbf{E}_k{}^*\times\mathbf{H}_m)_{\mathrm{t}}]$$
$$+\frac{\partial}{\partial z}[(\mathbf{E}_m\times\mathbf{H}_k{}^*)_z+(\mathbf{E}_k{}^*\times\mathbf{H}_m)_z], \tag{6}$$

where $(\mathbf{E}_m\times\mathbf{H}_k{}^*)_t$ represents the components transverse to the z axis and $(\mathbf{E}_m\times\mathbf{H}_k{}^*)_z$ is the axial component. For a lossless guide, the fields are

$$\mathbf{E}_i(\mathbf{r})\equiv\mathbf{e}_i(x,y)e^{-j\beta_i z},\qquad \mathbf{H}_i(\mathbf{r})\equiv\mathbf{h}_i(x,y)e^{-j\beta_i z}. \tag{7}$$

Then

$$\nabla\cdot(\mathbf{E}_m\times\mathbf{H}_k{}^*+\mathbf{E}_k{}^*\times\mathbf{H}_m)=\nabla\cdot\left[(\mathbf{e}_m\times\mathbf{h}_k+\mathbf{e}_k\times\mathbf{h}_m)e^{-j(\beta_m-\beta_k)z}\right]$$
$$=0$$

and

$$\nabla_{\mathrm{t}}\cdot[(\mathbf{e}_m\times\mathbf{h}_k)_{\mathrm{t}}+(\mathbf{e}_k\times\mathbf{h}_m)_{\mathrm{t}}]=j(\beta_m-\beta_k)[(\mathbf{e}_m\times\mathbf{h}_k)_z+(\mathbf{e}_k\times\mathbf{h}_m)_z]. \tag{8}$$

Integration of Eq. (8) over the cross section of the guide yields

$$\int_S\nabla_{\mathrm{t}}\cdot[(\mathbf{e}_m\times\mathbf{h}_k)_{\mathrm{t}}+(\mathbf{e}_k\times\mathbf{h}_m)_{\mathrm{t}}]\,da$$
$$=\int_S j(\beta_m-\beta_k)[(\mathbf{e}_m\times\mathbf{h}_k)_z+(\mathbf{e}_k\times\mathbf{h}_m)_z]\,da, \tag{9}$$

$$\oint_C[(\mathbf{e}_m\times\mathbf{h}_k)_{\mathrm{t}}+(\mathbf{e}_k\times\mathbf{h}_m)_{\mathrm{t}}]\cdot\hat{\mathbf{n}}\,dl$$
$$=j(\beta_m-\beta_k)\int_S[(\mathbf{e}_m\times\mathbf{h}_k)_z+(\mathbf{e}_k\times\mathbf{h}_m)_z]\,da. \tag{10}$$

The two-dimensional divergence theorem (Section AI.3) is used to transform the left side of Eq. (9) into a contour integral. The contour C bounds the surface S.

In the case of the dielectric slab guide, the contour C extends to infinity. If one or both of the fields are guided modes of the guide, its field strength vanishes at infinity and there is no contribution from the contour integral.

If both of the fields are radiation modes of the guide and these are harmonic functions of the transverse variables x and y, the eigenvalues have continuous spectrum for the radiation mode. For $m \neq k$, an integration over a small region around m or k sums up to zero. This is because the integrand contains an oscillatory factor with an infinitely large argument in x and y. Thus

$$\oint_C [(\mathbf{e}_m \times \mathbf{h}_k)_t + (\mathbf{e}_k \times \mathbf{h}_m)_t] \cdot \hat{\mathbf{n}}\, dl = 0. \tag{11}$$

For either guided modes, radiation modes, or mixed modes, Eq. (10) becomes

$$j(\beta_m - \beta_k) \int_S [(\mathbf{e}_m \times \mathbf{h}_k)_z + (\mathbf{e}_k \times \mathbf{h}_m)_z]\, da = 0$$

or

$$\int_S [(\mathbf{e}_m \times \mathbf{h}_k)_z + (\mathbf{e}_k \times \mathbf{h}_m)_z]\, da = 0 \qquad (\beta_m \neq \beta_k). \tag{12}$$

All uniform guides inhomogeneously filled with isotropic media have reflection symmetry with respect to the axial coordinate. Let the axis of the guide be the z axis. Then if the solution is $(\mathbf{E}_{it}, \mathbf{E}_{iz}, \mathbf{H}_{it}, \mathbf{H}_{iz}, \gamma_i)$ there exists a solution $(\mathbf{E}_{it}, -\mathbf{E}_{iz}, -\mathbf{H}_{it}, \mathbf{H}_{iz}, -\gamma_i)$. This is the definition of reflection symmetry, and it can be easily verified (by direct substitution into Maxwell's equations) that the above fields satisfy the same set of relations.

The above analysis will be carried out for the following modes:

$$\begin{aligned} \mathbf{E}_m(\mathbf{r}) &= \mathbf{e}_m(x, y) e^{-j\beta_m z} = (\hat{\mathbf{x}} e_{mx} + \hat{\mathbf{y}} e_{my} + \hat{\mathbf{z}} e_{mz}) e^{-j\beta_m z}, \\ \mathbf{H}_m(\mathbf{r}) &= \mathbf{h}_m(x, y) e^{-j\beta_m z} = (\hat{\mathbf{x}} h_{mx} + \hat{\mathbf{y}} h_{my} + \hat{\mathbf{z}} h_{mz}) e^{-j\beta_m z}, \end{aligned} \tag{13}$$

$$\begin{aligned} \mathbf{E}_k(\mathbf{r}) &= \mathbf{e}_k(x, y) e^{-j\beta_k z} = (\hat{\mathbf{x}} e_{kx} + \hat{\mathbf{y}} e_{ky} + \hat{\mathbf{z}} e_{kz}) e^{-j\beta_k z}, \\ \mathbf{H}_k(\mathbf{r}) &= \mathbf{h}_k(x, y) e^{-j\beta_k z} = (\hat{\mathbf{x}} h_{kx} + \hat{\mathbf{y}}_{ky} + \hat{\mathbf{z}} h_{kz}) e^{-j\beta_k z}. \end{aligned} \tag{14}$$

Let the kth mode be replaced by its reflection symmetric modes:

$$\begin{aligned} \mathbf{E}_k'(\mathbf{r}) &= \mathbf{e}_k'(x, y) e^{j\beta_k z} = (\hat{\mathbf{x}} e_{kx} + \hat{\mathbf{y}} e_{ky} - \hat{\mathbf{z}} e_{kz}) e^{j\beta_k z}, \\ \mathbf{H}_k'(\mathbf{r}) &= \mathbf{h}_k'(x, y) e^{j\beta_k z} = (-\hat{\mathbf{x}} h_{kx} - \hat{\mathbf{y}} h_{ky} + \hat{\mathbf{z}} h_{kz}) e^{j\beta_k z}. \end{aligned} \tag{15}$$

When the entire analysis is repeated for $\mathbf{E}_m$, $\mathbf{H}_m$, $\mathbf{E}_k'$ and $\mathbf{H}_k'$, the final result, from Eq. (12), is

$$\int_S [(\mathbf{e}_m \times \mathbf{h}_k')_z + (\mathbf{e}_k' \times \mathbf{h}_m)_z]\, da = 0. \tag{16}$$

It can be shown by direct expansion that

$$(\mathbf{e}_m \times \mathbf{h}_k')_z = -(\mathbf{e}_m \times \mathbf{h}_k)_z \qquad \text{and} \qquad (\mathbf{e}_k' \times \mathbf{h}_m)_z = (\mathbf{e}_k \times \mathbf{h}_m)_z. \tag{17}$$

Therefore, Eq. (16) becomes

$$-\int_S [(\mathbf{e}_m\times\mathbf{h}_k)_z-(\mathbf{e}_k\times\mathbf{h}_m)]\,da=0. \tag{18}$$

The sum and difference of Eq. (12) and Eq. (18) yield

$$\begin{aligned}&\int_S(\mathbf{e}_m\times\mathbf{h}_k)_z\,da=0, \qquad m\neq k,\\ &\int_S(\mathbf{e}_k\times\mathbf{h}_m)_z\,da=0, \qquad m\neq k.\end{aligned} \tag{19}$$

These relations imply that the Poynting vector in the axial direction is zero for fields of different modes. Since $\mathbf{e}_m$ and $\mathbf{h}_m$ are mutually perpendicular, their unit vectors obey

$$\begin{aligned}&\hat{\mathbf{e}}_m\times\hat{\mathbf{h}}_m=\hat{\gamma} \qquad \text{or} \qquad \hat{\mathbf{h}}_m=\hat{\gamma}\times\hat{\mathbf{e}}_m,\\ &\hat{\mathbf{e}}_m\times\hat{\mathbf{h}}_k=\hat{\mathbf{e}}_m\times(\hat{\gamma}\times\hat{\mathbf{e}}_k)\\ &\qquad=\hat{\gamma}(\hat{\mathbf{e}}_m\cdot\hat{\mathbf{e}}_k)-\hat{\mathbf{e}}_k(\hat{\gamma}\cdot\hat{\mathbf{e}}_m)\\ &\qquad=\hat{\gamma}(\hat{\mathbf{e}}_m\cdot\hat{\mathbf{e}}_k).\end{aligned} \tag{20}$$

Then

$$\int_S(\mathbf{e}_m\times\mathbf{h}_k)\,da=\int\hat{\gamma}(\mathbf{e}_m\cdot\mathbf{e}_k)\,da=0.$$

Therefore

$$\mathbf{e}_m\perp\mathbf{e}_k. \tag{21}$$

This proves that fields of different modes are orthogonal.

Example 2

A grounded dielectric sheet waveguide is made of a thin film of lossless dielectric (μ_1, ϵ_1) of thickness x_0. This film is protected by a second lossless dielectric coating (μ_2, ϵ_2). For convenience of analysis, the thickness of this coating is assumed to extend to infinity. The dielectric film is supported by a perfectly conducting surface. Both width and length extend to infinity. These conditions are illustrated in Figure 4. The fields are assumed to be independent of y. Determine

a. the TE fields in the guide;
b. the cutoff frequency and possible values of k_{1x} if $x_0=10^{-6}$ m, $\epsilon_1=1.500\epsilon_0$, $\epsilon_2=1.498\epsilon_0$, $\mu_1=\mu_2=\mu_0$, with cutoff at $k_{1x}x_0=3/2\pi$, and $f=6\times10^{15}$ Hz.

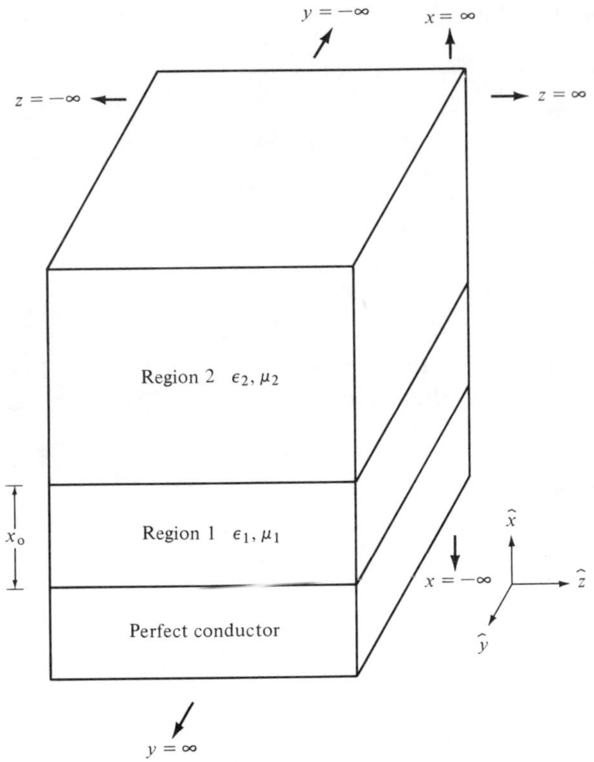

Figure 4. Example 2.

SOLUTION

a. The guided waves in the z direction have the general form

$$\tilde{\mathbf{E}}(\mathbf{r}, t) = \mathbf{E}(x) e^{j\omega t - \gamma_z z}. \tag{1}$$

For the TE modes, E_y is the only component of the **E** field and satisfies the wave equation (2.1.6),

$$\frac{d^2 E_y}{dx^2} + k_x{}^2 E_y = 0 \qquad k_x{}^2 \equiv \gamma_z{}^2 - \gamma^2, \quad \gamma^2 \equiv -\omega^2 \mu \epsilon. \tag{2}$$

The corresponding **H** field is given by Eq. (2.1.5),

$$H_x = \frac{-\gamma_z}{j\omega\mu} E_y, \qquad H_z = \frac{-1}{j\omega\mu} \frac{dE_y}{dx}. \tag{3}$$

The solution of Eq. (2) will now be obtained for each region. Region 1 of Figure 4 is bounded by the surfaces $x=0$ and $x=x_0$, and a standing wave solution is expected in the x direction. Consequently, the trigonometric form of solution is chosen for this region:

$$E_{1y}(x)=E^{e}\cos k_{1x}x+E^{o}\sin k_{1x}x. \tag{4}$$

The boundary condition at $x=0$ requires the vanishing of the tangential **E** field, so that

$$E_{1y}(x=0)=0=E^{e}. \tag{5}$$

Hence Eqs. (3) and (4) become

$$E_{1y}(x)=E^{o}\sin k_{1x}x, \qquad k_{1x}{}^{2}=\gamma_{1z}{}^{2}-\gamma_{1}{}^{2}, \quad \gamma_{1}{}^{2}=-\omega^{2}\mu_{1}\epsilon_{1}, \tag{6}$$

$$H_{1x}(x)=-\frac{\gamma_{1z}}{j\omega\mu_{1}}E_{1y}=\frac{-\gamma_{1z}}{j\omega\mu_{1}}E^{o}\sin k_{1x}x, \tag{7}$$

$$H_{1z}(x)=\frac{-1}{j\omega\mu_{1}}\frac{dE_{1y}}{dx}=\frac{-k_{1x}}{j\omega\mu_{1}}E^{o}\cos k_{1x}x. \tag{8}$$

The solution in exponential form will be taken for region 2. For an efficient guide, evanescent modes are required in the unbounded region. Accordingly,

$$E_{2y}(x)=E^{+}e^{-\alpha_{2x}(x-x_{0})}+E^{-}e^{\alpha_{2x}(x-x_{0})},$$
$$\alpha_{2x}\equiv\gamma_{2}{}^{2}-\gamma_{2z}{}^{2},$$
$$\gamma_{2}{}^{2}\equiv-\omega^{2}\mu_{2}\epsilon_{2}. \tag{9}$$

The second term in Eq. (9) can be extremely large as x increases. It is therefore an unacceptable solution and should be discarded.

$$E_{2y}(x)=E^{+}e^{-\alpha_{2x}(x-x_{0})}, \qquad x>x_{0}, \tag{10}$$

$$H_{2x}(x)=\frac{-\gamma_{2z}}{j\omega\mu_{2}}E^{+}e^{-\alpha_{2x}(x-x_{0})}, \tag{11}$$

$$H_{2z}(x)=\frac{\alpha_{2x}}{j\omega\mu_{2}}E^{+}e^{-\alpha_{2x}(x-x_{0})}. \tag{12}$$

The boundary condition at $x=x_0$ requires the continuity of all tangential field components.

$$E_{1y}(x=x_{0})=E_{2y}(x=x_{0}), \qquad E^{o}\sin k_{1x}x_{0}=E^{+}, \tag{13}$$

$$H_{1z}(x=x_{0})=H_{2z}(x=x_{0}), \qquad \frac{-k_{1x}}{j\omega\mu_{1}}E^{o}\cos k_{1x}x_{0}=\frac{\alpha_{2x}}{j\omega\mu_{2}}E^{+}. \tag{14}$$

Dividing Eq. (13) by Eq. (14) yields

$$\tan k_{1x}x_{0}=-k_{1x}\mu_{2}/\alpha_{2x}\mu_{1}. \tag{15}$$

In order to satisfy the boundary condition, the waves in both regions must travel with identical velocity in the z direction. This is equivalent to requiring

$$\gamma_{1z}{}^2 = \gamma_{2z}{}^2 \equiv \gamma_g{}^2, \qquad k_{1x}{}^2 + \gamma_1{}^2 = \gamma_2{}^2 - \alpha_{2x}{}^2$$

or

$$\begin{aligned}\alpha_{2x}{}^2 &= -\gamma_1{}^2 + \gamma_2{}^2 - k_{1x}{}^2 \\ &\equiv G^2 - k_{1x}{}^2, \qquad G^2 \equiv \omega^2(\mu_1\epsilon_1 - \mu_2\epsilon_2).\end{aligned} \tag{16}$$

Evanescent modes exist when $G^2 > k_{1x}{}^2$ or $\alpha_{2x}{}^2$ is real and positive, cutoff occurs when $G^2 = k_{1x}{}^2$ or $\alpha_{2x}{}^2 = 0$.

The cutoff frequency is given by

$$\omega_c = k_{1x}(\mu_1\epsilon_1 - \mu_2\epsilon_2)^{-1/2} = 2\pi f_c. \tag{17}$$

α_{2x} can be eliminated from Eq. (15) by the use of Eq. (16):

$$\tan k_{1x}x_0 = -\frac{\mu_2}{\mu_1}\frac{k_{1x}}{\sqrt{G^2 - k_{1x}{}^2}}$$

or

$$\cot k_{1x}x_0 = -\frac{\mu_1}{\mu_2}\left[\left(\frac{G}{k_{1x}}\right)^2 - 1\right]^{1/2}. \tag{18}$$

The graphical method may be used to find the solution of Eq. (18). This is accomplished by plotting each side of Eq. (18) as a function of k_{1x} and locating their intersections (see Figure 5). The fraction μ_1/μ_2 serves as a

Figure 5. Graphical solution to Example 2.

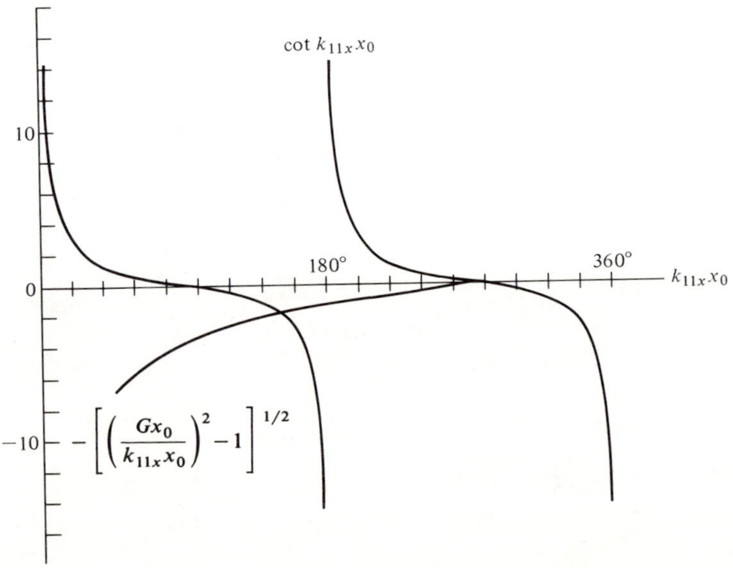

scale factor. When this is unity, $\mu_1/\mu_2=1$, one has

$$\cot^2 k_{1x}x_0=\left(\frac{G}{k_{1x}}\right)^2-1, \qquad \left(\frac{G}{k_{1x}}\right)^2=\cot^2 k_{1x}x_0+1=\frac{1}{\sin^2 k_{1x}x_0}. \tag{19}$$

At cutoff

$$\left(\frac{G}{k_{1x}}\right)^2=1=\frac{1}{\sin^2 k_{1x}x_0}, \qquad \text{or} \qquad |\sin k_{1x}x_0|=1$$

$$k_{1nx}x_0=\frac{2n+1}{2}\pi, \quad k_{1nx}=\frac{(2n+1)\pi}{2x_0}, \qquad n=0,1,\ldots. \tag{20}$$

The guide propagation constant is given by Eq. (6),

$$\begin{aligned}\gamma_g&=\gamma_{2nz}=\gamma_{1nz}=\left(k_{1nx}{}^2+\gamma_1{}^2\right)^{1/2}\\&=\left(k_{1nx}{}^2-\omega^2\mu_1\epsilon_1\right)^{1/2}.\end{aligned} \tag{21}$$

With the exception of E°, all parameters in the field expression, Eqs. (6)–(8) and (10)–(12), have been specified. The constant E° can be related to the incident power as shown in Example 1.

b. The cutoff frequency is determined by Eq. (17),

$$\omega_{nc}=k_{1nx}(\mu_1\epsilon_1-\mu_2\epsilon_2)^{-1/2}=2\pi f_{nc}.$$

For $k_{1nx}=(2n+1)\pi/2x_0=3\pi/2x_0$, this means $n=1$, and the cutoff frequency of $n=1$ mode is

$$\begin{aligned}\omega_{1c}&=k_{11x}\left[\mu_0\epsilon_0(\epsilon_{1r}-\epsilon_{2r})\right]^{-1/2}\\&=\frac{3\pi}{2\times10^{-6}}\times3\times10^8(1.500-1.498)^{-1/2}\\&=\frac{4.5\pi\times10^{15}}{\sqrt{0.2}}=2\pi f_{1c},\end{aligned}$$

$$f_{1c}=5.03\times10^{15}\ \text{Hz}.$$

At $f=6\times10^{15}$ Hz, the parameter G is

$$G=\frac{2\pi-6\times10^{15}}{3\times10^8}(1.500-1.498)^{1/2}-5.62\times10^6.$$

Equation (18) is modified as follows:

$$\cot k_{11x}x_0=-\left[(Gx_0/k_{11x}x_0)^2-1\right]^{1/2}. \tag{22}$$

Each side of Eq. (22) may now be plotted as a function of $k_{11x}x_0$. The values of the contangent function can be obtained easily from a standard

$k_{11x}x_0$	$A \equiv \frac{Gx_0}{k_{11x}x_0}$	$B \equiv A^2$	$C \equiv B-1$	$\sqrt{C}$
$3\pi/2$	1	1	0	0
π	1.789	3.200	2.200	1.483
$140° = 7\pi/9$	2.300	5.290	4.290	2.071
$150° = 5\pi/6$	2.147	4.608	3.608	1.900
$\pi/2$	3.578	12.800	11.800	3.435
$\pi/4$	7.155	51.197	50.197	7.085
$152° = 2.6529$	2.118	4.487	3.487	1.867

mathematical table. The values of the right-hand side are then calculated and are listed in the accompanying table. The intersection is found to be around $k_{11x}x_0 = 150°$. Additional calculations must be made at $k_{11x}x_0 = 150°$. Equation (22) is found to be satisfied within 1%.

References

1. J. Brown, The Types of Wave Which May Exist Near a Guiding Surface, *Proc. IEEE (London)* 100 (68), 363–364 (1953).
2. R. E. Collin, *Field Theory of Guided Waves*, McGraw-Hill, New York, 1960.
3. D. Hondors and P. Debye, Elektromagnetische Wellen an dielektrischen Drahten, *Ann. Phys.* 32, 465–476 (1910).
4. D. Marcuse, *Light Transmission Optics*, Van Nostrand Reinhold, Princeton, N.J., 1972.
5. D. Marcuse, *Theory of Dielectric Optical Waveguides*, Academic, New York, 1974.
6. O. Schriever, Elektromagnetische Wellen an dielektrischen Drahten, *Ann. Phys.* 63, 645–673 (1920).
7. M. S. Sodha and A. K. Ghatak, *Inhomogeneous Optical Waveguides*, Plenum, New York, 1977.
8. W. L. Weeks, *Electromagnetic Theory for Engineering Applications*, Wiley, New York, 1964.
9. R. M. Whitmer, Fields in Nonmetallic Waveguides, *Proc. IRE* 36, 1105–1109 (1948).
10. R. M. Whitmer, Radiation from a Dielectric Waveguide, *J. Appl. Phys.* 23, 949–953 (1952).
11. H. G. Unger, *Planar Optical Waveguides and Fibres*, Clarendon, Oxford, 1977.
12. H. Zahn, Uber den Nachweis elektromagnetischer Wellen an dielektrischen Drahten, *Ann. Phys.* 49, 907–933 (1916).

3

Eigenvalues and Eigenfunctions

Many problems in engineering science involve the solution of second-order partial differential equations. The mathematical techniques for treating the general homogeneous problem of this type will be reviewed in this chapter, followed by an investigation of the properties of eigenfunctions and eigenvalues. The Gram–Schmidt orthogonalization procedure will also be examined.

3.1. Homogeneous Problems

In many physical problems one seeks a solution of a partial differential equation of the general type

$$\nabla^2 U = MU + N\frac{\partial U}{\partial t} + P\frac{\partial^2 U}{\partial t^2} \tag{1}$$

subject to one of the following homogeneous boundary conditions:

$$\begin{aligned}
&\text{(a)} \qquad U=0,\\
&\text{(b)} \qquad \frac{\partial U}{\partial n}=0,\\
&\text{(c)} \qquad \frac{\partial U}{\partial n}+\xi U=0, \qquad \xi>0,
\end{aligned} \tag{2}$$

where $U \equiv \tilde{U}(\mathbf{r}, t)$, n is the variable normal to the boundary, and M, N, P, and ξ are constants. This type of problem is known as *homogeneous* [1, 4, 6, 8, 10] since both the differential equation and boundary conditions are homogeneous.

This problem can be subdivided into a spatial-domain and a time-domain problem. Let

$$\tilde{U}(\mathbf{r},t)\equiv R(\mathbf{r})T(t). \tag{3}$$

Then

$$\nabla^2(RT)=M(RT)+N\frac{\partial}{\partial t}(RT)+P\frac{\partial^2}{\partial t^2}(RT),$$

$$T\nabla^2 R=MRT+NR\frac{dT}{dt}+PR\frac{d^2T}{dt^2}, \tag{4}$$

$$\frac{1}{R}\nabla^2 R-M=\frac{1}{T}\left(N\frac{dT}{dt}+P\frac{d^2T}{dt^2}\right)\equiv -Q,$$

or

$$\nabla^2 R+\lambda R=0, \qquad \lambda\equiv Q-M, \tag{5}$$

$$QT+N\frac{dT}{dt}+P\frac{d^2T}{dt^2}=0. \tag{6}$$

The last part of Eq. (4) is obtained by the same reasoning as in Section 1.5.

Equation (6) is a *homogeneous second-order ordinary differential equation* and can be solved by one of the standard techniques. Let the trial solution be

$$T=T_0e^{Pt}. \tag{7}$$

Then

$$\left(p^2+\frac{N}{P}p+\frac{Q}{P}\right)T_0e^{pt}=0,$$

$$p_{1,2}=\tfrac{1}{2}\left[-N/P\pm\sqrt{(N/P)^2-4Q/P}\ \right],$$

and the general solution is

$$T=T_1e^{p_1t}+T_2e^{p_2t}, \tag{8}$$

where the constants of integration T_1 and T_2 can be evaluated from the initial conditions.

Equation (5) is known as *Helmholtz's equation*, which is still a partial differential equation. This equation has solutions which satisfy the specified boundary conditions for certain values of λ only. These particular values of λ are known as the *eigenvalues* (or characteristic values) of the problem. The corresponding solutions are known as the *eigenfunctions* (or characteristic functions).

Some of the properties of eigenvalues and eigenfunctions will be investigated in the following sections.

3.2. Nonnegativity of Eigenvalues

That eigenvalues are nonnegative [1, 3–6, 10] can be shown by using Green's theorem (A1.2.4),

$$\int_V f\nabla^2 g\, dv = \oint_S f\frac{\partial g}{\partial n}\, da - \int_V \nabla f\cdot\nabla g\, dv, \tag{1}$$

where $f \equiv f(\mathbf{r})$ and $g \equiv g(\mathbf{r})$. If $f = g = R$, then

$$\int_V R\,\nabla^2 R\, dv = \oint_S R\frac{\partial R}{\partial n}\, da - \int_V (\nabla R)\cdot(\nabla R)\, dv,$$

$$-\lambda\int_V R^2\, dv = \oint_S R\frac{\partial R}{\partial n}\, da - \int_V (\nabla R)^2\, dv, \tag{2}$$

$$(\nabla R)^2 = (\nabla R)\cdot(\nabla R) = \left(\frac{\partial R}{\partial x}\right)^2 + \left(\frac{\partial R}{\partial y}\right)^2 + \left(\frac{\partial R}{\partial z}\right)^2$$

in cartesian coordinates. Equation (3.1.5) is used to obtain the left-hand side of Eq. (2). The first term on the right of Eq. (2) vanishes for either of the boundary conditions, (a) or (b) in Eq. (3.1.2). For boundary condition (c), $\partial R/\partial n = -\xi R$, one has

$$\oint_S R\frac{\partial R}{\partial n}\, da = -\xi\oint_S R^2\, da,$$

and Eq. (2) becomes

$$\lambda\int_V R^2\, dv = \xi\oint_S R^2\, da + \int_V (\nabla R)^2\, dv. \tag{3}$$

It follows that λ is nonnegative, since each integrand in Eq. (3) is always positive. In fact, $\lambda > 0$ if R is not zero.

3.3. Orthogonality of Eigenfunctions

Let R_i and R_j be solutions of Eq. (3.1.5) with the homogeneous boundary conditions, Eq. (3.1.2). Then

$$\nabla^2 R_i + \lambda_i R_i = 0 \quad \text{or} \quad R_j\nabla^2 R_i + \lambda_i R_i R_j = 0, \tag{1}$$

$$\nabla^2 R_j + \lambda_j R_j = 0 \quad \text{or} \quad R_i\nabla^2 R_j + \lambda_j R_i R_j = 0. \tag{2}$$

The difference of Eqs. (1) and (2) is

$$R_j\nabla^2 R_i - R_i\nabla^2 R_j = (\lambda_j - \lambda_i)R_i R_j \tag{3}$$

and so

$$\int_V \left(R_j\nabla^2 R_i - R_i\nabla^2 R_j\right) dv = (\lambda_j - \lambda_i)\int_V R_i R_j\, dv. \tag{4}$$

Applying Green's theorem in symmetrical form [Eq. (A1.2.6)],

$$\int_V (f\nabla^2 g - g\nabla^2 f)\,dv = \oint_S \left(f\frac{\partial g}{\partial n} - g\frac{\partial f}{\partial n} \right) da, \tag{5}$$

to the left-hand side of Eq. (4) yields

$$\int_V (R_j \nabla^2 R_i - R_i \nabla^2 R_j)\,dv = \oint_S \left(R_j \frac{\partial R_i}{\partial n} - R_i \frac{\partial R_j}{\partial n} \right) da = 0. \tag{6}$$

The right-hand side of Eq. (6) vanishes owing to the boundary conditions, Eq. (3.1.2). Substituting Eq. (6) into Eq. (4) yields

$$\int_V R_i R_j\,dv = 0 \qquad \text{if} \quad \lambda_i \neq \lambda_j. \tag{7}$$

Thus eigensolutions having different eigenvalues are orthogonal [1, 3–6, 10].

The eigenfunctions may be normalized by dividing each by $(\int_V R_i^{\,2}\,dv)^{1/2}$,

$$R_{in} = R_i \bigg/ \left(\int_V R_i^{\,2}\,dv \right)^{1/2}, \tag{8}$$

where R_{in} is the normalized eigenfunction of R_i. Then

$$\int_V R_{in} R_{jn}\,dv = \delta_{ij}, \tag{9}$$

where δ_{ij} is the Kronecker delta function, defined as

$$\begin{aligned} \delta_{ij} &= 0 \qquad \text{for} \qquad i \neq j, \\ \delta_{ij} &= 1 \qquad \text{for} \qquad i = j. \end{aligned} \tag{10}$$

3.4. Gram–Schmidt Orthogonalization

In some cases there is more than one linearly independent solution with the same eigenvalue [1, 3, 6–8]; such cases are known as degenerate. Let $R_{i1}, R_{i2}, \ldots, R_{ip}$ be the eigensolutions with eigenvalue λ_i, each orthogonal to all other eigenfunctions R_j. An orthogonal set $r_1, r_2, \ldots, r_p$ can be constructed from $R_{i1}, R_{i2}, \ldots, R_{ip}$ as follows [1, 3, 6–8].

One starts by choosing

$$r_1 = R_{i1} \qquad \text{and} \qquad r_{1n} = r_1/|r_1|. \tag{1}$$

Next one picks the second function R_{i2} from the original set and subtracts a multiple of r_{1n} from it.

$$r_2 = R_{i2} - m_1 r_{1n}. \tag{2}$$

The factor m is chosen such that r_2 and r_{1n} are orthogonal.

$$r_{1n}\cdot r_2=0=R_{12}\cdot r_{1n}-m_1 r_{1n}\cdot r_{1n}$$

or

$$m_1=R_{i2}\cdot r_{1n}.$$

Then

$$r_2=R_{i2}-(R_{i2}\cdot r_{1n})r_{1n} \qquad \text{and} \qquad r_{2n}=r_2/|r_2|. \tag{3}$$

This construction would fail if and only if r_2 were identically zero. This would imply that R_{i1} and R_{i2} are linearly dependent, contradicting the original statement.

Next r_3 is formed by subtracting multiples of r_{1n} and r_{2n} from the third given function R_{i3}:

$$r_3=R_{i3}-m_2 r_{1n}-n_2 r_{2n}, \tag{4}$$

where m_2 and n_2 are determined such that r_3 is orthogonal to both r_{1n} and r_{2n}.

$$r_{1n}\cdot r_3=R_{i3}\cdot r_{1n}-m_2=0, \qquad \text{or} \qquad m_2=R_{i3}\cdot r_{1n}; \tag{5}$$

$$r_{2n}\cdot r_3=R_{i3}\cdot r_{2n}-n_2=0, \qquad \text{or} \qquad n_2=R_{i3}\cdot r_{2n}. \tag{6}$$

Thus

$$r_3=R_{i3}-(R_{i3}\cdot r_{1n})r_{1n}-(R_{i3}\cdot r_{2n})r_{2n}, \qquad \text{and} \qquad r_{3n}=r_3/|r_3|. \tag{7}$$

In general

$$\begin{aligned} r_k=R_{ik}&-(R_{ik}\cdot r_{1n})r_{1n}-(R_{1k}\cdot r_{2n})r_{2n}-\cdots \\ &-(R_{ik}\cdot r_{(k-1)n})r_{(k-1)n}. \end{aligned} \tag{8}$$

From the method of formation of r_k, it is clear that r_k is a linear combination of $R_{i1}, R_{i2},\ldots, R_{ip}$. This procedure could fail if r_k were identically zero for some k, which would mean that $R_{i1}, R_{i2},\ldots, R_{ir}$ were linearly dependent, contradicting the original assumption.

Example 3

Determine an orthonormal set within the interval $-1\le z\le 1$ from a group of functions defined by

$$g_n(z)=z^n, \qquad n=0,1,2,\ldots.$$

SOLUTION

Let the first function be

$$f_0(z)\equiv g_0(z)=1, \qquad M_0{}^2\equiv\int_{-1}^{1} f_0{}^2(z)\,dz=\int_{-1}^{1} dz=2.$$

The normalized function satisfies the relation

$$\int_{-1}^{1} F^2(z)\,dz=1.$$

Therefore

$$F_0(z)=f_0(z)/M_0=1/\sqrt{2}\,. \tag{1}$$

The second function is constructed according to the Gram–Schmidt orthogonalization process. First set

$$f_1(z)\equiv g_1(z)-a_0F_0(z). \tag{2}$$

Multiply Eq. (2) by $F_0(z)$ and integrate over the interval $-1\leq z\leq 1$:

$$\int_{-1}^{1} f_1(z)F_0(z)\,dz=\int_{-1}^{1} F_0(z)[g_1(z)-a_0F_0(z)]\,dz=0.$$

Because $f_1(z)$ and $F_0(z)$ are orthogonal,

$$a_0=\frac{1}{\sqrt{2}}\int_{-1}^{1} g_1(z)\,dz\Big/\int_{-1}^{1} F_0{}^2(z)\,dz=\frac{1}{\sqrt{2}}\int_{-1}^{1} z\,dz=0. \tag{3}$$

Therefore, Eqs. (2) and (3) give

$$f_1(z)=g_1(z)=z,\qquad M_1{}^2\equiv\int_{-1}^{1} f_1{}^2(z)\,dz=\tfrac{2}{3},$$

and

$$F_1(z)=f_1(z)/M_1=\sqrt{\tfrac{3}{2}}\;z. \tag{4}$$

The next function $f_2(z)$ is given by

$$f_2(z)=g_2(z)-b_1F_1(z)-b_0F_0(z), \tag{5}$$

$$\int_{-1}^{1} F_0(z)f_2(z)\,dz=0=\int_{-1}^{1}\frac{1}{\sqrt{2}}z^2\,dz-\int_{-1}^{1} b_1F_1(z)F_0(z)\,dz$$

$$-b_0\int_{-1}^{1} F_0{}^2(z)\,dz,$$

$$b_0=\frac{1}{\sqrt{2}}\int_{-1}^{1} z^2\,dz=\frac{1}{\sqrt{2}}\,\frac{1}{3}z^3\Big|_{-1}^{1}=\frac{\sqrt{2}}{3}, \tag{6}$$

$$\int_{-1}^{1} F_1(z)f_2(z)\,dz=0=\int_{-1}^{1}\sqrt{\tfrac{3}{2}}\;z^3\,dz-b_1\int_{-1}^{1} F_1{}^2(z)\,dz,$$

$$b_1=\sqrt{\tfrac{3}{2}}\;\tfrac{1}{4}z^4\Big|_{-1}^{1}=0. \tag{7}$$

Thus

$$f_2(z)=z^2-\frac{\sqrt{2}}{3}\frac{1}{\sqrt{2}}=z^2-\tfrac{1}{3},$$

$$M_2{}^2=\int_{-1}^{1} f_2{}^2(z)\,dz=\int_{-1}^{1}\left(z^4-\tfrac{2}{3}z^2+\tfrac{1}{9}\right)dz=\tfrac{8}{45}, \tag{8}$$

$$F_2(z)=f_2(z)/M_2=\sqrt{\tfrac{45}{8}}\left(z^2-\tfrac{1}{3}\right);$$

and this process continues.

3.5. Functionals and Eigenvalues

Consider the functional [1, 2, 6, 9]

$$M[f]\equiv-\int_V f\nabla^2 f\,dv. \tag{1}$$

A *functional* is defined as a quantity that depends on the entire measure of another function or functions, as a function is a quantity that depends on one or more discrete variables. In other words, the functional $M[f(s)]$ spans the space of the function $f(s)$, while a function $g(x, y, z)$ spans the space of the coordinates x, y, z. It is also a measure of a certain behavior of the function $f(\mathbf{r})$ within the region V. The normalization factor is defined to be

$$N[f]=\int_V f^2\,dv. \tag{2}$$

With this definition one can consider

$$\hat{M}[f]\equiv M[f]/N[f], \tag{3}$$

which may be interpreted as the value of $M[f]$ normalized with respect to $N[f]$.

The functional $\hat{M}[f]$ will next be minimized by the functions f satisfying the homogeneous boundary conditions. Let $g_1(\mathbf{r})$ be such a function which minimizes $\hat{M}[f]$ and p be the resulting minimum. The necessary condition to be satisfied by $g_1(\mathbf{r})$ is obtained by comparison of $\hat{M}[g_1]$ with $\hat{M}[h]$, where $h(\mathbf{r})$ is a function slightly different from $g_1(\mathbf{r})$,

$$h(\mathbf{r})\equiv g_1(\mathbf{r})+\epsilon\zeta(\mathbf{r}), \tag{4}$$

where ϵ is a small constant, and $\zeta(\mathbf{r})$ satisfies the same homogeneous boundary conditions as $g_1(\mathbf{r})$. By hypothesis

$$\hat{M}[h]=\hat{M}[g_1+\epsilon\zeta]\geq\hat{M}[g_1]=p_1, \tag{5}$$

$$\hat{M}[h]\equiv M[h]/N[h]\geq p_1,$$

$$M[g_1+\epsilon\zeta]\geq p_1 N[g_1+\epsilon\zeta],$$

$$-\int_V (g_1+\epsilon\zeta)\nabla^2(g_1+\epsilon\zeta)\,dv\geq p_1\int_V (g_1+\epsilon\zeta)^2\,dv,$$

$$-\int_V (g_1\nabla^2 g_1+\epsilon_1 g_1\nabla^2\zeta+\epsilon\zeta\nabla^2 g_1+\epsilon^2\zeta\nabla^2\zeta)\,dv$$

$$\geq p_1\int_V (g_1{}^2+2\epsilon g_1\zeta+\epsilon^2\zeta^2)\,dv \tag{6}$$

since

$$\hat{M}[g_1]=M[g_1]/N[g_1]=p_1.$$

Therefore,

$$-\int_V g_1\nabla^2 g_1\,dv=p\int_V g_1{}^2\,dv \quad\text{or}\quad \int_V (g_1\nabla^2 g_1+p_1 g_1{}^2)\,dv=0. \tag{7}$$

Since Eq. (7) is valid for any volume, its integrand must vanish, so that

$$g_1\nabla^2 g_1+p_1 g_1{}^2=0. \tag{8}$$

With Eq. (8) substituted into Eq. (6),

$$-\epsilon\left[2p_1\int_V g_1\zeta\,dv+\int_V g_1\nabla^2\zeta\,dv+\int_V \zeta\nabla^2 g_1\,dv\right]$$

$$-\epsilon^2\left[p_1\int_V \zeta^2\,dv+\int_V \zeta\nabla^2\zeta\,dv\right]\geq 0. \tag{9}$$

In order for this relation to be valid independent of the polarity of ϵ, it is necessary that the coefficient of ϵ vanish

$$2p_1\int_V g_1\zeta\,dv+\int_V (g_1\nabla^2\zeta+\zeta\nabla^2 g_1)\,dv=0. \tag{10}$$

By Green's theorem

$$\int_V (g_1\nabla^2\zeta-\zeta\nabla^2 g_1)\,dv=\oint_S\left(g_1\frac{\partial\zeta}{\partial n}-\zeta\frac{\partial g_1}{\partial n}\right)da=0 \tag{11}$$

or

$$\int_V g_1\nabla^2\zeta\,dv=\int_V \zeta\nabla^2 g_1\,dv. \tag{12}$$

The left side of Eq. (11) vanishes because both g_1 and ζ satisfy the homogeneous boundary condition. Substitution of Eq. (12) into Eq. (10) yields

$$2\int_V \zeta(\nabla^2 g_1 + p_1 g_1)\, dv = 0. \tag{13}$$

Since the function ζ is arbitrary, the above relation is valid if and only if

$$\nabla^2 g_1 + p_1 g_1 = 0 \tag{14}$$

throughout the region V. This implies that if $\hat{M}[f]$ can be minimized by a function g_1, then g_1 must satisfy Eq. (14), and the minimum value $\hat{M}[f]\|_{\min} = \hat{M}[g_1]$ is the eigenvalue p_1 of Eq. (14), under the same homogeneous conditions.

3.6. Nonnegativity of $\hat{M}[f]$

The functional $\hat{M}[f]$,

$$\hat{M}[f] \equiv M[f]/N[f], \tag{1}$$

where

$$M[f] = -\int_V f\, \nabla^2 f\, dv, \tag{2}$$

$$N[f] = -\int_V f^2\, dv \tag{3}$$

is nonnegative which can be shown as follows. For a real function f, $N[f] \geq 0$ because its integrand is always greater or equal to zero. Further, $M[f]$ can be rearranged by means of Green's first identity [Eq. (A.1.2.5)],

$$M[f] = -\oint_S f \frac{\partial f}{\partial n}\, da + \int_V (\nabla f)^2\, dv \geq 0. \tag{4}$$

The second term is never negative, since its integrand is nonnegative. The first term vanishes for either $f=0$ or $\partial f/\partial n = 0$ on the bounding surface S. For the third boundary condition, $\partial f/\partial n + \zeta f = 0$ on S, the first integrand becomes ζf^2, which is nonnegative for $\zeta > 0$. Hence $M[f]$ is never negative subject to the homogeneous boundary conditions. The nonnegativity of both $N[f]$ and $M[f]$ leads to the desired result, $\hat{M}[f] \geq 0$. This is consistent with the fact that all eigenvalues are nonnegative.

3.7. Successive Minima of $\hat{M}[f]$

It has been shown that the absolute minimum of $\hat{M}[f]$ is obtained when $f = g_1$ and g_1 satisfies the equation (Section 3.5)

$$\nabla^2 g_1 + p_1 g_1 = 0 \tag{1}$$

and the minimum value of $\hat{M}[g_1]$ is the eigenvalue p_1, the lowest eigenvalue of the above differential equation.

Next, consider the minimization of $\hat{M}[f]$ by a function g_2 which satisfies the following constraints:

$$\int_V g_2 g_1\, dv = 0. \tag{2}$$

That is, the function g_2 is orthogonal to g_1. The minimum thus obtained cannot be less than the absolute minimum p_1. Let this second minimum be p_2. By the same procedures used to obtain the absolute minimum p_1, let the comparison function be

$$h_1(\mathbf{r}) \equiv g_2(\mathbf{r}) + \epsilon_1 \zeta_1(\mathbf{r}), \tag{3}$$

where ϵ_1 is a small constant and $\zeta_1(\mathbf{r})$ is an arbitrary function which satisfies the homogeneous boundary conditions and the orthogonality relation

$$\int_V \zeta_1 g_1\, dv = 0. \tag{4}$$

The function ζ_1 which is orthogonal to the given function g_1 can be constructed by the Gram–Schmidt orthogonalization procedure (Section 3.4), that is, by subtracting from an arbitrary function $\eta(\mathbf{r})$ the portion not orthogonal to g_1:

$$\zeta_1(\mathbf{r}) = \eta(\mathbf{r}) - a g_1(\mathbf{r}) \qquad \text{or} \qquad \eta(\mathbf{r}) = \zeta_1(\mathbf{r}) + a g_1(\mathbf{r}). \tag{5}$$

The constant coefficient a is obtained as follows:

$$\int_V \eta g_1\, dv = \int_V \zeta_1 g_1\, dv + a\int_V g_1 g_1\, dv = a\int_V g_1 g_1\, dv.$$

Therefore,

$$a = \int_V \eta g_1\, dv \Big/ \int_V g_1{}^2\, dv, \tag{6}$$

since ζ_1 and g_1 are orthogonal.

The general comparison function then becomes

$$h_1(\mathbf{r}) = g_2(\mathbf{r}) + \epsilon_1[\eta(\mathbf{r}) - a g_1(\mathbf{r})], \tag{7}$$

where $\eta(\mathbf{r})$ must satisfy the boundary conditions.

Now by hypothesis

$$M[h_1] \geq p_2 N[h_1],$$

$$M[g_2 + \epsilon_1 \zeta_1] \geq p_2 N[g_2 + \epsilon_1 \zeta_1].$$

By a procedure similar to that followed in obtaining Eq. (3.5.13), one obtains

$$2\int_V \zeta_1(\nabla^2 g_2 + p_1 g_2)\, dv = 0. \tag{8}$$

Substituting ζ_1 from Eq. (5) into Eq. (8) yields

$$\int_V \eta(\nabla^2 g_2 + p_2 g_2)\, dv - a\int_V g_1(\nabla^2 g_2 + p_2 g_2)\, dv = 0. \tag{9}$$

The last term vanishes because of the orthogonality property. The third term can be transformed by Green's theorem into

$$\int_V g_1 \nabla^2 g_2\, dv = \int_V g_2 \nabla^2 g_1\, dv = p_1 \int_V g_2 g_1\, dv = 0. \tag{10}$$

Thus Eq. (9) becomes

$$\int_V \eta(\nabla^2 g_2 + p_2 g_2)\, dv = 0. \tag{11}$$

Since this must be valid for an arbitrary function $\eta(\mathbf{r})$, one concludes that

$$\nabla^2 g_2 + p_2 g_2 = 0 \tag{12}$$

throughout the volume V.

We have just shown that $\hat{M}[f]$ is minimized by a second function $g_2(\mathbf{r})$ which satisfies the eigenequation, Eq. (12). The value of $\hat{M}[g_2]$ is equal to the eigenvalue p_2 of Eq. (12).

This process can be continued indefinitely and the successive minima of $\hat{M}[g_i]$ can be identified with the corresponding eigenvalues, and

$$p_{n+1} \geq p_n \tag{13}$$

subject to the condition that each minimizing function $g_n(\mathbf{r})$ be orthogonal to all previous ones, $g_1, g_2, \ldots, g_{n-1}$.

It has been shown that p_n is the minimum value $M[f]$ for normalized functions $f(\mathbf{r})$ under the condition that $f(\mathbf{r})$ is orthogonal to $g_1(\mathbf{r}), g_2(\mathbf{r}), \ldots, g_{n-1}(\mathbf{r})$. These are the most severe conditions imposed on the function $f(\mathbf{r})$. If one requires $f(\mathbf{r})$ to be orthogonal to $n-1$ specified functions $u_1(\mathbf{r}), u_2(\mathbf{r}), \ldots, u_{n-1}(\mathbf{r})$, which may be different from $g_1(\mathbf{r}), \ldots, g_{n-1}(\mathbf{r})$, one finds that the minimum value of $M[f]$ must be less than p_n. Therefore p_n is the largest value that the minimum of $M[f]$ can assume, and this value occurs only when $u_i = g_i$, $i = 1, 2, \ldots$.

It will be shown that if a function $f(\mathbf{r})$ is orthogonal to an arbitrary sequence of functions $u_1(\mathbf{r}), u_2(\mathbf{r}), \ldots, u_{n-1}(\mathbf{r})$, then the value $M[f] \leq p_n$. Let the function be expressed

$$f(\mathbf{r}) = \sum_{k=1}^{n} C_k g_k(\mathbf{r}). \tag{14}$$

This function $f(\mathbf{r})$ can be made orthogonal to the sequence $\{u_m(\mathbf{r})\}$ by proper choice of the coefficients C_k. The orthogonality relations generate $n-1$ equations:

$$\int_V f(\mathbf{r})u_i(\mathbf{r})\,dv=0, \qquad i=1,2,\ldots,n-1. \tag{15}$$

The normalization condition

$$N[f]=\sum_{k=1}^{n} C_k^{\,2}=1 \tag{16}$$

provides the nth relation for determining the n coefficients. Now,

$$\begin{aligned}
M[f] &\equiv -\int_V f\nabla^2 f\,dv \\
&= -\int_V\left(\sum_{k=1}^{n} C_k g_k\right)\nabla^2\left(\sum_{q=1}^{n} C_q g_q\right)dv \\
&= -\int_V\left(\sum_{k=1}^{n} C_k g_k\right)\left(-\sum_{q=1}^{n} C_q p_q g_q\right)dv \\
&= \sum_{k=1}^{n} C_k^{\,2} p_k \\
&\le p_n \sum_{k=1}^{n} C_k^{\,2} \qquad (\text{since } p_n \ge p_k,\ k=1,2,\ldots) \\
&\le p_n \qquad \left(\text{since } \sum_{k=1}^{n} C_k^{\,2}=1\right).
\end{aligned}$$

Thus the p_i are not only successive minima of $M[f]$ but also the maximum values of these minima. Hence weakening the restriction on the successive comparison functions can only make the eigenvalues smaller, or at least unchanged. Conversely, no strengthening of the weak restrictions can lower the eigenvalues.

References

1. R. Courant and D. Hilbert, *Methods of Mathematical Physics*, Interscience, New York, 1953 and 1962.
2. J. W. Dettman, *Mathematical Methods in Physics and Engineering*, McGraw-Hill, New York, 1962.
3. D. G. B. Edelen and A. D. Kydoniefs, *An Introduction to Linear Algebra for Science and Engineering*, Elsevier, New York, 1972.
4. E. Kreyszig, *Advanced Engineering Mathematics*, Wiley, New York, 1968.

5. H. Lass, *Elements of Pure and Applied Mathematics*, McGraw-Hill, New York, 1957.
6. P. M. Morse and H. Feshbach, *Methods of Theoretical Physics*, McGraw-Hill, New York, 1953.
7. B. Noble, *Applied Linear Algebra*, Prentice-Hall, Englewood Cliffs, N.J., 1969.
8. K. F. Riley, *Mathematical Methods for the Physical Sciences*, Cambridge University Press, London, 1974.
9. H. Sagan, *Boundary and Eigenvalue Problems in Mathematical Physics*, Wiley, New York, 1961.
10. S. A. Schelkunoff, *Applied Mathematics for Engineers and Scientists*, Van Nostrand, Princeton, N.J., 1965.

4

Nonhomogeneous Problems

Problems described by homogeneous differential equations subject to homogeneous boundary conditions are known as homogeneous problems. This type was investigated in the last chapter.

Problems involving forcing functions or disturbances excited by initial distortion constitute the group known as nonhomogeneous. [1–6]. Nonhomogeneous problems are characterized by either nonhomogeneous differential equations or nonhomogeneous boundary conditions or both.

The nonhomogeneous problem will be analyzed in this chapter. The general problem is subdivided into simpler cases, and each case is then examined separately.

4.1. Nonhomogeneous Equation and Nonhomogeneous Boundary Condition

A general *nonhomogeneous boundary value problem* is described by the differential equation

$$\nabla^2 U = MU + N\frac{\partial U}{\partial t} + P\frac{\partial^2 U}{\partial t^2} - \tilde{F} \equiv L_t[U] - \tilde{F},$$
$$L_t[U] \equiv MU + N\frac{\partial U}{\partial t} + P\frac{\partial^2 U}{\partial t^2} \qquad (1)$$

in a region V, where the response is $U \equiv \tilde{U}(\mathbf{r}, t)$ and the forcing function $\tilde{F} = \tilde{F}(\mathbf{r}, t)$. The response is subject to the boundary condition

$$C_1\frac{\partial U}{\partial n} + C_2 U = \tilde{f}(\mathbf{r}_s, t), \qquad (2)$$

where C_1 and C_2 are constant coefficients, and $\tilde{f}(\mathbf{r}_s, t)$ is the specified function on the boundary surface indicated by the position vector $\mathbf{r}_s$.

This problem of a nonhomogeneous differential equation with nonhomogeneous boundary condition may be subdivided into three simpler cases.

I. Homogeneous equation with homogeneous boundary condition.

$$\nabla^2 U_0 = L_t[U_0], \tag{3}$$

$$C_1 \frac{\partial U_0}{\partial n} + C_2 U_0 = 0. \tag{4}$$

II. Nonhomogeneous equation with homogeneous boundary condition.

$$\nabla^2 U_1 = L_t[U_1] - \tilde{F}, \tag{5}$$

$$C_1 \frac{\partial U_1}{\partial n} + C_2 U_1 = 0. \tag{6}$$

III. Homogeneous equation with nonhomogeneous boundary condition.

$$\nabla^2 U_2 = L_t[U_2], \tag{7}$$

$$C_1 \frac{\partial U_2}{\partial n} + C_2 U_2 = \tilde{f}. \tag{8}$$

The general case is the sum of the above three cases, $U = U_0 + U_1 + U_2$, and we have

$$\nabla^2 U = L_t[U] - \tilde{F}, \tag{9}$$

$$C_1 \frac{\partial U}{\partial n} + C_2 U = \tilde{f}. \tag{10}$$

Each of the above cases will now be investigated separately.

4.2. Homogeneous Equation with Homogeneous Boundary Condition

Case I is the generalized homogeneous problem treated in Chapter 3 and can be summarized as follows. The problem is described by the equation

$$\nabla^2 U_0 = L_t[U_0], \qquad U_0 \equiv \tilde{U}_0(\mathbf{r}, t), \quad L_t = M + N\frac{\partial}{\partial t} + P\frac{\partial^2}{\partial t^2}, \tag{1}$$

in the region V and satisfies the boundary condition

$$C_1 \frac{\partial U_0}{\partial n} + C_2 U_0 = 0 \tag{2}$$

on the bounding surface S. The initial conditions are

$$U_0(\mathbf{r}, t=0) \equiv m(\mathbf{r}), \qquad \left(\frac{\partial U_0}{\partial t}\right)_{t=0} \equiv n(\mathbf{r}). \tag{3}$$

This problem can be solved by the method of separation of variables. Let

$$U_0(\mathbf{r}, t) \equiv R_0(\mathbf{r}) T_0(t). \tag{4}$$

Then Eq. (1) is decomposed into two equations,

$$\nabla^2 R_{0k} + \lambda_k R_{0k} = 0, \tag{5}$$

$$\frac{d^2T_{0k}}{dt^2} + 2\xi\omega\frac{dT_{0k}}{dt} + \omega^2 T_{0k} = 0, \qquad 2\xi\omega \equiv \frac{N}{P}, \qquad \omega^2 \equiv \frac{M+\lambda_k}{P}, \tag{6}$$

where λ_k is the constant of separation.

Equation (5) specifies the function $R_{0k}(\mathbf{r})$ in the spatial domain, subject to the boundary condition (2) on the bounding surface. This problem has discrete eigenvalues λ_k that increases without bound as $k \to \infty$, and the corresponding solution R_{0k} is known as the eigenfunction.

Equation (6) is an ordinary differential equation and its solution can be found by assuming a trial solution $T_{0k} = T_k e^{pt}$.

$$\begin{aligned} p_{1,2} &= -\xi\omega \pm j\omega\sqrt{1-\xi^2} \equiv \alpha_k \pm j\omega_k, \\ T_{0k} &= A_k e^{p_1 t} + B_k e^{p_2 t} = e^{-\alpha_k t}\left(A_k e^{j\omega_k t} + B_k e^{-j\omega_k t}\right). \end{aligned} \tag{7}$$

The general solution is

$$\begin{aligned} \tilde{U}_0(\mathbf{r}, t) &= \sum_{k=1}^{\infty} R_{0k}(\mathbf{r}) T_{0k}(t) \\ &= \sum_{k=1}^{\infty} R_{0k}(\mathbf{r}) e^{-\alpha_k t}\left(A_k e^{j\omega_k t} + B_k e^{-j\omega_k t}\right). \end{aligned} \tag{8}$$

The following relations are obtained from the initial conditions, Eq. (3):

$$m(\mathbf{r}) = \sum_{k=1}^{\infty} R_{0k}(\mathbf{r})(A_k + B_k), \tag{9}$$

$$n(\mathbf{r}) = \sum_{k=1}^{\infty} R_{0k}(\mathbf{r})\left[-\alpha_k(A_k + B_k) + j\omega_k(A_k - B_k)\right]. \tag{10}$$

If $m(\mathbf{r})$ and $n(\mathbf{r})$ are well-behaved functions, then the constants A_k and B_k may be evaluated as follows. Multiply both sides of Eq. (9) by $R_{0j}(\mathbf{r})$ and integrate over the region V:

$$\begin{aligned} \int_V m(\mathbf{r}) R_{0j}(\mathbf{r})\, dv &= \int_V R_{0j}(\mathbf{r}) \sum_{k=1}^{\infty} R_{0k}(\mathbf{r})(A_k + B_k)\, dv \\ &= \sum_{k=1}^{\infty} \left[(A_k + B_k) \int_V R_{0j}(\mathbf{r}) R_{0k}(\mathbf{r})\, dv \right], \end{aligned}$$

$$a_j \equiv A_j + B_j = \int_V m(\mathbf{r}) R_{0j}(\mathbf{r})\, dv \bigg/ \int_V R_{0j}(\mathbf{r}) R_{0j}(\mathbf{r})\, dv, \tag{11}$$

since $\int_V R_{0j} R_{0k}\, dv = 0$ for $j \neq k$. Similarly,

$$\int_V n(\mathbf{r}) R_{0j}(\mathbf{r})\, dv = \left[j\omega_j (A_j - B_j) - \alpha_j (A_j + B_j) \right] \int_V R_{0j}(\mathbf{r}) R_{0j}(\mathbf{r})\, dv,$$

$$b_j \equiv A_j - B_j = \frac{1}{j\omega_j} \left[\alpha_j a_j + \int_V n(\mathbf{r}) R_{0j}(\mathbf{r})\, dv \Big/ \int_V R_{0j}(\mathbf{r}) R_{0j}(\mathbf{r})\, dv \right], \tag{12}$$

and

$$A_j = \tfrac{1}{2}(a_j + b_j), \qquad B_j = \tfrac{1}{2}(a_j - b_j). \tag{13}$$

This completes the solution of the homogeneous problem.

4.3. Nonhomogeneous Equation with Homogeneous Boundary Condition

Case II deals with a nonhomogeneous differential equation,

$$\nabla^2 U_1 = L_t[U_1] - \tilde{F}, \qquad U_1 \equiv \tilde{U}_1(\mathbf{r}, t), \quad \tilde{F} = \tilde{F}(\mathbf{r}, t), \tag{1}$$

in region V, with the condition that

$$C_1 \frac{\partial U_1}{\partial n} + C_2 U_1 = 0 \tag{2}$$

on the boundary and the initial conditions

$$m(\mathbf{r}) = U_1(\mathbf{r}, 0) \qquad \text{and} \qquad n(\mathbf{r}) = \left(\frac{\partial U_1}{\partial t} \right)_{t=0}. \tag{3}$$

We consider three possibilities for $\tilde{F}(\mathbf{r}, t)$.

Case a: The forcing function $\tilde{F}(\mathbf{r}, t)$ is separable. We have

$$\tilde{F}(\mathbf{r}, t) \equiv F(\mathbf{r}) T_1(t). \tag{4}$$

Then let the response be

$$\tilde{U}_1(\mathbf{r}, t) \equiv R_1(\mathbf{r}) T_1(t). \tag{5}$$

Substituting of Eq. (5) into Eq. (1) yields

$$\frac{1}{R_1}\left[\nabla^2 R_1 + F(\mathbf{r}) \right] = \frac{1}{T_1} L_t[T_1] \equiv -\lambda \tag{6}$$

or

$$\nabla^2 R_1 + \lambda R_1 = -F(\mathbf{r}), \tag{7}$$

$$L_t[T_1] + \lambda T_1 = 0. \tag{8}$$

The problem now becomes that of solving Eqs. (7) and (8) subject to the conditions

$$C_1 \frac{\partial R_1}{\partial n} + C_2 R_1 = 0 \tag{9}$$

and

$$R_1(\mathbf{r})T_1(0) = m(\mathbf{r}), \qquad R_1(\mathbf{r})\left(\frac{dT_1}{dt}\right)_{t=0} = n(\mathbf{r}). \tag{10}$$

The homogeneous equation, Eq. (8), has already been treated in Section 4.2. The nonhomogeneous equation in the spatial domain will be treated in Chapter 5.

Case b: The forcing function $\tilde{F}(\mathbf{r}, t)$ is a periodic function in the time domain. If this is the case, a Fourier series may be used to represent the time function:

$$\tilde{F}(\mathbf{r}, t) = F(\mathbf{r}) \sum_{k=1}^{\infty} C_k g_k(t). \tag{11}$$

Let the trial solution be

$$\tilde{U}(\mathbf{r}, t) = R(\mathbf{r}) \sum_{k=1}^{\infty} C_k g_k(t). \tag{12}$$

Substituting Eq. (12) into (Eq. (1) yields

$$\nabla^2 R + \lambda R = -F(\mathbf{r}) \tag{13}$$

and

$$L_t[g_k] + \lambda_k g_k = 0. \tag{14}$$

The problem is to solve Eqs. (13) and (14) subject to the following conditions:

$$C_1 \frac{\partial R}{\partial n} + C_2 R = 0 \tag{15}$$

and

$$R(\mathbf{r}) \left[\sum_{k=1}^{\infty} C_k g_k(t) \right]_{t=0} = m(\mathbf{r}),$$

$$R(\mathbf{r}) \left[\sum_{k=1}^{\infty} C_k \frac{dg_k}{dt} \right]_{t=0} = n(\mathbf{r}). \tag{16}$$

Case c: A more general forcing function. Consider a forcing function given by

$$\tilde{F}(\mathbf{r}, t)=\sum_{k=1}^{\infty} F_k(\mathbf{r}) g_k(t). \tag{17}$$

Let the trial solution be

$$\tilde{U}(\mathbf{r}, t)=\sum_{k=1}^{\infty} R_k(\mathbf{r}) g_k(t). \tag{18}$$

Substituting Eq. (18) into Eq. (1) yields

$$\nabla^2 R_k+\lambda_k R_k=-F_k(\mathbf{r}) \tag{19}$$

$$L_t[g_k]+\lambda_k g_k=0. \tag{20}$$

The problem is to solve Eqs. (19) and (20) subject to the following conditions:

$$C_1 \frac{\partial R_k}{\partial n}+C_2 R_k=0, \tag{21}$$

$$m(\mathbf{r})=\left[\sum_{k=1}^{\infty} R_k C_k g_k\right]_{t=0} \quad \text{and} \quad n(\mathbf{r})=\left[\sum_{k=1}^{\infty} R_k C_k \frac{dg_k}{dt}\right]_{t=0}. \tag{22}$$

Example 4

Determine the solution of the following nonhomogeneous differential equation:

$$\frac{d^2F}{dx^2}+\pi^2 F=\pi^2 x, \qquad F \equiv F(x), \tag{1}$$

subject to the boundary conditions

$$F(0)=1 \quad \text{and} \quad F(1)=-1. \tag{2}$$

SOLUTION

The solution of this problem will be obtained by decomposing it into two simpler cases. Let

$$F(x) \equiv p(x)+q(x) \tag{3}$$

and

$$\frac{d^2p}{dx^2}+\pi^2 p=0, \qquad p(0)=0, \quad p(1)=0; \tag{4}$$

$$\frac{d^2q}{dx^2}+\pi^2 q=\pi^2 x, \qquad q(0)=1, \quad q(1)=-1. \tag{5}$$

The general solution of Eq. (4) is

$$p(x)=A\sin\pi x+B\cos\pi x, \qquad p(0)=0=B.$$

Thus

$$p(x)=A\sin\pi x. \tag{6}$$

Equation (6) also satisfies the boundary condition at $x=1$.

The general solution of Eq. (5) is

$$q(x)=Cx+D,$$

$$q(0)=1=D,$$

$$q(1)=-1=C+1, \qquad \text{or} \qquad C=-2,$$

and thus

$$q(x)=1-2x. \tag{7}$$

The complete solution of Eq. (1) is the sum of Eqs. (6) and (7),

$$F(x)\equiv p(x)+q(x)=A\sin\pi x-2x+1. \tag{8}$$

Because of the boundary conditions specified, the constant coefficient A cannot be determined. It will be shown in the next chapter that this example belongs to a class of problems having an infinite number of solutions.

4.4. Homogeneous Equation with Nonhomogeneous Boundary Condition

Case III deals with a homogeneous equation

$$\nabla^2 U_2=L_t[U_2] \tag{1}$$

in region V, subject to the boundary condition

$$C_1\frac{\partial U_2}{\partial n}+C_2U_2=\tilde{f}(\mathbf{r}_s,t) \tag{2}$$

on the bounding surface S, where $\mathbf{r}_s$ is the position vector on the surface. The initial conditions are

$$\tilde{U}_2(\mathbf{r},0)=m(\mathbf{r}) \qquad \text{and} \qquad \left(\frac{\partial\tilde{U}_2}{\partial t}\right)_{t=0}=n(\mathbf{r}). \tag{3}$$

Consider three simple forms of boundary condition.

Case a: The boundary condition is separable.
Thus

$$\tilde{f}(\mathbf{r}_s,t)=R(\mathbf{r}_s)T_2(t). \tag{4}$$

Then let the trial solution be

$$\tilde{U}(\mathbf{r}, t) \equiv R_2(\mathbf{r}) T_2(t). \tag{5}$$

Substituting Eq. (5) into Eq. (1) yields

$$\nabla^2 R_2 + \lambda R_2 = 0 \tag{6}$$

and

$$L_t[T_2] + \lambda T_2 = 0. \tag{7}$$

The problem is to solve Eqs. (6) and (7) subject to the following conditions:

$$C_1 \frac{\partial R_2}{\partial n} + C_2 R_2 = R_2(\mathbf{r}_s) \tag{8}$$

and

$$R_2(\mathbf{r}) \mathrm{T}_2(0) = m(\mathbf{r}), \qquad R_2(\mathbf{r}) \frac{dT_2}{dt}\bigg|_{t=0} = n(\mathbf{r}). \tag{9}$$

Case b: A more complicated periodic function. $T_2(t)$ can here be expanded in a Fourier series as

$$\tilde{f}(\mathbf{r}_s, t) = R(\mathbf{r}_s) \sum_{k=1}^{\infty} C_k g_k(t). \tag{10}$$

Then let

$$\tilde{U}(\mathbf{r}, t) = R_2(\mathbf{r}) \sum_{k=1}^{\infty} C_k g_k(t). \tag{11}$$

The problem is to solve

$$\nabla^2 R_2 + \lambda R_2 = 0 \tag{12}$$

and

$$L_t[g_k] + \lambda g_k = 0 \tag{13}$$

subject to the conditions

$$C_1 \frac{\partial R_2}{\partial n} + C_2 R_2 = R(\mathbf{r}_s) \tag{14}$$

and

$$m(\mathbf{r}) = R_2(\mathbf{r}) \left[\sum_{k=1}^{\infty} C_k g_k(t) \right]_{t=0}, \qquad n(\mathbf{r}) = R_2(\mathbf{r}) \left[\sum_{k=1}^{\infty} C_k \frac{g_k(t)}{dt} \right]_{t=0}. \tag{15}$$

Case c: A more general boundary condition. This may be given by

$$\tilde{f}(\mathbf{r}_s, t) = \sum_{k=1}^{\infty} f_k(\mathbf{r}_s) g_k(t). \tag{16}$$

Let the trial solution be

$$\tilde{U}(\mathbf{r}, t) = \sum_{k=1}^{\infty} R_k(\mathbf{r}) g_k(t). \tag{17}$$

The problem is then to solve

$$\nabla^2 R_k + \lambda_k R_k = 0 \tag{18}$$

and

$$L_t[g_k] + \lambda_k g_k = 0 \tag{19}$$

subject to the conditions

$$C_1 \frac{\partial R_k}{\partial n} + C_2 R_k = f_k(\mathbf{r}_s) \qquad \text{on } S \tag{20}$$

and

$$m(\mathbf{r}) = \sum_{k=1}^{\infty} R_k(\mathbf{r}) g_k(t)\bigg|_{t=0}, \qquad n(\mathbf{r}) = \sum_{k=1}^{\infty} R_k(\mathbf{r}) \frac{dg_k}{dt}\bigg|_{t=0}. \tag{21}$$

In all three cases, it is necessary to solve an equation of the form

$$\nabla^2 R + \lambda R = 0 \tag{22}$$

subject to the nonhomogeneous boundary condition

$$C_1 \frac{\partial R}{\partial n} + C_2 R = f(\mathbf{r}_s), \qquad \mathbf{r} = \mathbf{r}_s. \tag{23}$$

The combination of a homogeneous equation and a nonhomogeneous boundary condition is equivalent to a nonhomogeneous equation with a homogeneous boundary condition. This can be shown as follows.

Let $R_1(\mathbf{r})$ be an arbitrary differential function which satisfies the boundary condition

$$C_1 \frac{\partial R_1}{\partial n} + C_2 R_1 = f(\mathbf{r}_s) \qquad \text{at} \qquad \mathbf{r} = \mathbf{r}_s. \tag{24}$$

Consider the difference function $Q(\mathbf{r})$ defined by

$$Q(\mathbf{r}) \equiv R(\mathbf{r}) - R_1(\mathbf{r}). \tag{25}$$

At the boundary,

$$C_1 \frac{\partial Q}{\partial n} + C_2 Q = C_1 \frac{\partial (R - R_1)}{\partial n} + C_2 (R - R_1) = 0 \qquad (\mathbf{r} = \mathbf{r}_s). \tag{26}$$

That is, $Q(\mathbf{r})$ satisfies the homogeneous boundary condition, and

$$\begin{aligned}\nabla^2 Q+\lambda Q&=\nabla^2(R-R_1)+\lambda(R-R_1)\\&=(\nabla^2 R+\lambda R)-(\nabla^2 R_1+\lambda R_1)\\&=-(\nabla^2 R_1+\lambda R_1),\end{aligned}\tag{27}$$

which is a nonhomogeneous equation. Hence, solving Eq. (27) for $Q(\mathbf{r})$ with the homogeneous boundary condition, Eq. (26), is equivalent to solving Eq. (22) with the nonhomogeneous boundary condition, Eq. (23).

4.5. General Problem

In a general nonhomogeneous problem, one must solve an equation of the form

$$\nabla^2 R+\lambda R=F \qquad \text{within } V, \tag{1}$$

where $R\equiv R(\mathbf{r})$ and $F\equiv F(\mathbf{r})$, subject to the homogeneous boundary condition

$$C_1\frac{\partial R}{\partial n}+C_2 R=0 \qquad \text{on } S. \tag{2}$$

Let the eigenfunction for the given domain be $g_k(\mathbf{r})$; that is,

$$\nabla^2 g_k+\lambda_k g_k=0 \qquad \text{within } V \tag{3}$$

and

$$C_1\frac{\partial g_k}{\partial n}+C_2 g_k=0 \qquad \text{on } S. \tag{4}$$

$R(\mathbf{r})$ and $F(\mathbf{r})$ may be expressed in terms of these eigenfunctions as

$$F(\mathbf{r})=\sum_{k=1}^{n} a_k g_k, \tag{5}$$

$$R(\mathbf{r})=\sum_{k=1}^{n} b_k g_k. \tag{6}$$

Substituting Eqs. (5) and (6) into Eq. (1) yields

$$\nabla^2\left(\sum_{k=1}^{n} b_k g_k\right)+\lambda\left(\sum_{k=1}^{n} b_k g_k\right)=\sum_{k=1}^{n} a_k g_k,$$

$$\sum_{k=1}^{n}\left[b_k(\nabla^2 g_k+\lambda g_k)-a_k g_k\right]=0,$$

$$\sum_{k=1}^{n}\left[-b_k(\lambda_k-\lambda)-a_k\right]g_k=0, \tag{7}$$

$$b_k=a_k/(\lambda-\lambda_k). \tag{8}$$

Equation (3) is used to obtain Eq. (7).

When $\lambda=\lambda_k$, the coefficient b_k becomes infinite if $a_k \neq 0$. This is the case where the forcing function has a component whose frequency coincides with one of the eigenfunctions.

The homogeneous problem

$$\nabla^2 R+\lambda R=0 \qquad \text{within } V \tag{9}$$

and

$$C_1\frac{\partial R}{\partial n}+C_2 R=0 \qquad \text{on } S \tag{10}$$

has nontrivial solution only if λ is the eigenvalue of the domain and consequently $R(\mathbf{r})$ is the corresponding eigenfunction.

A finite solution for the nonhomogeneous problem [Eqs. (1) and (2)] does not exist when $\lambda=\lambda_k$ unless

$$\int_V F(\mathbf{r})g_k(\mathbf{r})\,dv=0. \tag{11}$$

On the other hand, if $\lambda \neq \lambda_k$, a unique solution exists for the nonhomogeneous problem.

The complete solution is given by the sum of the particular solution, Eq. (6), and the general solution of the homogeneous problem, Eqs. (9) and (10).

Example 5

Find the particular solution of the nonhomogeneous differential equation

$$\frac{d^2F}{dy^2}+\lambda F=\sin\frac{\pi y}{d}+3\sin\frac{3\pi y}{d}+5\sin\frac{5\pi y}{d}, \qquad F\equiv F(y), \tag{1}$$

that satisfies the boundary conditions

$$F(0)=0 \qquad \text{and} \qquad F(d)=0 \tag{2}$$

SOLUTION

Let the trial solution be

$$F(y)=A\sin\frac{\pi y}{d}+B\sin\frac{3\pi y}{d}+C\sin\frac{5\pi y}{d}, \tag{3}$$

where A, B, and C are some constants to be determined. The substitution of Eq. (3) into Eq. (1) yields the relation

$$A\left[-\left(\frac{\pi}{d}\right)^2+\lambda\right]\sin\frac{\pi y}{d}+B\left[-\left(\frac{3\pi}{d}\right)^2+\lambda\right]\sin\frac{3\pi y}{d}$$
$$+C\left[-\left(\frac{5\pi}{d}\right)^2+\lambda\right]\sin\frac{5\pi y}{d}=\sin\frac{\pi y}{d}+3\sin\frac{3\pi y}{d}+5\sin\frac{5\pi y}{d}$$

or

$$\left\{A\left[-\left(\frac{\pi}{d}\right)^2+\lambda\right]-1\right\}\sin\frac{\pi y}{d}+\left\{B\left[-\left(\frac{3\pi}{d}\right)+\lambda\right]-3\right\}\sin\frac{3\pi y}{d}$$

$$+\left\{C\left[-\left(\frac{5\pi}{d}\right)^2+\lambda\right]-5\right\}\sin\frac{5\pi y}{d}=0. \tag{4}$$

Equation (4) can be satisfied if and only if the coefficient of each term vanishes. Therefore

$$A=\frac{1}{\lambda-(\pi/d)^2},\qquad B=\frac{3}{\lambda-(3\pi/d)^2},\qquad C=\frac{5}{\lambda-(5\pi/d)^2}. \tag{5}$$

The complete solution of Eq. (1) is

$$F(y)=\frac{\sin(\pi y/d)}{\lambda-(\pi/d)^2}+\frac{3\sin(3\pi y/d)}{\lambda-(3\pi/d)^2}+\frac{5\sin(5\pi y/d)}{\lambda-(5\pi/d)^2}. \tag{6}$$

Equation (6) is the particular solution of Eq. (1) provided $\lambda\neq(n\pi/d)^2$ with $n=1$, 3, and 5.

References

1. R. Courant and D. Hilbert, *Methods of Mathematical Physics*, Interscience, New York, 1953 and 1962.
2. J. W. Dettman, *Mathematical Methods in Physics and Engineering*, McGraw-Hill, New York, 1962.
3. E. Kreyszig, *Advanced Engineering Mathematics*, Wiley, New York, 1968.
4. P. M. Morse and H. Feshbach, *Methods of Theoretical Physics*, McGraw-Hill, New York, 1953.
5. H. Sagan, *Boundary and Eigenvalue Problems in Mathematical Physics*, Wiley, New York, 1961.
6. S. A. Schelkunoff, *Applied Mathematics for Engineers and Scientists*, Van Nostrand, Princeton, N.J., 1965.

5

Green's Function

The previous two chapters analyzed the homogeneous and nonhomogeneous problems. The solution of nonhomogeneous differential equations will now be examined. The chapter begins with an investigation of the simple problem of excitation by a unit point source. The basic method of solution will then be generalized using Green's function techniques [1–9] to the problem of an arbitrary forcing function. The properties of Green's function and the existence of solutions will also be discussed.

5.1. Influence Function

The nonhomogeneous problem is described by

$$L[U]=f(x), \qquad L[U]\equiv m(x)U(x)+n(x)\dot{U}(x)+p(x)\ddot{U}(x) \quad (1)$$

in the region $a\leq x\leq b$, where $\dot{U}(x)\equiv dU/dx$ and $\ddot{U}(x)\equiv d^2U/dx^2$. The homogeneous boundary conditions at the endpoints are

$$\begin{aligned} &a_{11}\dot{U}(a)+a_{12}U(a)=0, \\ &a_{21}\dot{U}(b)+a_{22}U(b)=0. \end{aligned} \quad (2)$$

The continuous forcing function $f(x)$ may be viewed approximately as a sum of discrete point sources $\bar{f}_1, \bar{f}_2,\ldots, \bar{f}_n$ located at various points $x_1, x_2,\ldots, x_n$, respectively, within the interval. The point source may be obtained as follows.

For a given $f(x)$, let the interval $a<x<b$ be subdivided into n segments as shown in Figure 6.

The length Δl of each segment is then given by

$$\Delta l=(b-a)/n$$

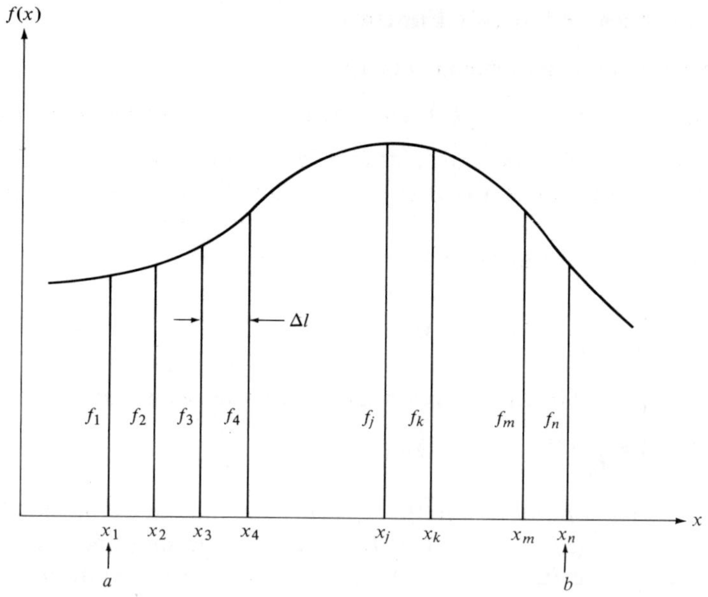

Figure 6. Driving function.

and the position x_k is

$$x_k = k\,\Delta l, \qquad k = 1, 2, \ldots.$$

Then the average value of the forcing function over the kth segment is

$$\bar{f}_k = \frac{1}{\Delta l}\int_{x_{k-1}}^{x_k} f(s)\,ds. \tag{3}$$

Let the response caused by a point source of unity strength located at x_k be denoted by $G(x, x_k)$. The response produced by a point source of strength $\bar{f}_k(x_k)$ at x_k will then be $G(x, x_k)\bar{f}_k(x_k)$. The total response produced by all sources $\bar{f}_1, \bar{f}_2, \ldots, \bar{f}_n$ is the linear sum of the responses due to the individual sources,

$$U(x) \simeq \sum_{k=1}^{n} G(x, x_k)\bar{f}_k(x_k). \tag{4}$$

This approximation can be improved by reducing the length of segment Δl, and in the limit $\Delta l \to 0$ one has

$$U(x) = \int_a^b G(x, x')f(x')\,dx'. \tag{5}$$

The influence function $G(x, x')$ is known as *Green's function*. Physically, Green's function is the response of a system to a point source of unit strength.

5.2. Properties of Green's Function

Consider the nonhomogeneous equation

$$L[U]=f(x), \qquad L[U]\equiv m(x)U(x)+n(x)\dot{U}(x)+p(x)\ddot{U}(x), \quad (1)$$

where again $\dot{U}\equiv dU/dx$ and $\ddot{U}\equiv d^2U/dx^2$. Equation (1) is valid in the interval $a\le x\le b$ and $U(x)$ satisfies the following boundary conditions:

$$\begin{aligned} &a_{11}\left.\frac{\partial U}{\partial x}\right|_{x=a}+a_{12}U(a)=0,\\ &a_{21}\left.\frac{\partial U}{\partial x}\right|_{x=b}+a_{22}U(b)=0. \end{aligned} \quad (2)$$

Assume that a Green's function $G(x, x')$ exists such that

$$U(x)=\int_a^b G(x,x')f(x')\,dx'. \quad (3)$$

The necessary conditions on the Green's function may be obtained by substituting Eq. (3) into Eq. (1). In so doing one must be careful when differentiating the function $U(x)$. Although $G(x, x')$ is continuous in the interval $a\le x\le b$, its derivative could be discontinuous at $x=x'$.

Equation (3) can be written

$$U(x)=\lim_{\epsilon\to 0}\left[\int_a^{x-\epsilon}G(x,x')f(x')\,dx'+\int_{x+\epsilon}^b G(x,x')f(x')\,dx'\right], \quad (4)$$

and the derivatives of $U(x)$ can be obtained by *Leibnitz's rule*:

$$\begin{aligned}\frac{d}{dt}\left[\int_{a(t)}^{b(t)}F(x,t)\,dx\right]&=F(b,t)\frac{db(t)}{dt}-F(a,t)\frac{da(t)}{dt}\\ &\quad+\int_{a(t)}^{b(t)}\frac{\partial F(x,t)}{\partial t}dx, \end{aligned} \quad (5)$$

$$\begin{aligned}\dot{U}(x)&=\lim_{\epsilon\to 0}\left[G(x,x-\epsilon)f(x-\epsilon)\right.\\ &\quad+\int_a^{x-\epsilon}\frac{\partial G(x,x')}{\partial x}f(x')\,dx'\\ &\quad\left.+\int_{x+\epsilon}^{b}\frac{\partial G(x,x')}{\partial x}f(x')\,dx'-G(x,x+\epsilon)f(x+\epsilon)\right]\\ &=\lim_{\epsilon\to 0}\left[\int_a^{x-\epsilon}\frac{\partial G(x,x')}{\partial x}f(x')\,dx'\right.\\ &\quad\left.+\int_{x+\epsilon}^{b}\frac{\partial G(x,x')}{\partial x}f(x')\,dx'\right]. \end{aligned} \quad (6)$$

Since $G(x, x')$ is a continuous function,

$$\lim_{\epsilon\to 0} G(x, x-\epsilon) = \lim_{\epsilon\to 0} G(x, x+\epsilon),$$

$$\begin{aligned}\ddot{U}(x) &= \frac{d}{dx}\left\{ \lim_{\epsilon\to 0}\left[\int_a^{x-\epsilon} \frac{\partial G(x,x')}{\partial x} f(x')\,dx' + \int_{x+\epsilon}^{b} \frac{\partial G(x,x')}{\partial x} f(x')\,dx' \right]\right\} \\ &= \lim_{\epsilon\to 0}\left\{ f(x)\left[\frac{\partial G(x,x-\epsilon)}{\partial x} - \frac{\partial G(x,x+\epsilon)}{\partial x}\right] + \int_a^{x-\epsilon} \frac{\partial G^2(x,x')}{\partial x^2} f(x')\,dx' + \int_{x+\epsilon}^{b} \frac{\partial^2 G(x,x')}{\partial x^2} f(x')\,dx' \right\}. \qquad (7)\end{aligned}$$

Substitution of Eqs. (4), (6), and (7) into Eq. (1) yields

$$\begin{aligned}L[U] &= \lim_{\epsilon\to 0}\left\{ m(x)\left[\int_a^{x-\epsilon} G(x,x')f(x')\,dx' + \int_{x+\epsilon}^{b} G(x,x')f(x')\,dx' \right] \right. \\ &\quad + n(x)\left[\int_a^{x-\epsilon} \frac{\partial G(x,x')}{\partial x} f(x')\,dx' + \int_{x+\epsilon}^{b} \frac{\partial G(x,x')}{\partial x} f(x')\,dx' \right] \\ &\quad + p(x)\left[f(x)\left[\frac{\partial G(x,x-\epsilon)}{\partial x} - \frac{\partial G(x,x+\epsilon)}{\partial x}\right] \right. \\ &\quad \left.\left. + \int_a^{x-\epsilon} \frac{\partial^2 G(x,x')}{\partial x^2} f(x')\,dx' + \int_{x+\epsilon}^{b} \frac{\partial^2 G(x,x')}{\partial x^2} f(x')\,dx' \right]\right\} \\ &= \lim_{\epsilon\to 0}\left\{ p(x)f(x)\left[\frac{\partial G(x,x-\epsilon)}{\partial x} - \frac{\partial G(x,x+\epsilon)}{\partial x}\right] \right. \\ &\quad + \int_a^{x-\epsilon} L[G(x,x')]\,f(x')\,dx \\ &\quad \left. + \int_{x+\epsilon}^{b} L[G(x,x')]\,f(x')\,dx' \right\} = f(x). \qquad (8)\end{aligned}$$

Equation (8) will be valid—that is, Eq. (3) will be a solution of Eq. (1)—if $G(x, x')$ satisfies the following conditions:

a. The Green's function is a solution of the associated homogeneous equation

$$L[G(x,x')]=0 \tag{9}$$

everywhere in the interval $a \leq x \leq b$ except at $x=x'$.

b. At $x=x'$, the derivative of Green's function satisfies the relation

$$\lim_{\epsilon\to 0}\left[\left.\frac{\partial G(x,x')}{\partial x}\right|_{x'=x-\epsilon}-\left.\frac{\partial G(x,x')}{\partial x}\right|_{x'=x+\epsilon}\right]=p(x)^{-1}$$

or

$$\lim_{\epsilon\to 0}\left[\left.\frac{\partial G(x,x')}{\partial x}\right|_{x'=x+\epsilon}^{x'=x-\epsilon}\right]=p(x)^{-1}. \tag{10}$$

The boundary conditions for Green's function can be determined as follows. The value of $U(x)$ at $x=a$ is given by Eq. (4):

$$U(x=a)=\lim_{\epsilon\to 0}\left[\int_a^{a-\epsilon}G(a,x')f(x')\,dx'+\int_{a+\epsilon}^{b}G(a,x')f(x')\,dx\right]$$

$$=\lim_{\epsilon\to 0}\int_{a+\epsilon}^{b}G(a,x')f(x')\,dx'. \tag{11}$$

Similarly, at $x=b$,

$$U(b)=\lim_{\epsilon\to 0}\int_a^{b-\epsilon}G(b,x')f(x')\,dx'. \tag{12}$$

The value of $\dot{U}(x)$ is given by Eq. (6) and the corresponding values at the endpoints are

$$\dot{U}(a)=\lim_{\epsilon\to 0}\int_{a+\epsilon}^{b}\left.\frac{\partial G(x,x')}{\partial x}\right|_{x=a}f(x')\,dx', \tag{13}$$

$$\dot{U}(b)=\lim_{\epsilon\to 0}\int_a^{b-\epsilon}\left.\frac{\partial G(x,x')}{\partial x}\right|_{x=b}f(x')\,dx'. \tag{14}$$

Substituting Eqs. (11)–(14) into Eq. (2) yields

$$\lim_{\epsilon\to 0}\int_{a+\epsilon}^{b}\left[a_{11}\left.\frac{\partial G(x,x')}{\partial x}\right|_{x=a}+a_{12}G(a,x')\right]f(x')\,dx'=0, \tag{15}$$

$$\lim_{\epsilon\to 0}\int_a^{b-\epsilon}\left[a_{21}\left.\frac{\partial G(x,x')}{\partial x}\right|_{x=b}+a_{22}G(b,x')\right]f(x')\,dx'=0. \tag{16}$$

Equations (15) and (16) are satisfied if the Green's function satisfies the same boundary conditions as U [Eq. (2)], that is, if

$$\left[a_{11}\dot{G}(x,x')+a_{12}G(x,x')\right]_{x=a}=0,$$
$$\left[a_{21}\dot{G}(x,x')+a_{22}G(x,x')\right]_{x=b}=0. \tag{17}$$

Summary

The solution of the nonhomogeneous equation

$$L[U]=f(x) \tag{18}$$

satisfying the homogeneous boundary conditions

$$\left[a_{11}\dot{U}+a_{12}U\right]_{x=a}=0,$$
$$\left[a_{21}\dot{U}+a_{22}U\right]_{x=b}=0 \tag{19}$$

is given by

$$U(x)=\int_a^b G(x,x')f(x')\,dx'$$
$$=\lim_{\epsilon\to 0}\left[\int_a^{x-\epsilon} G(x,x')f(x')\,dx'+\int_{x+\epsilon}^{b} G(x,x')f(x')\,dx'\right],$$

where $G(x,x')$, the Green's function of the problem, satisfies the following conditions.

a. $L[G(x,x')]=0$ everywhere except at $x=x'$.
b. At $x=x'$,

$$\lim_{\epsilon\to 0}\left[\dot{G}(x,x')\big|_{x'=x-\epsilon}-\dot{G}(x,x')\big|_{x'=x+\epsilon}\right]=1/p(x).$$

c. The Green's function satisfies the boundary conditions of Eq. (2),

$$\left[a_{11}\dot{G}(x,x')+a_{12}G(x,x')\right]_{x=a}=0,$$
$$\left[a_{21}\dot{G}(x,x')+a_{22}G(x,x')\right]_{x=b}=0.$$

Example 6

Find the solution of the following differential equation by the Green's function method:

$$\frac{d^2U(x)}{dx^2}+U(x)=x \tag{1}$$

with boundary conditions

$$U(0)=0 \qquad \text{and} \qquad U(\pi/2)=0. \tag{2}$$

SOLUTION

The Green's function $G(x, x')$ of this problem satisfies the homogeneous equation

$$\frac{d^2G(x,x')}{dx^2}+G(x,x')=0 \tag{3}$$

and the boundary conditions

$$G(0,x')=0 \qquad \text{and} \qquad G(\pi/2,x')=0. \tag{4}$$

It is convenient to choose the Green's function to have the form

$$G(x,x')=f(x)g(x'). \tag{5}$$

The function $F(x)$ is determined by substituting Eq. (5) into Eq. (3):

$$g(x')\left[\frac{d^2f(x)}{dx^2}+f(x)\right]=0. \tag{6}$$

Either $\sin x$ or $\cos x$ satisfies Eqs. (4) and (6). The Green's function is chosen as follows:

$$G(x,x')=\begin{cases} g_1(x')\sin x\equiv G(x'>x), & 0\le x\le x', \\ g_2(x')\cos x\equiv G(x'<x), & x'<x\le \pi/2. \end{cases} \tag{7}$$

The functions $g_1(x')$ and $g_2(x')$ are determined by imposing the condition (5.2.10).

$$\lim_{\epsilon\to 0}\left[\frac{\partial G(x,x')}{\partial x}\Bigg|_{x'=x+\epsilon}^{x'=x-\epsilon}\right]=p(x)^{-1}, \tag{8}$$

$$\lim_{\epsilon\to 0}\left[\frac{\partial G(x'<x)}{\partial x}\Bigg|_{x'=x-\epsilon}-\frac{\partial G(x'>x)}{\partial x}\Bigg|_{x'=x+\epsilon}\right]=1, \qquad p(x)=1$$

$$-g_2(x)\sin x-g_1(x)\cos x=1. \tag{9}$$

Equation (9) will be satisfied if one chooses

$$g_2(x)=-\sin x \qquad \text{and} \qquad g_1(x)=-\cos x. \tag{10}$$

The Green's function [Eq. (7)] then becomes

$$G(x,x')=\begin{cases} G(x'>x)=-\cos x'\sin x, & 0\le x<x', \\ G(x'<x)=-\sin x'\cos x, & x'<x\le \pi/2. \end{cases} \tag{11}$$

The solution of the problem is given by Eq. (5.2.4),

$$U(x)=\lim_{\epsilon\to 0}\left[\int_0^{x-\epsilon}G(x'<x)f(x')\,dx'+\int_{x+\epsilon}^{\pi/2}G(x'>x)f(x')\,dx'\right]$$

$$=\lim_{\epsilon\to 0}\left(\int_0^{x-\epsilon}-x'\cos x\sin x'\,dx'+\int_{x+\epsilon}^{\pi/2}-x'\cos x'\sin x\,dx'\right)$$

$$=x-(\pi/2)\sin x. \tag{12}$$

It can be shown by direct substitution that Eq. (12) satisfies Eqs. (1) and (2).

5.3. Method of Variation of Parameters

Consider the second-order nonhomogeneous differential equation

$$L[U]=f(x), \qquad L[U]\equiv m(x)U(x)+n(x)\dot{U}(x)+p(x)\ddot{U}(x) \quad (1)$$

in the interval $a \le x \le b$, and assume that $f(x)$, $m(x)$, $n(x)$, and $p(x)$ are continuous functions within this interval.

Let $U_1(x)$ and $U_2(x)$ be two independent solutions of the corresponding homogeneous equation $L[U]=0$. Then the solution to the nonhomogeneous equation (1) may be assumed to be

$$U(x)=v_1(x)U_1(x)+v_2(x)U_2(x), \quad (2)$$

where the parameters $v_1(x)$ and $v_2(x)$ are to be determined by substituting Eq. (2) back into Eq. (1). The first and second derivatives of $U(x)$ are

$$\dot{U}(x)=(v_1\dot{U}_1+v_2\dot{U}_2)+(\dot{v}_1U_1+\dot{v}_2U_2), \quad (3)$$

$$\ddot{U}(x)=\frac{d}{dx}(v_1\dot{U}_1+v_2\dot{U}_2)+\frac{d}{dx}(\dot{v}_1U_1+\dot{v}_2U_2). \quad (4)$$

Now, v_1 and v_2 are some auxiliary functions that in combination with U_1 and U_2 give the particular solution. In order to simplify the determination of these auxiliary functions we impose the condition that the second expression in parentheses in Eq. (3) must vanish; thus

$$\dot{v}_1U_1+\dot{v}_2U_2=0. \quad (5)$$

Any other choice for these terms would lead to second derivatives of $v_1(x)$ and $v_2(x)$ in Eq. (4); as a consequence, determining $v_1(x)$ and $v_2(x)$ would require solving a second-order differential equation, which is no simpler than the original problem of solving Eq. (1).

However, with the restriction (5), Eqs. (3) and (4) become

$$\dot{U}(x)=v_1\dot{U}_1+v_2\dot{U}_2, \quad (6)$$

$$\ddot{U}(x)=v_1\ddot{U}_1+v_2\ddot{U}_2+\dot{v}_1\dot{U}_1+\dot{v}_2\dot{U}_2, \quad (7)$$

respectively. Substituting Eqs. (6) and (7) into Eq. (1) yields

$$v_1L[U_1]+v_2L[U_2]+\dot{v}_1\dot{U}_1+\dot{v}_2\dot{U}_2=f(x)$$

or

$$\dot{v}_1\dot{U}_1+\dot{v}_2\dot{U}_2=f(x). \quad (8)$$

Since both $U_1(x)$ and $U_2(x)$ satisfy the homogeneous equation $L[U]=0$, $\dot{v}_1(x)$ and $\dot{v}_2(x)$, and consequently $v_1(x)$ and $v_2(x)$, may be determined from Eqs. (5) and (8).

$$\dot{v}_1(x)=\frac{1}{W(U_1,U_2)}\begin{vmatrix}0 & U_2\\ f(x) & \dot{U}_2\end{vmatrix}=\frac{-fU_2}{W(U_1,U_2)},$$
$$v_1(x)=\int\frac{-f(s)U_2(s)}{W(U_1,U_2)}\,ds, \tag{9}$$

$$\dot{v}_2(x)=\frac{1}{W(U_1,U_2)}\begin{vmatrix}U_1 & 0\\ \dot{U}_1 & f(x)\end{vmatrix}=\frac{fU_1}{W(U_1,U_2)},$$
$$v_2(x)=\int\frac{f(s)U_1(s)}{W(U_1,U_2)}\,ds, \tag{10}$$

where

$$W(U_1,U_2)\equiv\begin{vmatrix}U_1 & U_2\\ \dot{U}_1 & \dot{U}_2\end{vmatrix}=U_1\dot{U}_2-\dot{U}_1U_2. \tag{11}$$

The particular solution of Eq. (1) is obtained by substituting Eqs. (9) and (10) into Eq. (2):

$$U(x)=U_1(x)\int\frac{-f(s)U_2(s)}{W(U_1,U_2)}\,ds+U_2(x)\int\frac{f(s)U_1(s)}{W(U_1,U_2)}\,ds. \tag{12}$$

Equation (12) contains not only the particular integrals but also the complementary solutions, since the constants of integration produce terms of the form $C_1U_1(x)+C_2U_2(x)$. Equation (12) is therefore actually the complete solution of Eq. (1).

Equation (12) can be expressed in terms of definite integrals as

$$\begin{aligned}U(x)&=U_1(x)\int_{x_0}^{x}\frac{-f(s)U_2(s)}{W(U_1,U_2)}\,ds+U_2(x)\int_{x_0}^{x}\frac{f(s)U_1(s)}{W(U_1,U_2)}\,ds\\&=\int_{x_0}^{x}\frac{U_1(s)U_2(x)}{W(U_1,U_2)}f(s)\,ds-\int_{x_0}^{x}\frac{U_1(x)U_2(s)}{W(U_1,U_2)}f(s)\,ds\\&=\int_{x_0}^{x}\frac{1}{W(U_1,U_2)}\left[U_1(s)U_2(x)-U_1(x)U_2(s)\right]f(s)\,ds,\\W(U_1,U_2)&\equiv W\left[U_1(s),U_2(s)\right],\end{aligned} \tag{13}$$

where x_0 is some reference point in the interval $a\le x\le b$. The endpoints of the interval can be used as the references; that is, one can take

$$\begin{aligned}U(x)&=\int_a^x\frac{U_1(s)U_2(x)}{W(U_1,U_2)}f(s)\,ds+\int_x^b\frac{U_1(x)U_2(s)}{W(U_1,U_2)}f(s)\,ds\\&=\int_a^x G(x,s)f(s)\,ds+\int_x^b G(x,s)f(s)\,ds\\&=\int_a^b G(x,s)f(s)\,ds,\end{aligned} \tag{14}$$

where

$$G(x,s)=\frac{U_1(s)U_2(x)}{W(U_1,U_2)}, \qquad a<s<x,$$
$$=\frac{U_1(x)U_2(s)}{W(U_1,U_2)}, \qquad x<s<b. \tag{15}$$

5.4. Green's Function Method

The solution of the nonhomogeneous equation

$$\nabla^2 U(\mathbf{r})+\lambda U(\mathbf{r})=F(\mathbf{r}) \tag{1}$$

in a finite region V may be obtained by means of the Green's function. The homogeneous boundary condition for the problem is

$$C_1\frac{\partial U}{\partial n}+C_2U=0 \qquad \text{or} \qquad \frac{\partial U}{\partial n}+\xi U=0, \quad \xi\equiv\frac{C_2}{C_1} \tag{2}$$

on the bounding surface S.

The method of Green's function involves finding the Green's function $G(\mathbf{r},\mathbf{r}')$ corresponding to the equation to be solved. The Green's function satisfies

$$\nabla^2 G(\mathbf{r},\mathbf{r}')+G(\mathbf{r},\mathbf{r}')=\delta(\mathbf{r}-\mathbf{r}'), \tag{3}$$

where $\delta(\mathbf{r}-\mathbf{r}')$ is the Dirac delta function (Appendix 2), with properties

$$\int_V \delta(\mathbf{r}-\mathbf{r}')\,dv=1, \qquad \mathbf{r}' \text{ within } V,$$
$$=0, \qquad \mathbf{r}' \text{ exterior to } V. \tag{4}$$

Multiply Eq. (1) by $G(\mathbf{r},\mathbf{r}')$ and integrate over the volume V of the source points.

$$\int_V\left[G(\mathbf{r},\mathbf{r}')\nabla^2U(\mathbf{r})+\lambda G(\mathbf{r},\mathbf{r}')U(\mathbf{r})\right]dv'=\int_V G(\mathbf{r},\mathbf{r}')F(\mathbf{r}')\,dv'. \tag{5}$$

Next multiply Eq. (3) by $U(\mathbf{r}')$ and integrate over the volume V of the source points.

$$\int_V\left[U(\mathbf{r}')\,\nabla^2G(\mathbf{r},\mathbf{r}')+\lambda G(\mathbf{r},\mathbf{r}')U(\mathbf{r}')\right]dv'=\int_V \delta(\mathbf{r}-\mathbf{r}')U(\mathbf{r}')\,dv'. \tag{6}$$

Subtracting Eq. (6) from Eq. (5) yields

$$\int_V\left[G(\mathbf{r},\mathbf{r}')\,\nabla^2U(\mathbf{r})-U(\mathbf{r}')\,\nabla^2G(\mathbf{r},\mathbf{r}')\right]dv'$$
$$=\int_V G(\mathbf{r},\mathbf{r}')F(\mathbf{r}')\,dv'-\int_V\delta(\mathbf{r}-\mathbf{r}')U(\mathbf{r}')\,dv', \tag{7}$$

$$U(\mathbf{r})=\int_V G(\mathbf{r},\mathbf{r}')F(\mathbf{r}')\,dv'-\oint_S\left[G(\mathbf{r},\mathbf{r}')\frac{\partial U(\mathbf{r}')}{\partial n}-U(\mathbf{r}')\frac{\partial G(\mathbf{r},\mathbf{r}')}{\partial n}\right]da'$$

$$=\int_V G(\mathbf{r},\mathbf{r}')F(\mathbf{r}')\,dv'+\oint_S U(\mathbf{r}')\left(\frac{\partial G}{\partial n}+\xi G\right)da'. \tag{8}$$

Green's theorem (Section A1.2),

$$\int_V (G\nabla^2 U - U\nabla^2 G)\, dv' = \int_S \left(G\frac{\partial U}{\partial n} - U\frac{\partial G}{\partial n} \right) da', \tag{9}$$

is used to obtain Eq. (7).

The boundary condition, Eq. (2), is used to obtain Eq. (8). Equation (8) implies that the solution of Eq. (1) is given by

$$U(\mathbf{r}) = \int_V G(\mathbf{r}, \mathbf{r}') F(\mathbf{r})\, dv' \tag{10}$$

if the Green's function $G(\mathbf{r}, \mathbf{r}')$ satisfies the same boundary condition as U, that is, if

$$\frac{\partial G(\mathbf{r}, \mathbf{r}')}{\partial n} + \xi G(\mathbf{r}, \mathbf{r}') = 0 \tag{11}$$

on the bounding surface S.

Summary

The solution of the nonhomogeneous equation, Eq. (1), may be obtained by solving the same equation with an impulse forcing function, Eq. (3). The solution of Eq. (3) is known as the Green's function of the problem. If the Green's function is made to satisfy the boundary condition for the original problem, Eq. (2), then the solution of Eq. (1) is given by Eq. (10). Equation (10) may be interpreted as the summation of the impulse response weighted by the actual forcing function.

The Green's function $G(\mathbf{r}, \mathbf{r}')$ is symmetrical with respect to its variables.

$$G(\mathbf{r}, \mathbf{r}') = G(\mathbf{r}', \mathbf{r}). \tag{12}$$

The Green's function $G(\mathbf{r}, \mathbf{r}_1')$ for an impulse forcing function $\delta(\mathbf{r} - \mathbf{r}_1')$ located at $\mathbf{r}_1'$ is determined by

$$\nabla^2 G(\mathbf{r}, \mathbf{r}_1') + \lambda G(\mathbf{r}, \mathbf{r}_1') = \delta(\mathbf{r} - \mathbf{r}_1'). \tag{13}$$

The corresponding Green's function $G(\mathbf{r}, \mathbf{r}_2')$ for an impulse $\delta(\mathbf{r} - \mathbf{r}_2')$ satisfies

$$\nabla^2 G(\mathbf{r}, \mathbf{r}_2') + \lambda G(\mathbf{r}, \mathbf{r}_2') = \delta(\mathbf{r} - \mathbf{r}_2'). \tag{14}$$

Both Green's functions satisfy the same homogeneous boundary condition, Eq. (11). The application of Green's theorem to $G(\mathbf{r}, \mathbf{r}_1')$ and $G(\mathbf{r}, \mathbf{r}_2')$ yields

$$\begin{aligned}
&\int_V \left[G(\mathbf{r}, \mathbf{r}_1') \nabla^2 G(\mathbf{r}, \mathbf{r}_2') - G(\mathbf{r}, \mathbf{r}_2') \nabla^2 G(\mathbf{r}, \mathbf{r}_1') \right] dv' \\
&= \oint_S \left[G(\mathbf{r}, \mathbf{r}_1') \frac{\partial G(\mathbf{r}, \mathbf{r}_2')}{\partial n} - G(\mathbf{r}, \mathbf{r}_2') \frac{\partial G(\mathbf{r}, \mathbf{r}_1')}{\partial n} \right] da' \\
&= \oint_S \{ G(\mathbf{r}, \mathbf{r}_1')[-\xi G(\mathbf{r}, \mathbf{r}_2')] - G(\mathbf{r}, \mathbf{r}_2')[-\xi G(\mathbf{r}, \mathbf{r}_1')] \}\, da' \\
&= 0.
\end{aligned} \tag{15}$$

Substituting Eq. (13) and (14) into Eq. (15) yields

$$\int_V \{G(\mathbf{r},\mathbf{r}_1')[\delta(\mathbf{r}-\mathbf{r}_2')-\lambda G(\mathbf{r},\mathbf{r}_2')]$$

$$-G(\mathbf{r},\mathbf{r}_2')[\delta(\mathbf{r}-\mathbf{r}_1')-\lambda G(\mathbf{r},\mathbf{r}_1')]\}\,dv'=0$$

$$\int_V [G(\mathbf{r},\mathbf{r}_1')\delta(\mathbf{r}-\mathbf{r}_2')-G(\mathbf{r},\mathbf{r}_2')\delta(\mathbf{r}-\mathbf{r}_1')]\,dv'=0$$

$$G(\mathbf{r}_2',\mathbf{r}_1')-G(\mathbf{r}_1',\mathbf{r}_2')=0.$$

In general,

$$G(\mathbf{r}_j',\mathbf{r}_k')=G(\mathbf{r}_k',\mathbf{r}_j'), \tag{16}$$

which is identical to Eq. (12) with different notation. Equation (16) implies that the response at position $\mathbf{r}_j'$, caused by an impulse at $\mathbf{r}_k'$ is identical to the response at position $\mathbf{r}_k'$ produced by an impulse located at $\mathbf{r}_j'$.

5.5. Integral Equation Method

The general solution of the equation

$$\nabla^2 f(\mathbf{r})+\lambda f(\mathbf{r})=F(\mathbf{r}) \tag{1}$$

has been obtained by means of the Green's function, which satisfies the equation

$$\nabla^2 g(\mathbf{r},\mathbf{r}')+\lambda g(\mathbf{r},\mathbf{r}')=\delta(\mathbf{r}-\mathbf{r}'). \tag{2}$$

The boundary condition is

$$\frac{\partial f}{\partial n}+\xi f=0. \tag{3}$$

For the special case when $\lambda=0$, Eq. (1) becomes

$$\nabla^2 f_0(\mathbf{r})=F(\mathbf{r}) \tag{4}$$

and the corresponding equation for the Green's function is

$$\nabla^2 g_0(\mathbf{r},\mathbf{r})=\delta(\mathbf{r}-\mathbf{r}'), \tag{5}$$

where the subscript 0 indicates $\lambda=0$. The application of Green's theorem to the functions $f_0(\mathbf{r})$ and $g_0(\mathbf{r},\mathbf{r}')$ yields

$$\int_V (f_0\nabla^2 g_0-g_0\nabla^2 f_0)\,dv'=\oint_S\left(f_0\frac{\partial g_0}{\partial n}-g_0\frac{\partial f_0}{\partial n}\right)da'=0,$$

$$\int_V [f_0\delta(\mathbf{r}-\mathbf{r}')-g_0F(\mathbf{r}')]\,dv'=0,$$

$$f_0(\mathbf{r})=\int_V g_0(\mathbf{r},\mathbf{r}')F(\mathbf{r}')\,dv'. \tag{6}$$

Now consider the Helmholtz equation

$$\nabla^2 h(\mathbf{r}) + \lambda h(\mathbf{r}) = 0$$

or

$$\nabla^2 h(\mathbf{r}) = -\lambda h(\mathbf{r}). \tag{7}$$

Then by virtue of Eq. (6), one has

$$h(\mathbf{r}) = -\lambda \int_V g_0(\mathbf{r},\mathbf{r}')h(\mathbf{r})\, dv'. \tag{8}$$

This is an integral equation for the unknown function $h(\mathbf{r})$. Two properties of the solution will now be shown.

To see the first, let there be a solution $h_j(\mathbf{r})$ corresponding to a value λ_j. Then

$$\nabla^2 h_j(\mathbf{r}) + \lambda_j h_j(\mathbf{r}) = 0, \qquad h_j(\mathbf{r}) = -\lambda_j \int_V g_0(\mathbf{r},\mathbf{r}')h_j(\mathbf{r}')\, dv'. \tag{9}$$

For $\lambda = \lambda_k$ the solution is $h_k(\mathbf{r})$:

$$\nabla^2 h_k(\mathbf{r}) + \lambda_k h_k(\mathbf{r}) = 0, \qquad h_k(\mathbf{r}) = -\lambda_k \int_V g_0(\mathbf{r},\mathbf{r}')h_k(\mathbf{r}')\, dv'. \tag{10}$$

Multiplying Eq. (9) by $\lambda_k h_k(\mathbf{r})$ and integrate to obtain

$$\begin{aligned}
\int_V \lambda_k h_k(\mathbf{r}) h_j(\mathbf{r})\, dv &= -\lambda_j \lambda_k \int_V h_k(\mathbf{r})\, dv \int_{V'} g_0(\mathbf{r},\mathbf{r}')h_j(\mathbf{r}')\, dv' \\
&= -\lambda_j \lambda_k \int_{V'} h_j(\mathbf{r}')\, dv' \int_V g_0(\mathbf{r},\mathbf{r}')h_k(\mathbf{r})\, dv \\
&= -\lambda_j \lambda_k \int_{V'} h_j(\mathbf{r}')\, dv' \left[\frac{-1}{\lambda_k} h_k(\mathbf{r}') \right],
\end{aligned}$$

$$\lambda_k \int_V h_k(\mathbf{r}) h_j(\mathbf{r})\, dv = \lambda_j \int_V h_j(\mathbf{r}) h_k(\mathbf{r})\, dv, \tag{11}$$

where a change of notation has been introduced in the integral on the right of Eq. (11). Finally,

$$(\lambda_j - \lambda_k) \int_V h_j(\mathbf{r}) h_k(\mathbf{r})\, dv = 0. \tag{12}$$

Therefore if $\lambda_j \neq \lambda_k$, the solutions $h_j(\mathbf{r})$ and $h_k(\mathbf{r})$ are orthogonal to each other.

Second, it can be shown that all eigenvalues are real. If the eigenvalues were complex,

$$\lambda \equiv \alpha + j\beta, \tag{13}$$

then the corresponding eigenfunction would also be complex.

$$h(\mathbf{r}) \equiv q(\mathbf{r}) + jp(\mathbf{r}). \tag{14}$$

Then

$$\begin{aligned} h(\mathbf{r}) &= -\lambda \int_V g_0(\mathbf{r},\mathbf{r}') h(\mathbf{r}')\, dv' \\ &= -(\alpha + j\beta) \int_V g_0(\mathbf{r},\mathbf{r}') [q(\mathbf{r}') + jp(\mathbf{r}')]\, dv'. \end{aligned} \tag{15}$$

For the corresponding complex conjugate eigenvalue

$$\lambda^* = \alpha - j\beta \qquad \text{and} \qquad h^*(\mathbf{r}) = q(\mathbf{r}) - jp(\mathbf{r}) \tag{16}$$

and

$$h^*(\mathbf{r}) = -(\alpha - j\beta) \int_V g_0(\mathbf{r},\mathbf{r}') [q(\mathbf{r}') - jp(\mathbf{r}')]\, dv'. \tag{17}$$

By virtue of the orthogonality relation, Eq. (12), one has

$$\begin{aligned} 0 = \int_V h(\mathbf{r}) h^*(\mathbf{r})\, dv = \int_V (\alpha + j\beta)(\alpha - j\beta) \int_V g_0(\mathbf{r},\mathbf{r}')(q + jp)\, dv' \\ \times \int_V g_0(\mathbf{r},\mathbf{r}')(q - jp)\, dv'\, dv. \end{aligned}$$

Therefore

$$\int_V \int_V \int_V [g_0(\mathbf{r},\mathbf{r}')]^2 [q^2(\mathbf{r}') + p^2(\mathbf{r}')]\, dv'\, dv'\, dv = 0. \tag{18}$$

Since Green's function has a finite nonzero value, one concludes that

$$q^2(\mathbf{r}) + p^2(\mathbf{r}') = 0,$$

or $q^2(\mathbf{r}) = 0$ and $p^2(\mathbf{r}) = 0$, which implies that the eigenfunction for a complex eigenvalue is nonexistent.

5.6. Nonhomogeneous and Associated Homogeneous Problems

A nonhomogeneous problem is described by the following differential equation and boundary conditions:

$$L[U(x)] = F(x), \quad L[U(x)] \equiv m(x)U(x) + n(x)\dot{U}(x) + p(x)\ddot{U}(x) \tag{1}$$

in the region $a \le x \le b$. The boundary conditions are

$$\begin{aligned} B_a[U] &\equiv B_a[U(a)] = \left[\dot{U}(x) + \xi_a U(x)\right]_{x=a} = C_a, \\ B_b[U] &\equiv B_b[U(b)] = \left[\dot{U}(x) + \xi_b U(x)\right]_{x=b} = C_b, \end{aligned} \tag{2}$$

where ξ_a, ξ_b, C_a, and C_b are constants.

The associated homogeneous boundary value problem is defined by the following relations:

$$L[G(x,x')]=0 \tag{3}$$

within $a \le x \le b$ and

$$\begin{aligned} B_a[G(a,x')] &\equiv B_a[G]=\left[\dot{G}(x,x')+\xi_a G(x,x')\right]_{x=a}=0,\\ B_b[G(b,x')] &\equiv B_b[G]=\left[\dot{G}(x,x')+\xi_b G(x,x')\right]_{x=b}=0 \end{aligned} \tag{4}$$

at the endpoints.

Assume Eq. (3) has two independent solutions $G_1(x, x')$ and $G_2(x, x')$. The general solution is then given by

$$G(x,x')=A_1G_1(x,x')+A_2G_2(x,x'). \tag{5}$$

The general solution of the nonhomogeneous equation is the sum of the solutions of the associated homogeneous equation and the particular solution of Eq. (1),

$$U(x)=U_{\mathrm{p}}(x)+A_1G_1(x,x')+A_2G_2(x,x'), \tag{6}$$

where $U_{\mathrm{p}}(x)$ is the particular solution. The constants of integration A_1 and A_2 can be determined by imposing the boundary conditions, Eq. (2).

$$\begin{aligned} B_a[U]&=B_a[U_{\mathrm{p}}]+A_1B_a[G_1]+A_2B_a[G_2]=C_a,\\ B_b[U]&=B_b[U_{\mathrm{p}}]+A_1B_b[G_1]+A_2B_b[G_2]=C_b, \end{aligned} \tag{7}$$

or

$$\begin{aligned} A_1B_a[G_1]+A_2B_a[G_2]&=C_a-B_a[U_{\mathrm{p}}],\\ A_1B_b[G_1]+A_2B_b[G_2]&=C_b-B_b[U_{\mathrm{p}}]. \end{aligned} \tag{8}$$

Equation (8) may be expressed in matrix notation as

$$\Delta\begin{pmatrix} A_1\\ A_2\end{pmatrix}=\begin{pmatrix} C_a-B_a[U_{\mathrm{p}}]\\ C_b-B_b[U_{\mathrm{p}}]\end{pmatrix}, \tag{9}$$

where

$$\Delta\equiv\begin{pmatrix} B_a[G_1] & B_a[G_2]\\ B_b[G_1] & B_b[G_2]\end{pmatrix}. \tag{10}$$

Case a: $\Delta \neq 0$. In this case the values of A_1 and A_2 can be determined uniquely from Eq. (9). The solution of the nonhomogeneous problem, Eqs. (1) and (2), then exists and is uniquely determined.

For the associated homogeneous problem, Eqs. (3) and (4), the solution is Eq. (5). The corresponding boundary condition, Eq. (4), supplies the following relations:

$$B_a[G]=A_1B_a[G_1]+A_2B_a[G_2]=0,$$
$$B_b[G]=A_1B_b[G_1]+A_2B_b[G_2]=0$$

or

$$\Delta\begin{bmatrix} A_1 \\ A_2 \end{bmatrix}=\begin{bmatrix} 0 \\ 0 \end{bmatrix}, \tag{11}$$

where Δ is defined by Eq. (10).

Since $\Delta \neq 0$, the associated homogeneous problem, Eqs. (3) and (4), has the trivial solution

$$G(x,x')=0. \tag{12}$$

Case b: $\Delta=0$. In this case Eq. (11) has a nontrivial solution, and hence the associated homogeneous problem, Eqs. (3) and (4), a uniquely defined solution.

However, Eq. (9) has no solution in general and consequently the solution to the nonhomogeneous problem does not exist. However, for certain values of $C_a - B_a[U_p]$ and $C_b - B_b[U_p]$ such that the two relations in Eq. (8) are linearly dependent, nontrivial solutions do exist. In this case one has only a single equation to determine two unknowns, A_1 and A_2. One of the constants remains arbitrary and there are thus infinitely many solutions.

Summary

If the homogeneous problem, Eqs. (3) and (4), has trivial solution only, then the nonhomogeneous problem, Eqs. (1) and (2), has a unique solution for any forcing function and nonhomogeneous boundary conditions.

On the other hand, if the homogeneous problem has a nontrivial solution, then the nonhomogeneous problem has a solution only if $F(x)$, C_a, and C_b satisfy certain conditions. If this is the case, then Eqs. (1) and (2) have infinitely many solutions.

Example 7

Investigate the possible solutions for the differential equation

$$\frac{d^2F}{dx^2}+\pi^2F=\pi^2x \tag{1}$$

subject to each of the following sets of boundary conditions:

a. $F(0)=1$ and $F(\frac{1}{2})=0$; (2)

b. $F(0)=1$ and $F(1)=0$. (3)

SOLUTION

a. The solution of Eq. (1) can be obtained by the method of variation of parameters. Let

$$F(x)=U_1(x)\cos\pi x+U_2(x)\sin\pi x,$$

$$\dot F(x)=-U_1\pi\sin\pi x+U_2\pi\cos\pi x+\left(\dot U_1\cos\pi x+\dot U_2\sin\pi x\right). \tag{4}$$

Set

$$\dot U_1\cos\pi x+\dot U_2\sin\pi x=0. \tag{5}$$

Then

$$\ddot F(x)=-U_1\pi^2\cos\pi x-U_2\pi^2\sin\pi x-\dot U_1\pi\sin\pi x+\dot U_2\pi\cos\pi x. \tag{6}$$

Substitute Eqs. (4) and (6) into Eq. (1):

$$-\dot U_1\pi\sin\pi x+\dot U_2\pi\cos\pi x=\pi^2 x. \tag{7}$$

$\dot U_1(x)$ and $\dot U_2(x)$ can be found from Eqs. (5) and (7),

$$\dot U_1(x)=-\pi x\sin\pi x,\qquad U_1(x)=(-1/\pi)(\sin\pi x-x\pi\cos\pi x)+C_1, \tag{8}$$

$$\dot U_2(x)=\pi x\cos\pi x,\qquad U_2(x)=(1/\pi)(\cos\pi x+\pi x\sin\pi x)+C_2. \tag{9}$$

The general solution, from Eq. (4), is

$$F(x)=C_1\cos\pi x+C_2\sin\pi x+x. \tag{10}$$

The coefficients C_1 and C_2 are evaluated by applying the boundary conditions, Eq. (2):

$$F(0)=1=C_1,\qquad F\left(\tfrac{1}{2}\right)=0=C_2+\tfrac{1}{2},\quad C_2=-\tfrac{1}{2}$$

Therefore,

$$F(x)=x+\cos\pi x-\tfrac{1}{2}\sin\pi x. \tag{11}$$

The associated homogeneous problem

$$\frac{d^2F_0}{dx^2}+\pi^2F_0=0 \tag{12}$$

has the general solution

$$F_0(x)=C_3\cos\pi x+C_4\sin\pi x. \tag{13}$$

The solution of the homogeneous equation (12) subject to boundary condition (2) is therefore

$$F_0(x)=\cos\pi x. \tag{14}$$

b. Consider the same differential equation (1), but subject to the boundary conditions specified by Eq. (3). The general solution is

$$F_1(x)=d_1\cos\pi x+d_2\sin\pi x+x. \tag{15}$$

The additional subscript has been used to distinguish the problem from that in part a. The boundary conditions yield

$$F_1(0)=1=d_1, \qquad F_1(1)=0=-d_1+1. \tag{16}$$

Therefore,

$$F_1(x)=x+\cos \pi x+d_2 \sin \pi x. \tag{17}$$

The coefficient d_2 is not specified and can have an infinite number of different values. Consequently, problem 6 has an infinite number of solutions.

The associated homogeneous equation

$$\frac{d^2F_{10}}{dx^2}+\pi^2 F_{10}=0 \tag{18}$$

has the general solution

$$F_{10}(x)=d_3 \cos \pi x+d_4 \sin \pi x, \tag{19}$$

$$F_{10}(0)=1=d_3, \tag{20}$$

$$F_{10}(1)=0=-d_3. \tag{21}$$

It is impossible to satisfy both Eqs. (20) and (21), and therefore this homogeneous problem does not have a solution.

References

1. G. Arfken, *Mathematical Methods for Physicists*, Academic, New York, 1966.
2. R. Courant and D. Hilbert, *Methods of Mathematical Physics*, Interscience, New York, 1953 and 1962.
3. P. Dennery and A. Krzywicki, *Mathematics for Physicists*, Harper & Row, New York, 1967.
4. J. W. Dettman, *Mathematical Methods in Physics and Engineering*, McGraw-Hill, New York, 1962.
5. P. M. Morse and H. Feshbach, *Methods of Theoretical Physics*, McGraw-Hill, New York, 1953.
6. T. Myint-U, *Ordinary Differential Equations*, Elsevier North Holland, New York, 1978.
7. K. F. Riley, *Mathematical Methods for the Physical Sciences*, Cambridge University Press, London, 1974.
8. H. Sagan, *Boundary and Eigenvalue Problems in Mathematical Physics*, Wiley, New York, 1961.
9. S. A. Schelkunoff, *Applied Mathematics for Engineers and Scientists*, Van Nostrand, Princeton, N.J., 1965.

6

Imperfect Waveguides

In previous chapters solutions were obtained for idealized waveguides; that is, the dimensions, shapes, media, and so on were assumed to be exactly as specified. However, in practice, it is impossible to construct such mathematically perfect structures, and the deviations from specifications become more critical as the wavelength is decreased. Knowing the effects of these imperfections is important to understanding waveguides at optical frequencies. Since imperfections can be numerous and complex [1–6], only one type will be treated here.

6.1. Imperfect Boundaries

The effect of an imperfect boundary of a dielectric sheet waveguide will now be investigated. To simplify the problem, it will be assumed that the imperfection in the thickness (the x direction) of the dielectric slab is a function of z only and is independent of y. It is further assumed that these imperfections are small deviations from the ideal situation, so that perturbation techniques can be applied.

The mode solutions of a perfect dielectric sheet waveguide were obtained (Chapter 2) by solving Maxwell's equations for regions filled with different dielectrics. Boundary conditions were then imposed to evaluate the constants of separation.

The permittivity of the medium may be considered to be a continuous function of space. The only boundary conditions then involved are those at infinity. An abrupt change in permittivity can be approximated by a continuous function and a solution obtained for this approximate permittivity. A limiting process is then applied to allow the variation to approach arbitrary steepness.

By treating the perfect case with discrete permittivities as the limiting situation of the imperfect case with continuous variation of permittivity, one can express the solutions of the imperfect waveguide as a sum of solutions for the perfect waveguide.

6.2. Imperfect Dielectric Sheet Waveguides

A defective dielectric sheet waveguide with a slight variation in thickness is shown in Figure 7. The upper and lower surfaces of the waveguide are defined by

$$x_u(z)=\delta+x_1(z), \qquad x_\ell(z)=-\delta+x_2(z). \tag{1}$$

The permittivity $\hat{\varepsilon}(x)$ for the perfect waveguide is

$$\begin{aligned}\hat{\varepsilon}(x)&=\varepsilon_1 \quad \text{for} \quad |x|<\delta,\\ &=\varepsilon_2 \quad \text{for} \quad |x|>\delta.\end{aligned} \tag{2}$$

The permittivity $\varepsilon(x, z)$ for the imperfect waveguide is

$$\varepsilon(x, z)=\hat{\varepsilon}(x)+\Delta\varepsilon(x, z), \tag{3}$$

Figure 7. Imperfect dielectric sheet waveguide.

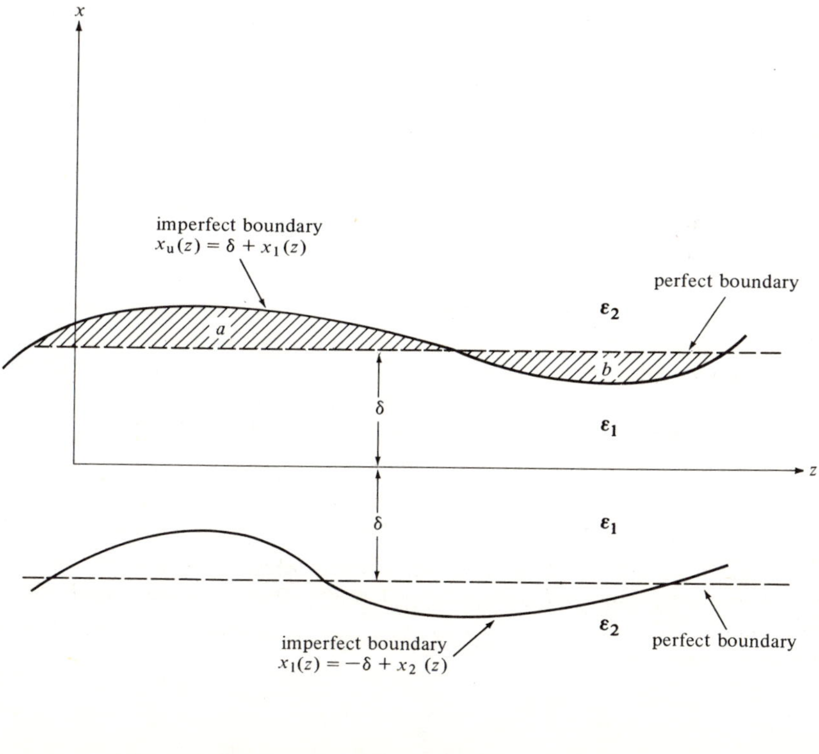

where $\Delta\varepsilon(x, z)$ is the increment in permittivity due to blemishes at the boundary. In region *a* of Figure 7, where the actual interface between two media is above that of the perfect guide, that is, $x_u(z) > \delta$, the increment in permittivity is

$$\Delta\varepsilon(x, z) = \varepsilon_1 - \varepsilon_2 \qquad \text{for} \qquad \delta < x < x_u. \tag{4}$$

In region *b*, on the other hand,

$$\Delta\varepsilon(x, z) = \varepsilon_2 - \varepsilon_1 \qquad \text{for} \qquad x_u < x < \delta.$$

The values of $\Delta\varepsilon(x, z)$ in the different regions can be summarized as follows. For $x_1(z) \geq 0$ and $x_2(z) \geq 0$,

$$\Delta\varepsilon(x, z) = \begin{cases} 0, & x > x_u(z), \\ \varepsilon_1 - \varepsilon_2, & \delta < x < x_u(z), \\ 0, & x_\ell(z) < x < \delta, \\ \varepsilon_1 - \varepsilon_2, & -\delta < x < x_\ell(z), \\ 0, & x < -\delta. \end{cases} \tag{5}$$

For $x_1(z) < 0$ and $x_2(z) < 0$,

$$\Delta\varepsilon(x, z) = \begin{cases} 0, & x > \delta, \\ \varepsilon_2 - \varepsilon_1, & x_u(z) < x < \delta, \\ 0, & -\delta < x < x_u(z), \\ \varepsilon_1 - \varepsilon_2, & x_\ell(z) < x < -\delta, \\ 0, & x < x_\ell(z). \end{cases} \tag{6}$$

In this analysis, it is not necessary to have an insignificant increment $\Delta\varepsilon$ in permittivity. The perturbation is still valid if the extent of the region where $\Delta\varepsilon \neq 0$ is very small.

An arbitrary field distribution of a dielectric sheet waveguide can always be expressed as a sum of guided modes and radiation modes (see Appendix 3),

$$\begin{aligned} E_y &= \sum_k g_k E_{ky}{}^{\mathrm{g}} + \sum \int_0^\infty r(\beta_\rho, z) E_y{}^{\mathrm{r}}\, d\beta_\rho \\ &\equiv E_1 + E_2, \end{aligned} \tag{7}$$

where the kth guided mode and the radiation mode are, respectively,

$$\begin{aligned} \tilde{E}_{ky}{}^{\mathrm{g}} &= e_{ky}{}^{\mathrm{g}}(x) e^{j(\omega t - \beta_{gk} z)}, \\ \tilde{E}_y{}^{\mathrm{r}} &= e^{\mathrm{r}}(\beta_\rho x) e^{j(\omega t - \beta_r z)}. \end{aligned} \tag{8}$$

The summations include both even and odd modes. $g_k(z)$ and $r(\beta_\rho, z)$ are weight factors for the guided mode and the radiation mode, respectively. These weight factors may be evaluated by applying the orthogonality property of the mode functions, Eq. (A3.4.8).

$$g_k=\int_{-\infty}^{\infty} E_y E_{ky}{}^{g*}\,dx \Big/ \sum_m \int_{-\infty}^{\infty} E_{my}{}^{g} E_{ky}{}^{g*}\,dx$$

$$=\frac{\beta_{gk}}{2\omega\mu_o P^g}\int_{-\infty}^{\infty} E_y E_{ky}{}^{g*}\,dx, \tag{9}$$

$$r(\beta_\rho, z)=\int_{-\infty}^{\infty} E_y E_y{}^{r*}(\beta_\rho)\,dx \Big/ \int_{-\infty}^{\infty} E_y{}^{r} E_y{}^{r*}(\beta_\rho)\,dx. \tag{10}$$

E_y is a solution of the wave equation, Eq. (1.6.3),

$$\nabla^2\mathbf{E}-\gamma^2\mathbf{E}=0, \qquad \gamma^2\equiv-\omega^2\mu\varepsilon. \tag{11}$$

In rectangular coordinates, this becomes

$$\frac{\partial^2 E_y}{\partial x^2}+\frac{\partial^2 E_y}{\partial z^2}+\omega^2\mu\varepsilon E_y=0, \qquad \varepsilon\equiv\hat{\varepsilon}+\Delta\varepsilon, \tag{12}$$

where it is assumed that the field is independent of y. Substituting Eq. (7) into Eq. (12) yields

$$\frac{\partial^2(E_1+E_2)}{\partial x^2}+\frac{\partial^2(E_1+E_2)}{\partial z^2}+\omega^2\mu\varepsilon(E_1+E_2)=0,$$

$$\left(\frac{\partial^2 E_1}{\partial x^2}+\frac{\partial^2 E_1}{\partial z^2}+\omega^2\mu\varepsilon E_1\right)+\left(\frac{\partial^2 E_2}{\partial x^2}+\frac{\partial^2 E_2}{\partial z^2}+\omega^2\mu\varepsilon E_2\right)=0, \tag{13}$$

$$\frac{\partial E_1}{\partial x}=\sum_k g_k\frac{\partial E_{ky}{}^{g}}{\partial x},$$

$$\frac{\partial^2 E_1}{\partial x^2}=\sum_k g_k\frac{\partial^2 E_{ky}{}^{g}}{\partial x^2},$$

$$\frac{\partial E_1}{\partial z}=\sum_k\left(\frac{\partial g_k}{\partial z}E_{ky}{}^{g}-j\beta_{gk}g_k E_{ky}{}^{g}\right),$$

$$\frac{\partial^2 E_1}{\partial z^2}=\sum_k\left(\frac{\partial^2 g_k}{\partial z^2}E_{ky}{}^{g}-2j\beta_{gk}\frac{\partial g_k}{\partial z}E_{ky}{}^{g}+g_k\frac{\partial^2 E_{ky}{}^{g}}{\partial z^2}\right),$$

$$\frac{\partial^2 E_1}{\partial x^2}+\frac{\partial^2 E_1}{\partial z^2}+\omega^2\mu\varepsilon E_1=\sum_k g_k\left(\frac{\partial^2 E_{ky}{}^{g}}{\partial x^2}+\frac{\partial^2 E_{ky}{}^{g}}{\partial z^2}+\omega^2\mu\hat{\varepsilon}E_{ky}{}^{g}\right)$$

$$+\sum_k\left(\frac{\partial^2 g_k}{\partial z^2}-2j\beta_{gk}\frac{\partial g_k}{\partial z}+g_k\omega^2\mu\,\Delta\varepsilon\right)E_{ky}{}^{g}$$

$$=\sum_k\left(\frac{\partial^2 g_k}{\partial z^2}-2j\beta_{gk}\frac{\partial g_k}{\partial z}+\omega^2\mu\,\Delta\varepsilon g_k\right)E_{ky}{}^{g}. \tag{14}$$

The final form of Eq. (14) is obtained by virtue of the fact that $E_{ky}{}^{g}$ is a fundamental mode. Similarly,

$$\frac{\partial E_2}{\partial x}=\sum\int_0^\infty r(\beta_\rho,z)\frac{\partial E_y{}^{r}}{\partial x}\,d\beta_\rho,$$

$$\frac{\partial^2 E_2}{\partial x^2}=\sum\int_0^\infty r(\beta_\rho,z)\frac{\partial^2 E_y{}^{r}}{\partial x^2}\,d\beta_\rho,$$

$$\frac{\partial E_2}{\partial z}=\sum\int_0^\infty\left[\frac{\partial r(\beta_\rho,z)}{\partial z}E_y{}^{r}+r(\beta_\rho,z)\frac{\partial E_y{}^{r}}{\partial z}\right]d\beta_\rho,$$

$$\frac{\partial^2 E_2}{\partial z^2}=\sum\int_0^\infty\left[\frac{\partial^2 r(\beta_\rho,z)}{\partial z^2}E_y{}^{r}+2\frac{\partial r(\beta_\rho,z)}{\partial z}\frac{\partial E_y{}^{r}}{\partial z}\right.$$
$$\left.+r(\beta_\rho,z)\frac{\partial^2 E_y{}^{r}}{\partial z^2}\right]d\beta_\rho,$$

$$\frac{\partial^2 E_2}{\partial x^2}+\frac{\partial^2 E_2}{\partial z^2}+\omega^2\mu\varepsilon E_2=\sum\int_0^\infty r(\beta_\rho,z)\left(\frac{\partial^2 E_y{}^{r}}{\partial x^2}+\frac{\partial^2 E_y{}^{r}}{\partial z^2}+\omega^2\mu\varepsilon E_y{}^{r}\right)d\beta_\rho$$
$$+\sum\int_0^\infty\left[\frac{\partial^2 r(\beta_\rho,z)}{\partial z^2}-2j\beta_r\frac{\partial r(\beta_\rho,z)}{\partial z}\right.$$
$$\left.+\omega^2\mu\,\Delta\varepsilon\, r(\beta_\rho,z)\right]E_y{}^{r}\,d\beta_\rho$$
$$=\sum\int_0^\infty\left[\frac{\partial^2 r(\beta_\rho,z)}{\partial z^2}-2j\beta_r\frac{\partial r(\beta_\rho,z)}{\partial z}\right.$$
$$\left.+\omega^2\mu\,\Delta\varepsilon\, r(\beta_\rho,z)\right]E_y{}^{r}\,d\beta_\rho. \tag{15}$$

Substitution of Eqs. (14) and (15) into Eq. (13) yields

$$\sum_k\left(\frac{\partial^2 g_k}{\partial z^2}-2j\beta_{gk}\frac{\partial g_k}{\partial z}+\omega^2\mu\,\Delta\epsilon\, g_k\right)E_{ky}{}^{g}$$
$$+\sum\int_0^\infty\left(\frac{\partial^2 r}{\partial z^2}-2j\beta_r\frac{\partial r}{\partial z}+\omega^2\mu\,\Delta\varepsilon\, r\right)E_y{}^{r}\,d\beta_\rho=0. \tag{16}$$

For small perturbation it is assumed that the fields in the imperfect waveguide have components similar to those in the perfect guides; that is, for TE modes the fields have nonvanishing E_y, H_x, and H_z. E_y is determined from Eq. (16); H_x and H_z are related to E_y by Eq. (2.1.5),

$$H_x=\frac{-\gamma_g}{j\omega\mu}E_y \quad\text{and}\quad H_z=\frac{-1}{j\omega\mu}\frac{\partial E_y}{\partial x}. \tag{17}$$

6.3. Mode Conversion

Assume that a perfect dielectric sheet waveguide is connected to a piece of imperfect guide of length L. The imperfect guide is followed by another perfect guide. The dominant TE guided mode of the perfect guide is propagating in the perfect guide on the left and is incident upon the imperfect section. We shall investigate the fields in all three sections (see Figure 8).

The equation to be solved is Eq. (6.2.16),

$$\sum_k \left(\frac{\partial^2 g_k}{\partial z^2} - 2j\beta_{gk} \frac{\partial g_k}{\partial z} + \omega^2 \mu \, \Delta\varepsilon \, g_k \right) E_{ky}{}^{g}$$
$$+ \sum \int_0^\infty \left(\frac{\partial^2 r}{\partial z^2} - 2j\beta_r \frac{\partial r}{\partial z} + \omega^2 \mu \, \Delta\varepsilon \, r \right) E_y{}^{r} \, d\beta_\rho = 0. \qquad (1)$$

This equation involves both functions $g_k(z)$ and $r(\beta_\rho, z)$. The orthogonality relation will be used to investigate the possibility of separating Eq. (1) into equations containing only one of these functions. Multiply Eq. (1) by $(\beta_{gm}/2\omega\mu)E_{my}{}^{g*}$ and integrate over the cross-sectional area of the guide.

$$\int_{x=-\infty}^{\infty} \int_{y=y_0}^{y_0+1} \sum_k \left(\frac{\partial^2 g_k}{\partial z^2} - 2j\beta_{gk} \frac{\partial g_k}{\partial z} + \omega^2 \mu \, \Delta\varepsilon \, g_k \right)$$
$$\times E_{ky}{}^{g} \frac{\beta_{gm}}{2\omega\mu} E_{my}{}^{g*} \, dx \, dy$$
$$+ \int_{x=-\infty}^{\infty} \int_{y=y_0}^{y_0+1} \sum \int_0^\infty \left(\frac{\partial^2 r}{\partial z^2} - 2j\beta_r \frac{\partial r}{\partial z} + \omega^2 \mu \, \Delta\varepsilon \, r \right)$$
$$\times E_y{}^{r} \frac{\beta_{gm}}{2\omega\mu} E_{mg}{}^{g*} \, dx \, dy \, d\beta_\rho = 0,$$

$$P^{g}\left(\frac{\partial^2 g_m}{\partial z^2} - 2j\beta_{gm} \frac{\partial g_m}{\partial z} \right) + \sum_k \frac{\omega \beta_{gm} g_k}{2} \int_{-\infty}^{\infty} \Delta\varepsilon \, E_{ky}{}^{g} E_{my}{}^{g*} \, dx$$
$$+ \sum \frac{\omega \beta_{gm}}{2} \int_0^\infty r \, d\beta_\rho \int_{-\infty}^{\infty} \Delta\varepsilon \, E_y{}^{r} E_{my}{}^{g*} \, dx = 0,$$

$$\frac{\partial^2 g_m}{\partial z^2} - 2j\beta_{gm} \frac{\partial g_m}{\partial z} + \sum_k g_k G_{km} + \sum \int_0^\infty r R_m \, d\beta_\rho = 0, \qquad (2)$$

where

$$G_{km}(z) \equiv \frac{\omega \beta_{gm}}{2P^{g}} \int_{-\infty}^{\infty} \Delta\varepsilon \, E_{ky}{}^{g} E_{my}{}^{g*} \, dx, \qquad (3)$$

$$R_m(\beta_\rho, z) \equiv \frac{\omega \beta_{gm}}{2P^{g}} \int_{-\infty}^{\infty} \Delta\varepsilon \, E_y{}^{r} E_{mg}{}^{g*} \, dx. \qquad (4)$$

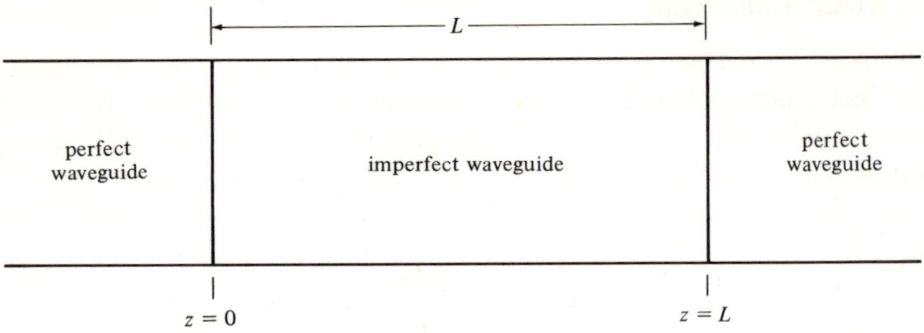

Figure 8. Arrangement for investigating mode conversion.

The power relation, Eq. (2.2b.12), is used in defining P^{g}.

$$\int_{-\infty}^{\infty} E_{ky}{}^{g} E_{my}{}^{g*}\, dx \equiv \delta_{km} \frac{2P^{g}\omega\mu}{\beta_{gm}}. \tag{5}$$

Similarly, multiplying Eq. (1) by $(\beta_{r}/2\omega\mu_{o})E_{y}{}^{r*}$ and integrating over the cross-sectional area of the guide, one obtains

$$\frac{\partial^2 r}{\partial z^2} - 2j\beta_{r}\frac{\partial r}{\partial z} + \sum \int_{0}^{\infty} \overline{R} r\, d\beta_{\rho} + \sum_{k} g_{k}\overline{G}_{k} = 0, \tag{6}$$

where

$$\overline{G}_{k}(\beta_{r}, z) \equiv \frac{\omega\beta_{r}}{2P^{r}} \int_{-\infty}^{\infty} \Delta\varepsilon\, E_{y}{}^{r*} E_{ky}{}^{g}\, dx, \tag{7}$$

$$\overline{R}(\beta_{\rho}, \beta_{\rho}') = \frac{\omega\beta_{r}}{2P^{r}} \int_{-\infty}^{\infty} \Delta\varepsilon\, E_{y}{}^{r}(\beta_{\rho}) E_{y}{}^{r*}(\beta_{\rho}')\, dx, \tag{8}$$

and P^{r} is defined by Eq. (2.5a.27),

$$\int_{-\infty}^{\infty} E_{y}{}^{r}(\beta_{\rho}) E_{y}{}^{r*}(\beta_{\rho}')\, dx = \frac{2\omega\mu_{o}}{\beta_{\rho}'} P^{r}\delta(\beta_{\rho} - \beta_{\rho}'). \tag{9}$$

Equations (2) and (6) are two coupled equations to be solved. Note that these equations become uncoupled for the case $\Delta\varepsilon = 0$. This simpler case will be solved first and the imperfect case $\Delta\varepsilon \neq 0$ will be treated as a nonhomogeneous problem.

6.4. Perfect Guide: $\Delta\varepsilon = 0$

For the case where $\Delta\varepsilon = 0$, the two coupled equations (6.3.2) and (6.3.6) become independent:

$$\frac{\partial^2 \hat{g}_m}{\partial z^2} - 2j\beta_{gm}\frac{\partial \hat{g}_m}{\partial z} = 0, \qquad \hat{g}_m \equiv g_m(\Delta\varepsilon = 0), \tag{1}$$

$$\frac{\partial^2 \hat{r}}{\partial z^2} - 2j\beta_{r}\frac{\partial \hat{r}}{\partial z} = 0, \qquad \hat{r} \equiv r(\Delta\varepsilon = 0). \tag{2}$$

The solution can easily be determined by using a trial solution $\hat{g}_m = ge^{pt}$, $p=0$ or $2j\beta_{gm}$:

$$\hat{g}_m(z)=a_1+a_2e^{2j\beta_{gm}z} \qquad \text{and} \qquad \hat{r}(z)=b_1+b_2e^{2j\beta_r z}. \tag{3}$$

The time and the z dependence of the field are given by $e^{j(\omega t-\beta_{gk}z)}$ for the guided mode. This is a traveling wave in the positive z direction ($\beta_{gk}>0$). In the expansion of Eq. (6.2.7),

$$E_y=\sum_k g_k E_{ky}{}^{g}+\sum\int_0^\infty r(\beta_\rho, z)E_y{}^{r}\,d\beta_\rho, \tag{4}$$

the normal modes are weighted by the factor $\hat{g}_m$ and $\hat{r}$. Thus

$$\begin{aligned}\hat{g}_m e^{j(\omega t-\beta_{gk}z)}&=a_1e^{j(\omega t-\beta_{gk}z)}+a_2e^{j(\omega t+\beta_{gk}z)},\\ \hat{r}e^{j(\omega t-\beta_r z)}&=b_1e^{j(\omega t-\beta_r z)}+b_2e^{j(\omega t+\beta_r z)}.\end{aligned} \tag{5}$$

The response due to a positive traveling wave consists of traveling waves in both the positive and negative directions.

6.5. Perturbed Case: $\Delta\varepsilon\neq 0$

Consider the nonhomogeneous equations (6.3.2) and (6.3.6),

$$\frac{\partial^2 g_m}{\partial z^2}-2j\beta_{gm}\frac{\partial g_m}{\partial z}=M_m, \tag{1}$$

$$M_m\equiv-\sum_k g_k G_{km}-\sum\int_0^\infty rR_m\,d\beta_\rho;$$

$$\frac{\partial^2 r}{\partial z^2}-2j\beta_r\frac{\partial r}{\partial z}=N(\beta_r), \tag{2}$$

$$N(\beta_r)\equiv-\sum_k g_k\bar{G}_k-\sum\int_0^\infty r\bar{R}\,d\beta_\rho;$$

where G_{km}, R_m, $\bar{G}_k$, and $\bar{R}$ are defined by Eqs. (6.3.3), (6.3.4), (6.3.7), and (6.3.8), respectively. The particular solutions of the above equations can be obtained by the method of *variation of parameters*. Since both equations are of the same form, consider the general equation,

$$\frac{\partial^2 q}{\partial z^2}-p\frac{\partial q}{\partial z}=Q. \tag{3}$$

Let the solution of the associated homogeneous equation be

$$q^0=q_1+q_2. \tag{4}$$

Then the particular solution is assumed to be

$$q^{\mathrm{P}}=U_1q_1+U_2q_2, \tag{5}$$

where U_1 and U_2 are parameters to be determined.

$$\frac{\partial q^{\mathrm{P}}}{\partial z}=U_1\frac{\partial q_1}{\partial z}+U_2\frac{\partial q_2}{\partial z}+q_1\frac{\partial U_1}{\partial z}+q_2\frac{\partial U_2}{\partial z}. \tag{6}$$

To avoid higher-order derivatives of U_1 and U_2, impose the restriction

$$q_1\frac{\partial U_1}{\partial z}+q_2\frac{\partial U_2}{\partial z}=0. \tag{7}$$

Then

$$\frac{\partial q^{\mathrm{P}}}{\partial z}=U_1\frac{\partial q_1}{\partial z}+U_2\frac{\partial q_2}{\partial z}, \tag{8}$$

$$\frac{\partial^2 q^{\mathrm{P}}}{\partial z^2}=U_1\frac{\partial^2 q_1}{\partial z^2}+U_2\frac{\partial^2 q_2}{\partial z^2}+\frac{\partial U_1}{\partial z}\frac{\partial q_1}{\partial z}+\frac{\partial U_2}{\partial z}\frac{\partial q_2}{\partial z}. \tag{9}$$

Substituting Eqs. (8) and (9) into Eq. (3) yields

$$\frac{\partial q_1}{\partial z}\frac{\partial U_1}{\partial z}+\frac{\partial q_2}{\partial z}\frac{\partial U_2}{\partial z}=Q. \tag{10}$$

Equations (7) and (10) may be written in matrix form

$$\begin{pmatrix} q_1 & q_2 \\ \partial q_1/\partial z & \partial q_2/\partial z \end{pmatrix}\begin{pmatrix} \partial U_1/\partial z \\ \partial U_2/\partial z \end{pmatrix}=\begin{pmatrix} 0 \\ Q \end{pmatrix}. \tag{11}$$

Let

$$W\equiv\begin{vmatrix} q_1 & q_2 \\ \partial q_1/\partial z & \partial q_2/\partial z \end{vmatrix}=q_1\frac{\partial q_2}{\partial z}-q_2\frac{\partial q_1}{\partial z}. \tag{12}$$

The solution of Eq. (11) is

$$\frac{\partial U_1}{\partial z}=\frac{1}{W}\begin{vmatrix} 0 & q_2 \\ Q & \partial q_2/\partial z \end{vmatrix}=\frac{-Qq_2}{W}, \qquad U_1=\int_0^z-\frac{Qq_2}{W}\,ds, \tag{13}$$

$$\frac{\partial U_2}{\partial z}=\frac{1}{W}\begin{vmatrix} q_1 & 0 \\ \partial U_1/\partial z & Q \end{vmatrix}=\frac{Qq_1}{W}, \qquad U_2=\int_0^z\frac{Qq_1}{W}\,ds. \tag{14}$$

The solution of Eq. (1) may now be obtained by identifying

$$q=g_m,\qquad p=2j\beta_{gm},\qquad q_1=a_1,\qquad q_2=a_2e^{2j\beta_{gm}z},\qquad \frac{\partial q_1}{\partial z}=0,$$

$$\frac{\partial q_2}{\partial z}=j2\beta_{gm}a_2e^{2j\beta_{gm}z},\qquad W=2ja_1a_2\beta_{gm}e^{2j\beta_{gm}z},\qquad Q=M_m,$$

$$U_1=\int_0^z-\frac{Qq_2}{W}\,ds=\frac{-1}{2j\beta_{gm}a_1}\int_0^z M_m\,ds, \tag{15}$$

$$U_2=\int_0^z\frac{Qq_1}{W}\,ds=\frac{1}{2j\beta_{gm}a_2}\int_0^z M_me^{-2j\beta_{gm}s}\,ds, \tag{16}$$

$$\begin{aligned} g_m{}^{\mathrm{P}}&=U_1q_1+U_2q_2 \\ &=\frac{1}{2j\beta_{gm}}\left(-\int_0^z M_m\,ds+e^{2j\beta_{gm}z}\int_0^z M_me^{-2j\beta_{gm}s}\,ds\right). \end{aligned} \tag{17}$$

The complete solution of Eq. (1) is the sum of the solution of the associated homogeneous equation and the particular solution:

$$
\begin{aligned}
g_m &= g_m{}^0 + g_m{}^{\mathrm{p}} \\
&= a_m + b_m e^{2j\beta_{gm}z} + \frac{1}{2j\beta_{gm}}\left(-\int_0^z M_m\,ds + e^{2j\beta_{gm}z}\int_0^z M_m e^{-2j\beta_{gm}s}\,ds\right) \\
&= \left(a_m - \frac{1}{2j\beta_{gm}}\int_0^z M_m\,ds\right) \\
&\quad + e^{2j\beta_{gm}z}\left(b_m + \frac{1}{2j\beta_{gm}}\int_0^z M_m e^{-2j\beta_{gm}s}\,ds\right) \\
&\equiv g_m{}^+ + g_m{}^- .
\end{aligned} \tag{18}
$$

The solution of Eq. (2) may be obtained similarly.

$$
\begin{aligned}
r(z) &= r^0 + r^{\mathrm{p}} \\
&= \left[c - \frac{1}{2j\beta_r'}\int_0^z N(\beta_r)\,ds\right] + e^{2j\beta_r z}\left(d + \frac{1}{2j\beta_r}\int_0^z N e^{-2j\beta_r s}\,ds\right) \\
&\equiv r^+ + r^- .
\end{aligned} \tag{19}
$$

The constant coefficients a, b, c, and d can be evaluated by the boundary conditions. It has assumed that only the lowest TE mode is incident from the perfect guide on the left, and therefore

$$
g_m{}^+(z=0) = 0 \qquad \text{for} \quad m \neq 0,
$$

$$
g_0{}^+(z=0) = 1 = a_0 - \frac{1}{2j\beta_{g0}}\int_0^{z=0} M_m\,ds = a_0 .
$$

Therefore, $a_0 = 1$. In general

$$
g_m{}^+(0) = \delta_{0m}, \qquad a_0 = \delta_{00} = 1. \tag{20}
$$

At $z = L$, the imperfect guide is connected to a perfect guide, and there should be no reflected waves for the normal modes of the perfect guide.

$$
\begin{aligned}
g_m{}^-(L) &= 0 \qquad \text{for all } m \\
&= e^{2j\beta_{gm}L}\left(b_m + \frac{1}{2j\beta_{gm}}\int_0^{z=L} M_m e^{-2j\beta_{gm}s}\,ds\right), \\
b_m &= \frac{-1}{2j\beta_{gm}}\int_0^L M_m e^{-2j\beta_{gm}s}\,ds .
\end{aligned} \tag{21}
$$

Similar reasoning applies to the radiation modes:

$$
\begin{aligned}
r^+(0) &= 0, \qquad c = 0, \\
r^-(L) &= 0, \qquad d = \frac{-1}{2j\beta_r}\int_0^L N e^{-2j\beta_r s}\,ds .
\end{aligned} \tag{22}
$$

With these constants, one has

$$g_m{}^+(z)=\delta_{0m}-\frac{1}{2j\beta_{gm}}\int_0^z M_m\,ds, \tag{23}$$

$$g_m{}^-(z)=\frac{e^{2j\beta_{gm}z}}{2j\beta_{gm}}\left(-\int_0^L M_m e^{-2j\beta_{gm}s}\,ds+\int_0^z M_m e^{-2j\beta_{gm}s}\,ds\right), \tag{24}$$

$$r^+(z)=\frac{-1}{2j\beta_r}\int_0^z N\,ds, \tag{25}$$

$$r^-(z)=\frac{e^{2j\beta_r z}}{2j\beta_r}\left(-\int_0^L Ne^{-2j\beta_r s}\,ds+\int_0^z Ne^{-2j\beta_r s}\,ds\right). \tag{26}$$

6.6. Mode-Conversion Losses

The power transmitted along the guide filled with lossless medium is given by

$$P=\tfrac{1}{2}\int_S \mathrm{Re}\left[\tilde{\mathbf{E}}\times\tilde{\mathbf{H}}^*\right]\cdot\hat{\mathbf{z}}\,da$$

$$=-\frac{\beta_g}{\omega\mu}\int_0^\infty |E_y|^2\,dx. \tag{1}$$

For TE mode, $\mathbf{E}=\hat{\mathbf{y}}E_y$ and $H_x=(-\beta_g/\omega\mu)E_y$. E_y is expressed in terms of normal modes by Eq. (6.2.7),

$$E_y=\sum_k g_k E_{ky}{}^{g}+\sum\int_0^\infty r(\beta_\rho,z)E_y{}^{r}\,d\beta_\rho. \tag{2}$$

Let the fields of each mode be normalized such that the power carried by each mode is equal to some reference value, say $\hat{P}$, such that

$$\hat{P}=-\frac{\beta_{gk}}{\omega\mu}\int_0^\infty |E_{ky}|^2\,dx,\qquad k=1,2,\ldots. \tag{3}$$

Then

$$P=\hat{P}\left(\sum_k |g_k|^2+\sum\int_0^\infty |r|^2\,d\beta_\rho\right)$$

$$\equiv P^g+P^r, \tag{4}$$

where

$$P^g\equiv\hat{P}\sum_k |g_k|^2 \qquad \text{(power of guided modes)},$$

$$P^r\equiv\hat{P}\sum\int_0^\infty |r|^2\,d\beta_\rho \qquad \text{(power of radiation modes)}.$$

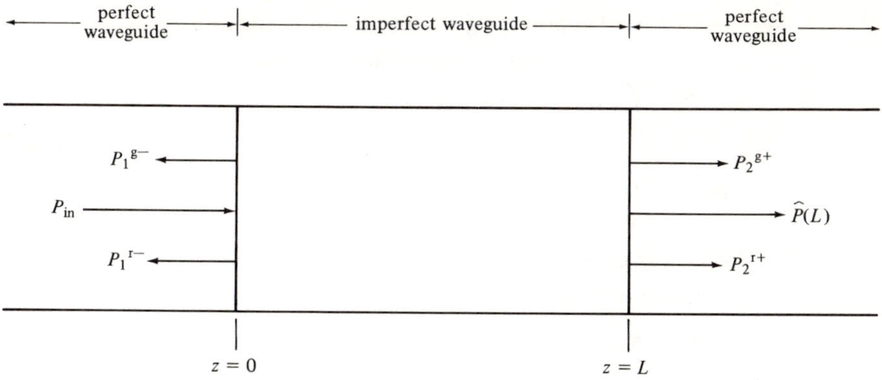

Figure 9. Power balance diagram.

The mode of the incident wave of $m=0$ is taken as the desired field and all other modes generated by the presence of imperfect guides are unwanted. The power carried by these unwanted modes are losses to the system. The *power loss* ΔP *due to mode conversion* (Figure 9) is given by

$$\Delta P=(P_1^{g-}+P_1^{r-})+(P_2^{g+}+P_2^{r+})-\hat{P}(L), \tag{5}$$

where

$$P_1^{g-}=P^{g-}(z=0)= \text{power of the guided modes propagating in the } -z \text{ direction at } z=0$$

$$=\hat{P}\sum_k |g_k^{\,-}(z=0)|^2$$

$$=\hat{P}\sum_k \left|\frac{1}{-2j\beta_{gk}}\int_0^L M_k e^{-2j\beta_{gk}s}\,ds\right|^2, \tag{6}$$

$$P_2^{g+}\equiv P^{g+}(z=L)= \text{power of the guided modes propagating in the } +z \text{ direction at } z=L$$

$$=\hat{P}\sum_k |g_k^{\,+}(z=L)|^2$$

$$=\hat{P}\sum_k \left|\delta_{0k}-\frac{1}{2j\beta_{gk}}\int_0^L M_k\,ds\right|^2, \tag{7}$$

$$P_1^{r-}\equiv P^{r}(z=0)= \text{power of the radiation modes propagating in the } -z \text{ direction at } z=0$$

$$=\hat{P}\sum\int_0^\infty |r^-(\beta_\rho,z=0)|^2\,d\beta_\rho$$

$$=\hat{P}\sum\int_0^\infty d\beta_\rho\left|\frac{-1}{2j\beta_r}\int_0^L N e^{-2j\beta_r s}\,ds\right|^2, \tag{8}$$

$$P_2^{\mathrm{r}+} = p^{\mathrm{r}}(z=L) = \text{power of the radiation modes propagating in the } +z \text{ direction at } z=L$$

$$= \hat{P} \sum \int_0^\infty |r^+(\beta_\rho, z=L)|^2 \, d\beta_\rho$$

$$= \hat{P} \sum \int_0^\infty \left| \frac{-1}{2 j\beta_{\mathrm{r}}} \int_0^L N\,ds \right|^2 d\beta_\rho, \tag{9}$$

$$\hat{P}(L) \equiv \text{power of the lowest mode propagating in the } +z \text{ direction at } z=L$$

$$= \hat{P} \sum_k |\delta_{0k} g_k{}^+(L)|^2 = \hat{P}. \tag{10}$$

Substituting Eqs. (6)–(10) into Eq. (5) yields

$$\Delta P/\hat{P} = \sum_k \left[(1+\delta_{0k}) |g_k{}^+(L)|^2 + |g_k{}^-(0)|^2 \right]$$
$$+ \sum \int_0^\infty \left[|r^+(\beta_\rho, L)|^2 + |r^-(\beta_\rho, 0)|^2 \right] d\beta_\rho. \tag{11}$$

This is the relative power loss due to mode conversion.

6.7. Functions M_m and N

The solutions of the coupled equations for coefficients g_m and $r(\beta_\rho, z)$ have been formally obtained in terms of the functions M_m and N, Eqs. (6.5.1) and (6.5.2). These are repeated here:

$$M_m \equiv -\sum_k g_k G_{km} - \sum \int_0^\infty r R_m \, d\beta_\rho$$
$$= -\sum_k (g_k{}^+ + g_k{}^-) G_{km} - \sum \int_0^\infty (r^+ + r^-) R_m \, d\beta_\rho, \tag{1}$$

$$N = -\sum_k g_k \bar{G}_k - \sum \int_0^\infty r \bar{R} \, d\beta_\rho$$
$$= -\sum_k (g_k{}^+ + g_k{}^-) \bar{G}_k - \sum \int_0^\infty (r^+ + r^-) \bar{R} \, d\beta_\rho, \tag{2}$$

where

$$g_k{}^+(z) \equiv \delta_{0k} - \frac{1}{2 j\beta_{gk}} \int_0^z M_k \, ds, \tag{3}$$

$$g_k{}^-(z) \equiv \frac{1}{2 j\beta_{gk}} e^{2 j\beta_{gk} z} \left(-\int_0^L M_k e^{-2 j\beta_{gk} s} ds + \int_0^z M_k e^{-2 j\beta_{gk} s} ds \right), \tag{4}$$

$$r^+(z) \equiv \frac{-1}{2 j\beta_{\mathrm{r}}} \int_0^z N\,ds, \tag{5}$$

$$r^-(z) \equiv \frac{1}{2 j\beta_{\mathrm{r}}} e^{2 j\beta_{\mathrm{r}} z} \left(-\int_0^L N e^{-2 j\beta_{\mathrm{r}} s} ds + \int_0^z N e^{-2 j\beta_{\mathrm{r}} s} ds \right), \tag{6}$$

$$G_{km}(z)\equiv\frac{\omega\beta_{gm}}{2\hat{P}}\int_{-\infty}^{\infty}\Delta\varepsilon E_{ky}{}^{g}E_{my}{}^{g^*}\,dx, \tag{7}$$

$$R_m(\beta_\rho)=\frac{\omega\beta_{gm}}{2\hat{P}}\int_{-\infty}^{\infty}\Delta\varepsilon E_y{}^{r}E_{my}{}^{g^*}\,dx, \tag{8}$$

$$\overline{G}_k=\frac{\omega\beta_r}{2\hat{P}}\int_{-\infty}^{\infty}\Delta\varepsilon E_{ky}{}^{g}E_y{}^{r^*}\,dx, \tag{9}$$

$$\overline{R}(\beta_\rho,\beta_\rho')=\frac{\omega\beta_r}{2\hat{P}}\int_{-\infty}^{\infty}\Delta\varepsilon E_y{}^{r}(\beta_\rho)E_y{}^{r^*}(\beta_\rho')\,dx. \tag{10}$$

An approximate expression for the functions M_m and N can be obtained by perturbation. For small imperfections in the waveguide, the fields in the imperfect guide should be predominately those of the incident wave from the perfect guide. This is equivalent to saying that the unwanted modes are very small compared to the incident modes. Substituting Eqs. (3)–(10) into Eqs. (1) and (2) and keeping only the first-order terms, one obtains

$$\begin{aligned}M_m&=-\sum_k\delta_{0k}G_{km}=-G_{0m}\\&=-\frac{\omega\beta_{gm}}{2\hat{P}}\int_{-\infty}^{\infty}\Delta\varepsilon E_{0y}{}^{g}E_{my}{}^{g^*}\,dx,\end{aligned} \tag{11}$$

$$\begin{aligned}N&=-\sum_k\delta_{0k}\overline{G}_k\\&=-\frac{\omega\beta_r}{2\hat{P}}\int_{-\infty}^{\infty}\Delta\varepsilon E_{0y}{}^{g}(\beta_\rho)E_y{}^{r^*}(\beta_\rho')\,dx.\end{aligned} \tag{12}$$

For the present perturbation theory to be valid, either $\Delta\varepsilon$ is very small or $\Delta\varepsilon$ exists within a very narrow region. In either case, the contribution to the integral in Eqs. (11) and (12) should be very small. It is therefore reasonable to assume that the fields remain practically constant over the region where $\Delta\varepsilon$ contributes and that these fields can be removed from the integrals. Thus

$$\begin{aligned}M_m&\simeq-\frac{\omega\beta_{gm}}{2\hat{P}}\left(\int_{-\delta}^{-\delta+x_2}\Delta\varepsilon E_{0y}^{g}E_{my}{}^{g^*}\,dx+\int_{\delta}^{\delta+x_1}\Delta\varepsilon E_{0y}{}^{g}E_{my}{}^{g^*}\,dx\right)\\&\simeq-\frac{\omega\beta_{gm}}{2\hat{P}}\left[\Delta\varepsilon(-\delta)E_{0y}{}^{g}(-\delta)E_{my}{}^{g^*}(-\delta)\int_{-\delta}^{-\delta+x_2}dx\right.\\&\qquad\left.+\Delta\varepsilon(\delta)E_{0y}{}^{g}(\delta)E_{my}{}^{g^*}(\delta)\int_{\delta}^{\delta+x_1}dx\right]\\&=-\frac{\omega\beta_{gm}}{2\hat{P}}\left[\Delta\varepsilon(-\delta)E_{0y}{}^{g}(-\delta)E_{my}{}^{g^*}(-\delta)x_2\right.\\&\qquad\left.+\Delta\varepsilon(\delta)E_{0y}{}^{g}(\delta)E_{my}{}^{g^*}(\delta)x_1\right],\end{aligned} \tag{13}$$

$$N \simeq -\frac{\omega\beta_r}{2\hat{P}}\left[\Delta\varepsilon(-\delta)E_{0y}{}^{g}(-\delta)E_y{}^{r*}(-\delta)x_2 + \Delta\varepsilon(\delta)E_{0y}{}^{g}(\delta)E_y{}^{r*}(\delta)x_1\right]. \quad (14)$$

These functions will be evaluated for TE even and odd modes separately.

TE even modes. The fields for the TE even guided modes are (see Sections 2.2 and 2.2b),

$$E_{my}{}^{eg}(x,y) = E_m^{eg}\cos k_m x\, e^{-j\beta_{gm}z}, \qquad k_m = \sqrt{\omega^2\mu_1\epsilon_1 - \beta_{gm}{}^2},$$

$$E_m{}^{eg} = \left[\frac{2\omega\mu_1 P_m{}^e}{\beta_{gm}(\delta + 1/\alpha_{2m})}\right]^{1/2}, \qquad \alpha_{2m} = \sqrt{\beta_{gm}{}^2 - \omega^2\mu_2\epsilon_2}. \quad (15)$$

The radiation field is given in Section 2.5:

$$E_y{}^{er} = E^{er}\cos\beta_\rho x\, e^{-j\beta_r z}, \qquad \beta_\rho{}^2 = \beta_r{}^2 - \omega^2\mu\epsilon,$$

$$E^{er} = \left[2\omega\mu_2 P^e/\pi\beta_r(f^2+g^2)\right]^{1/2}, \qquad f \equiv \cos\beta_r\delta,$$

$$g \equiv (\beta_r/\beta_\rho)\sin\beta_r\delta. \quad (16)$$

Substituting Eqs. (15) and (16) into Eqs. (13) and (14) yields

$$\begin{aligned} M_m{}^e = -\frac{\omega\beta_{gm}}{2\hat{P}}\Bigg\{ &-(\epsilon_1-\epsilon_2)\left[\frac{2\omega\mu P_0{}^e}{\beta_{g0}(\delta+1/\alpha_{20})}\right]^{1/2}\cos k_0\delta \\ &\times e^{-j\beta_{g0}z}\left[\frac{2\omega\mu P_m{}^e}{\beta_{gm}(\delta+1/\alpha_{2m})}\right]^{1/2}\cos k_m{}^*\delta\, e^{j\beta_{gm}z}x_2 \\ &+(\epsilon_1-\epsilon_2)\left[\frac{2\omega\mu P_0{}^e}{\beta_{g0}(\delta+1/\alpha_{20})}\right]^{1/2}\cos k_0\delta \\ &\times e^{-j\beta_{g0}z}\left[\frac{2\omega\mu P_m{}^e}{\beta_{gm}(\delta+1/\alpha_{2m})}\right]^{1/2}\cos k_m{}^*\delta\, e^{j\beta_{gm}z}x_1\Bigg\} \\ = \frac{-\omega\beta_{gm}}{2\hat{P}}&(\epsilon_1-\epsilon_2)\frac{2\omega\mu(P_0{}^eP_m{}^e)^{1/2}}{\left[\beta_{g0}(\delta+1/\alpha_{20})\beta_{gm}(\delta+1/\alpha_{2m})\right]^{1/2}} \\ &\times\cos k_0\delta\cos k_m{}^*\delta\, e^{j(\beta_{gm}-\beta_{g0})z}(x_1-x_2). \end{aligned}$$

For normalized fields, $P_m{}^{\mathrm{e}} = P_0{}^{\mathrm{e}} = \hat{P}$, one has

$$M_m{}^{\mathrm{e}} = \left\{ -\frac{\omega^2 \mu \beta_{gm}(\epsilon_1 - \epsilon_2)}{\left[\beta_{g0}(\delta + 1/\alpha_{20})\beta_{gm}(\delta + 1/\alpha_{2m})\right]^{1/2}} \cos k_0\delta \cos k_m{}^*\delta \right\}$$
$$\times (x_1 - x_2) e^{j(\beta_{gm} - \beta_{g0})z}$$
$$\equiv m_m{}^{\mathrm{e}}(x_1 - x_2) e^{j(\beta_{gm} - \beta_{g0})z}, \tag{17}$$

$$N^{\mathrm{e}} = \left\{ -\omega^2 \mu \beta_r (\epsilon_1 - \epsilon_2) \left[\frac{\beta_r}{2\beta_{g0}(\delta + 1/\alpha_{20})(f^2 + g^2)} \right]^{1/2} \right.$$
$$\left. \times \cos k_0\delta \cos \beta_r \delta \right\} (x_1 - x_2) e^{j(\beta_r - \beta_{g0})z}$$
$$\equiv n^{\mathrm{e}}(x_1 - x_2) e^{j(\beta_r - \beta_{g0})z}. \tag{18}$$

Substituting Eqs. (17) and (18) into Eqs. (3)–(6) yields

$$g_k{}^{\mathrm{e}+}(z) = \delta_{0k} - \frac{1}{2j\beta_{gk}} \int_0^z M_k{}^{\mathrm{e}}\, ds$$
$$= \delta_{0k} - \frac{m_k{}^{\mathrm{e}}}{2j\beta_{gk}} \int_0^z [x_1(s) - x_2(s)] e^{j(\beta_{gk} - \beta_{g0})s}\, ds, \tag{19}$$

$$g_k{}^{\mathrm{e}-}(z) = \frac{1}{2j\beta_{gk}} e^{2j\beta_{gk}z} \left(-\int_0^L M_k{}^{\mathrm{e}} e^{-2j\beta_{gk}s}\, ds + \int_0^z M_k{}^{\mathrm{e}} e^{-2j\beta_{gk}s}\, ds \right)$$
$$= \frac{m_k{}^{\mathrm{e}}}{2j\beta_{gk}} e^{2j\beta_{gk}z} \left\{ -\int_0^L [x_1(s) - x_2(s)] e^{-j(\beta_{gk} + \beta_{g0})s}\, ds \right.$$
$$\left. + \int_0^z [x_1(s) - x_2(s)] e^{-j(\beta_{gk} + \beta_{g0})s}\, ds \right\}, \tag{20}$$

$$r^{\mathrm{e}+}(z) = \frac{1}{2j\beta_r} \int_0^z N\, ds$$
$$= \frac{-n^{\mathrm{e}}}{2j\beta_r} \int_0^z [x_1(s) - x_2(s)] e^{j(\beta_r - \beta_{g0})s}\, ds, \tag{21}$$

$$r^{\mathrm{e}-}(z) = \frac{1}{2j\beta_r} e^{2j\beta_r z} \left(-\int_0^L N e^{-2j\beta_r s}\, ds + \int_0^z N e^{-2j\beta_r s}\, ds \right)$$
$$= \frac{n^{\mathrm{e}}}{2j\beta_r} e^{2j\beta_r z} \left\{ -\int_0^L [x_1(s) - x_2(s)] e^{-j(\beta_r + \beta_{g0})s}\, ds \right.$$
$$\left. + \int_0^z [x_1(s) - x_2(s)] e^{-j(\beta_r + \beta_{g0})s}\, ds \right\}. \tag{22}$$

TE odd modes. The fields for the guided modes are (see Sections 2.2 and 2.2b),

$$E_{my}{}^{og}=E_m{}^{og}\sin k_m x\, e^{-j\beta_{gm}z},\qquad k_m{}^2=\omega^2\mu_1\epsilon_1-\beta_{gm}{}^2, \tag{23}$$

$$E_m{}^{og}=\left[\frac{2\omega\mu_1P_m{}^{o}}{\beta_{gm}(\delta-1/\alpha_{2m})}\right]^{1/2},\qquad \alpha_{2m}{}^2=\beta_{gm}{}^2-\omega^2\mu_2\epsilon_2.$$

The radiation modes are (see Sections 2.5 and 2.5a)

$$E_y{}^{or}=E^{or}\sin\beta_\rho x\, e^{-j\beta_r z},\qquad \beta_\rho{}^2=\beta_r{}^2-\omega^2\mu\epsilon, \tag{24}$$

$$E^{or}=\left[2\omega\mu_2P^{o}/\pi\beta_r(f_0{}^2+g_0{}^2)\right]^{1/2}$$

$$f_0\equiv\sin\beta_r\delta,\qquad g_0\equiv(\beta_r/\beta_\rho)\cos\beta_r\delta.$$

Substituting Eqs. (23) and (24) into Eqs. (13) and (14) yields ($\mu=\mu_1=\mu_2$)

$$\begin{aligned}
M_m{}^{o}&=\frac{-\omega\beta_{gm}}{2\hat{P}}\left[\Delta\epsilon(-\delta)E_{0y}{}^{og}(-\delta)E_{my}{}^{og*}(-\delta)x_2\right.\\
&\qquad\left.+\Delta\epsilon(\delta)E_{0y}{}^{og}(\delta)E_{my}{}^{og*}(\delta)x_1\right]\\
&=\left\{-\omega^2\mu\beta_{gm}(\epsilon_1-\epsilon_2)\frac{\sin k_m\delta\sin k_0\delta}{\left[\beta_{g0}(\delta-1/\alpha_{20})\beta_{gm}(\delta-1/\alpha_{2m})\right]^{1/2}}\right\}\\
&\qquad\times(x_1-x_2)e^{j(\beta_{gm}-\beta_{g0})z}\\
&\equiv m_m{}^{o}\left[x_1(z)-x_2(z)\right]e^{j(\beta_{gm}-\beta_{g0})z},
\end{aligned} \tag{25}$$

$$\begin{aligned}
N^{o}&=\frac{-\omega\beta_r}{2\hat{P}}\left[\Delta\epsilon(-\delta)E_{0y}{}^{og}(-\delta)E_y{}^{or*}(-\delta)x_2\right.\\
&\qquad\left.+\Delta\epsilon(\delta)E_{0y}{}^{og}(\delta)E_y{}^{or*}(\delta)x_1\right]\\
&=\left\{\frac{-\omega^2\mu\beta_r}{2}(\epsilon_1-\epsilon_2)\frac{\sqrt{2}\,\sin\beta_r\delta\sin k_0\delta}{\left[\beta_{g0}(\delta-1/\alpha_{20})\beta_r(f_0{}^2+g_0{}^2)\right]^{1/2}}\right\}\\
&\qquad\times(x_1-x_2)e^{j(\beta_r-\beta_{g0})z}\\
&\equiv n^{o}\left[x_1(z)-x_2(z)\right]e^{j(\beta_r-\beta_{g0})z}.
\end{aligned} \tag{26}$$

The fields are normalized such that $P_m{}^{o}=P_0{}^{o}=\hat{P}$. Then substituting Eqs. (25) and (26) into Eqs. (3)–(6) yields

$$\begin{aligned}
g_k{}^{o+}(z)&=\delta_{0k}-\frac{1}{2j\beta_{gk}}\int_0^z M_k{}^{o}\,ds\\
&=\delta_{0k}-\frac{m_k{}^{o}}{2j\beta_{gk}}\int_0^z\left[x_1(s)-x_2(s)\right]e^{j(\beta_{gk}-\beta_{g0})s}\,ds,
\end{aligned} \tag{27}$$

$$g_k^{\,o-}(z)=\frac{m_k^{\,o}}{2j\beta_{gk}}e^{2j\beta_{gk}}\left\{-\int_0^L[x_1(s)-x_2(s)]e^{j(\beta_{gk}+\beta_{go})s}\,ds\right.$$

$$\left.+\int_0^z[x_1(s)-x_2(s)]e^{j(\beta_{gk}+\beta_{go})s}\,ds\right\}, \qquad (28)$$

$$r^{o+}(z)=\frac{-n^o}{2j\beta_r}\int_0^z[x_1(s)-x_2(s)]e^{j(\beta_r-\beta_{go})s}\,ds, \qquad (29)$$

$$r^{o-}(z)=\frac{n^o}{2j\beta_r}\left\{-\int_0^L[x_1(s)-x_2(s)]e^{-j(\beta_r+\beta_{go})s}\,ds\right.$$

$$\left.+\int_0^z[x_1(s)-x_2(s)]e^{-j(\beta_r+\beta_{go})s}\,ds\right\}e^{2j\beta_r z}. \qquad (30)$$

References

1. D. Marcuse, Mode Conversion Caused by Surface Imperfections of a Dielectric Slab Waveguide, *Bell Syst. Tech. J.* 48, 3187–3215 (1969).
2. D. Marcuse, Mode Conversion Caused by Diameter Changes of a Round Dielectric Waveguide, *Bell Syst. Tech. J.* 48, 3217–3232 (1969).
3. D. Marcuse, *Light Transmission Optics*, Van Nostrand Reinhold, Princeton, N.J., 1972.
4. D. Marcuse, *Theory of Dielectric Optical Waveguides*, Academic, New York, 1974.
5. A. W. Snyder, Radiation Losses Due to Variations of Radius on Dielectric or Optical Fibers, *IEEE Trans. Microwave Theory Tech.* MTT-18, 608–615 (1970).
6. H. G. Unger, *Planar Optical Waveguides and Fibres*, Clarendon, Oxford, 1977.

7

Inhomogeneous Waveguides

One of the most popular types of waveguide in optical communications consists of glass or plastic filaments with diameter of the order of that of human hair. A fiber of uniform dielectric constant does not by itself make a practical waveguide. Any scratches on it would cause radiation losses, while foreign objects next to the bare fiber would alter the boundary conditions and consequently affect the fields within.

One way to circumvent these drawbacks is to employ a protective coating, known as *cladding*. Cladded fiber will be discussed in the next chapter. Another method uses fiber with dielectric constant nonuniform in the radial direction. In *graded-index fiber* the dielectric constant decreases radially toward the surface. A properly designed graded-index fiber confines the fields within a region near the center (or, in the case of a dielectric sheet whose thickness is measured in the x direction, near $x=0$). Since the fields are practically zero near the surface of this type of waveguide, variation in the boundary conditions has little effect on the fields within the guide. The dielectric sheet with inhomogeneous dielectric constant will be analyzed in this chapter.

7.1. Inhomogeneous Dielectric Sheet

Dielectric waveguides characterized by continuous variation of the dielectric constant [4, 8, 9] have some interesting properties. These guides may be designed to give periodic focusing effects such that the energy of the beam is concentrated within a desired region. Imperfections at the peripheries of the guide may thus be made to have little effect on the performance of the guide.

In a medium characterized by an inhomogeneous permittivity $\varepsilon(\mathbf{r})$ and a uniform permeability μ, the spatial dependence of the fields satisfies Eqs. (1.3.10) and (1.3.11),

$$\nabla^2\mathbf{E}+\nabla\left[\frac{1}{\varepsilon_r(\mathbf{r})}\nabla\,\varepsilon_r(\mathbf{r})\cdot\mathbf{E}\right]-\gamma_0^2\varepsilon_r(\mathbf{r})\mathbf{E}=0,\qquad \gamma_0^2\equiv-\omega^2\mu\epsilon_0, \tag{1}$$

$$\nabla^2\mathbf{H}+\frac{1}{\varepsilon_r(\mathbf{r})}\nabla\,\varepsilon_r(\mathbf{r})\times(\nabla\times\mathbf{H})-\gamma_0^2\varepsilon_r(\mathbf{r})\mathbf{H}=0. \tag{2}$$

These equations are in general very complicated. If the relative variation $[1/\varepsilon_r(\mathbf{r})]\nabla\,\varepsilon_r(\mathbf{r})$ is very small, then the second term may be neglected to yield approximate relations.

For the special case when the permittivity is a function of only one of the spatial variables, say x, that is, $\varepsilon_r(\mathbf{r})\equiv\varepsilon_r(x)$, Eqs. (1) and (2) may be decomposed into the scalar equations

$$\nabla^2E_x+\frac{\partial}{\partial x}\left[\frac{1}{\varepsilon_r(x)}\frac{d\varepsilon_r(x)}{dx}E_x\right]-\gamma_0^2\varepsilon_r(x)E_x=0, \tag{3a}$$

$$\nabla^2E_y+\frac{\partial}{\partial y}\left[\frac{1}{\varepsilon_r(x)}\frac{d\varepsilon_r(x)}{dx}E_x\right]-\gamma_0^2\varepsilon_r(x)E_y=0, \tag{3b}$$

$$\nabla^2E_z+\frac{\partial}{\partial z}\left[\frac{1}{\varepsilon_r(x)}\frac{d\varepsilon_r(x)}{dx}E_x\right]-\gamma_0^2\varepsilon_r(x)E_z=0; \tag{3c}$$

$$\nabla\,\varepsilon_r(x)\times(\nabla\times\mathbf{H})=-\hat{\mathbf{y}}\frac{d\varepsilon_r(x)}{dx}\left(\frac{\partial H_y}{\partial x}-\frac{\partial H_x}{\partial y}\right)+\hat{\mathbf{z}}\frac{d\varepsilon_r(x)}{dx}\left(\frac{\partial H_x}{\partial z}-\frac{\partial H_z}{\partial x}\right),$$

$$\nabla^2H_x-\gamma_0^2\varepsilon_r(x)H_x=0, \tag{4a}$$

$$\nabla^2H_y-\frac{1}{\varepsilon_r(x)}\frac{d\varepsilon_r(x)}{dx}\left(\frac{\partial H_y}{\partial x}-\frac{\partial H_x}{\partial y}\right)-\gamma_0^2\varepsilon_r(x)H_y=0, \tag{4b}$$

$$\nabla^2H_z+\frac{1}{\varepsilon_r(x)}\frac{d\varepsilon_r(x)}{dx}\left(\frac{\partial H_x}{\partial z}-\frac{\partial H_z}{\partial x}\right)-\gamma_0^2\varepsilon_r(x)H_z=0. \tag{4c}$$

These equations point to the fact that when the permittivity is a function of x, $\varepsilon(\mathbf{r})\equiv\varepsilon(x)$, then the x components of the fields, E_x and H_x, satisfy two uncoupled equations, while equations for other components are coupled with E_x or H_x. It is therefore easier to determine E_x and H_x from Eqs. (3a) and (4a), respectively. Other field components are related to E_x and H_x through Maxwell's equations (see Appendix 4).

Equation (3a) may be solved by the method of separation of variables. Let the trial solution be

$$E_x(\mathbf{r}) \equiv e_x(x)Y(y)Z(z). \tag{5}$$

Substituting Eq. (5) into Eq. (3a) yields

$$\left[\frac{1}{e_x}\frac{d^2e_x}{dx^2}+\frac{1}{e_x}\frac{d}{dx}\left(\frac{1}{\varepsilon_r}\frac{d\varepsilon_r}{dx}e_x\right)-\gamma_0^2\varepsilon_r\right]+\frac{1}{Y}\frac{d^2Y}{dy^2}+\frac{1}{Z}\frac{d^2Z}{dz^2}=0,$$

$$\frac{1}{Y}\frac{d^2Y}{dy^2}=\gamma_y{}^2, \qquad Y(y)=A_1e^{\gamma_y y}+A_2e^{-\gamma_y y},$$

$$\frac{1}{Z}\frac{d^2Z}{dz^2}=\gamma_z{}^2, \qquad Z(z)=B_1e^{\gamma_z z}+B_2e^{-\gamma_z z},$$

$$\frac{d^2e_x}{dx^2}+\frac{d}{dx}\left(\frac{1}{\varepsilon_r}\frac{d\varepsilon_r}{dx}e_x\right)+\left(\gamma_y{}^2+\gamma_z{}^2-\gamma_0^2\varepsilon_r\right)e_x=0. \tag{6}$$

This is the equation to be solved. Once e_x is determined, the full field has the form

$$\tilde{E}_x(\mathbf{r},t)=E_0e_x(x)e^{j\omega t\pm\gamma_y y\pm\gamma_z z}, \tag{7}$$

where E_0 is the magnitude.

Equation (4a) may be solved similarly.

$$H_x(\mathbf{r})=h_x(x)Y(y)Z(z), \tag{8}$$

$$\frac{1}{h_x(x)}\frac{d^2h_x}{dx^2}-\gamma_0^2\varepsilon_r+\frac{1}{Y}\frac{d^2Y}{dy^2}+\frac{1}{Z}\frac{d^2Z}{dz^2}=0,$$

$$\frac{1}{Y}\frac{d^2Y}{dy^2}=\gamma_y{}^2, \qquad Y(y)=a_1e^{\gamma_y y}+a_2e^{-\gamma_y y},$$

$$\frac{1}{Z}\frac{d^2Z}{dz^2}=\gamma_z{}^2, \qquad Z(z)=b_1e^{\gamma_z z}+b_2e^{-\gamma_z z},$$

$$\frac{d^2h_x}{dx^2}+\left(\gamma_y{}^2+\gamma_z{}^2-\gamma_0^2\varepsilon_r\right)h_x=0. \tag{9}$$

This equation determines $h_x(x)$; $\tilde{H}_x(\mathbf{r},t)$ thus has the form

$$\tilde{H}_x(\mathbf{r},t)=H_0h_x(x)e^{j\omega t\pm\gamma_y y\pm\gamma_z z}, \tag{10}$$

where H_0 is the magnitude.

Equation (6) would be easier to solve if it did not contain the first derivative of $e_x(x)$. This simplification may be accomplished by a change of variable. Let

$$e_x(x) \equiv U(x)\bar{e}_x(x), \tag{11}$$

$$\frac{de_x}{dx} = U\frac{d\bar{e}_x}{dx} + \bar{e}_x\frac{dU}{dx},$$

$$\frac{d^2e_x}{dx} = U\frac{d^2\bar{e}_x}{dx^2} + \bar{e}_x\frac{d^2U}{dx^2} + 2\frac{dU}{dx}\frac{d\bar{e}_x}{dx}.$$

Substituting Eq. (11) into Eq. (6) yields

$$U\frac{d^2\bar{e}_x}{dx^2} + \left(2\frac{dU}{dx} + U\frac{1}{\varepsilon_r}\frac{d\varepsilon_r}{dx}\right)\frac{d\bar{e}_x}{dx} + \bar{e}_x\left[\frac{1}{\varepsilon_r}\frac{d\varepsilon_r}{dx}\frac{dU}{dx} + \frac{d^2U}{dx^2}\right.$$

$$\left. + \left(\gamma_y{}^2 + \gamma_z{}^2 - \gamma_0^2\varepsilon_r\right)U + U\frac{d}{dx}\left(\frac{1}{\varepsilon_r}\frac{d\varepsilon_r}{dx}\right)\right] = 0. \tag{12}$$

To obtain a relation without the first derivative $d\bar{e}_x/dx$, one must set its coefficient equal to zero, so that

$$\frac{2}{U}\frac{dU}{dx} = -\frac{1}{\varepsilon_r}\frac{d\varepsilon_r}{dx},$$

$$\frac{d(\ln U^2)}{dx} = \frac{d\ln(1/\varepsilon_r)}{dx},$$

$$U(x) = 1/\sqrt{\varepsilon_r(x)}\,. \tag{13}$$

Now

$$\frac{dU}{dx} = -\frac{1}{2}\varepsilon_r{}^{-3/2}\frac{d\varepsilon_r}{dx},$$

$$\frac{d^2U}{dx^2} = \frac{3}{4}\varepsilon_r{}^{-5/2}\left(\frac{d\varepsilon_r}{dx}\right)^2 - \frac{1}{2}\varepsilon_r{}^{-3/2}\frac{d^2\varepsilon_r}{dx^2},$$

$$\frac{d^2U}{dx^2} + \frac{1}{\varepsilon_r}\frac{d\varepsilon_r}{dx}\frac{dU}{dx} + U\frac{d}{dx}\left(\frac{1}{\varepsilon_r}\frac{d\varepsilon_r}{dx}\right) = \frac{1}{2}\varepsilon_r{}^{-3/2}\frac{d^2\varepsilon_r}{dx^2} - \frac{3}{4}\varepsilon_r{}^{-5/2}\left(\frac{d\varepsilon_r}{dx}\right)^2.$$

Substituting Eq. (13) into Eq. (12) yields

$$\frac{d^2\bar{e}_x}{dx^2} + \left[\frac{1}{2\varepsilon_r}\frac{d^2\varepsilon_r}{dx^2} - \frac{3}{4\varepsilon_r{}^2}\left(\frac{d\varepsilon_r}{dx}\right)^2 + \gamma_y{}^2 + \gamma_z{}^2 - \gamma_0^2\varepsilon_r\right]\bar{e}_x = 0 \tag{14}$$

and

$$\bar{e}_x(x) = \sqrt{\varepsilon_r(x)}\; e_x(x). \tag{15}$$

Note that Eq. (14) may be simplified to

$$\frac{d^2\bar{e}_x}{dx^2} + \left(\gamma_y{}^2 + \gamma_z{}^2 - \gamma_0^2\varepsilon_r\right)\bar{e}_x = 0, \tag{16}$$

if

$$\frac{1}{\varepsilon_r}\left[\frac{1}{2}\frac{d^2\varepsilon_r}{dx^2}-\frac{3}{4\varepsilon_r}\left(\frac{d\varepsilon_r}{dx}\right)^2\right]\ll|\gamma_0^2|=\omega^2\mu\epsilon_0=\left(\frac{2\pi}{\lambda}\right)^2.$$

This implies the variation of permittivity per unit wavelength is very small.

The problem of the inhomogeneous dielectric sheet is to solve Eqs. (9) and (14). Since these are independent equations, it is convenient to consider two separate cases: (a) $H_x=0$ and (b) $E_x=0$. The general case is the superposition of these special cases.

7.2. Fields in Terms of E_x and H_x

It has been shown that E_x and H_x satisfy two independent equations, (7.1.3a) and (7.1.4a). It would be useful to express all other field components in terms of E_x and H_x. Consequently, once E_x and H_x are known, the fields are completely determined.

It is shown in Appendix 4 that the tranverse field components are related to the parallel components [Eqs. (A4.17) and (A4.18)] by

$$-\nabla_\perp^2\mathbf{H}_\perp=j\omega\epsilon_0\varepsilon_r\nabla_\perp\times\mathbf{E}_\parallel+\nabla_\perp(\nabla_\parallel\cdot\mathbf{H}_\parallel),\tag{1}$$

$$-\nabla_\perp^2\mathbf{E}_\perp=\nabla_\perp\left(\frac{1}{\varepsilon_r}\nabla_\parallel\epsilon_r\cdot\mathbf{E}_\parallel\right)+\nabla_\perp(\nabla_\parallel\cdot\mathbf{E}_\parallel)-j\omega\mu\nabla_\perp\times\mathbf{H}_\parallel,\tag{2}$$

where for the present case

$$\begin{aligned}&\mathbf{E}_\parallel\equiv\hat{\mathbf{x}}E_x, &&\mathbf{E}_\perp\equiv\hat{\mathbf{y}}E_y+\hat{\mathbf{z}}E_z,\\ &\mathbf{H}_\parallel\equiv\hat{\mathbf{x}}H_x, &&\mathbf{H}_\perp\equiv\hat{\mathbf{y}}H_y+\hat{\mathbf{z}}H_z,\\ &\nabla_\parallel\equiv\hat{\mathbf{x}}\frac{\partial}{\partial x}, &&\nabla_\perp\equiv\hat{\mathbf{y}}\frac{\partial}{\partial y}+\hat{\mathbf{z}}\frac{\partial}{\partial z}.\end{aligned}\tag{3}$$

If the field's dependence on the variables in tranverse directions is of the form $e^{-\gamma_y y-\gamma_z z}$, then

$$\nabla_\perp^2\mathbf{H}_\perp=({\gamma_y}^2+{\gamma_z}^2)\mathbf{H}_\perp.\tag{4}$$

With (3) and (4), Eqs. (1) and (2) become

$$\mathbf{H}_\perp=\frac{-1}{{\gamma_y}^2+{\gamma_z}^2}\left[j\omega\epsilon_0\varepsilon_r(-\hat{\mathbf{y}}\gamma_z E_x+\hat{\mathbf{z}}\gamma_y E_x)+(-\hat{\mathbf{y}}\gamma_y-\hat{\mathbf{z}}\gamma_z)\frac{\partial H_x}{\partial x}\right]\tag{5}$$

or

$$H_y=\frac{1}{{\gamma_y}^2+{\gamma_z}^2}\left(j\omega\epsilon_0\varepsilon_r\gamma_z E_x-\gamma_y\frac{\partial H_x}{\partial x}\right),\tag{6}$$

$$H_z=\frac{-1}{{\gamma_y}^2+{\gamma_z}^2}\left(j\omega\epsilon_0\varepsilon_r\gamma_y E_x-\gamma_z\frac{\partial H_x}{\partial x}\right);\tag{7}$$

$$\mathbf{E}_\perp = \frac{-1}{\gamma_y{}^2+\gamma_z{}^2}\left[\nabla_\perp\left(\frac{1}{\varepsilon_r}\frac{\partial\varepsilon_r}{\partial x}E_x\right)+\nabla_\perp\frac{\partial E_x}{\partial x} - j\omega\mu\nabla_\perp \times \mathbf{H}_\parallel\right]$$

$$= \frac{-1}{\gamma_y{}^2+\gamma_z{}^2}\left[\frac{1}{\varepsilon_r}\frac{\partial\varepsilon_r}{\partial x}(-\hat{\mathbf{y}}\gamma_y-\hat{\mathbf{z}}\gamma_z)E_x\right.$$

$$\left. +(-\hat{\mathbf{y}}\gamma_y-\hat{\mathbf{z}}\gamma_z)\frac{\partial E_x}{\partial x} - j\omega\mu\nabla_\perp \times \mathbf{H}_\parallel\right]$$

$$= \frac{-1}{\gamma_y{}^2+\gamma_z{}^2}\left[(-\hat{\mathbf{y}}\gamma_y-\hat{\mathbf{z}}\gamma_z)\frac{1}{\varepsilon_r}\left(E_x\frac{\partial\varepsilon_r}{\partial x}+\varepsilon_r\frac{\partial E_x}{\partial x}\right)\right.$$

$$\left. - j\omega\mu\nabla_\perp \times (\hat{\mathbf{x}}H_x)\right]$$

$$= \frac{1}{\gamma_y{}^2+\gamma_z{}^2}\left[(\hat{\mathbf{y}}\gamma_y+\hat{\mathbf{z}}\gamma_z)\frac{1}{\varepsilon_r}\frac{\partial(\varepsilon_r E_x)}{\partial x} + j\omega\mu(-\hat{\mathbf{y}}\gamma_z+\hat{\mathbf{z}}\gamma_y)H_x\right],$$

$$E_y = \frac{1}{\gamma_y{}^2+\gamma_z{}^2}\left[\frac{\gamma_y}{\varepsilon_r}\frac{\partial(\varepsilon_r E_x)}{\partial x} - j\omega\mu\gamma_z H_x\right], \tag{8}$$

$$E_z = \frac{1}{\gamma_y{}^2+\gamma_z{}^2}\left[\frac{\gamma_z}{\varepsilon_r}\frac{\partial(\varepsilon_r E_x)}{\partial x} + j\omega\mu\gamma_y H_x\right]. \tag{9}$$

7.3. Guided Modes

To simplify mathematical analysis, waveguide problems are usually subdivided into special cases according to the vanishing of a certain field component. (Usually the axial component is chosen.) In this analysis it is assumed that the fields are invariant in the y direction, that is, $\partial/\partial y=0$, and that the z dependency is of the form $e^{-\gamma_z z}$.

TE Modes. In this case, the **E** field has no component in the axial direction. $E_z=0$. Maxwell's equations are $\nabla\times\mathbf{E}=-j\omega\mu\mathbf{H}$, or

$$\gamma_z E_y = -j\omega\mu H_x, \tag{1}$$

$$-\gamma_z E_x = -j\omega\mu H_y, \tag{2}$$

$$\frac{\partial E_y}{\partial x} = -j\omega\mu H_z; \tag{3}$$

and $\nabla\times\mathbf{H}=j\omega\epsilon_0\varepsilon_r(x)\mathbf{E}$; or

$$\gamma_z H_y = j\omega\epsilon_0\varepsilon_r E_x, \tag{4}$$

$$-\frac{\partial H_z}{\partial x} - \gamma_z H_x = j\omega\epsilon_0\varepsilon_r E_y, \tag{5}$$

$$\frac{\partial H_y}{\partial x} = 0. \tag{6}$$

Equation (6) implies that H_y is constant with respect to x. Since the addition of a constant field will not affect the general behavior of a varying field, it may be set equal to zero. Then Eq. (2) becomes

$$-\gamma_z E_x = 0 \qquad \text{or} \qquad E_x = \text{const} = 0;$$

that is, E_x is constant with respect to z and like H_y may therefore be set equal to zero. Equations (1)–(6) thus reduce to

$$\gamma_z E_y = -j\omega\mu H_x \qquad \text{or} \qquad E_y = \frac{-j\omega\mu}{\gamma_z} H_x, \tag{7}$$

$$\frac{\partial E_y}{\partial x} = -j\omega\mu H_z \qquad \text{or} \qquad H_z = \frac{-1}{j\omega\mu}\frac{\partial}{\partial x}\left(\frac{-j\omega\mu}{\gamma_z} H_x\right) = \frac{1}{\gamma_z}\frac{\partial H_x}{\partial x}. \tag{8}$$

H_x is given by Eq. (7.1.10) and h_x is solved for from Eq. (7.1.9).

TM Modes. The **H** field does not contain the axial component in this case; that is, $H_z = 0$. Maxwell's equations are $\nabla\times\mathbf{E} = -j\omega\mu\mathbf{H}$, or

$$\gamma_z E_y = -j\omega\mu H_x, \tag{9}$$

$$-\gamma_z E_x - \frac{\partial E_z}{\partial x} = -j\omega\mu H_y, \tag{10}$$

$$\frac{\partial E_y}{\partial x} = 0; \tag{11}$$

and $\nabla\times\mathbf{H}=j\omega\epsilon_0\varepsilon_r\mathbf{E}$, or

$$\gamma_z H_y = j\omega\epsilon_0\varepsilon_r E_x, \tag{12}$$

$$-\gamma_z H_x = j\omega\epsilon_0\varepsilon_r E_y, \tag{13}$$

$$\frac{\partial H_y}{\partial x} = j\omega\epsilon_0\varepsilon_r E_z. \tag{14}$$

Equation (11) leads to $E_y = 0$ and $H_x = 0$. Equations (12) and (14) yield

$$H_y = \frac{j\omega\epsilon_0\varepsilon_r}{\gamma_z} E_x, \tag{15}$$

$$E_z = \frac{1}{\gamma_z\varepsilon_r}\frac{\partial(E_x\varepsilon_r)}{\partial x}, \tag{16}$$

where E_x is given by Eq. (7.1.7) and $\bar{e}_x$ is to be determined from Eq. (7.1.14).

Summary

For the TE modes, $E_z=0$, $E_x=H_y=0$, and

$$\nabla^2 H_x - \gamma_0^2 \varepsilon_r H_x = 0, \qquad \gamma_0^2 \equiv -\omega^2 \mu \epsilon_0,$$
$$E_y = -\frac{j\omega\mu}{\gamma_z} H_x, \qquad H_z = \frac{1}{\gamma_z}\frac{\partial H_x}{\partial x}. \tag{17}$$

For the TM modes, $H_z=0$, $H_x=E_y=0$, and

$$\nabla^2 E_x + \frac{\partial}{\partial x}\left(\frac{1}{\varepsilon_r}\frac{d\varepsilon_r}{dx}E_x\right) - \gamma_0^2 \varepsilon_r E_x = 0,$$
$$E_z = \frac{1}{\gamma_z \varepsilon_r}\frac{\partial}{\partial x}(\varepsilon_r E_x), \qquad H_y = \frac{j\omega\epsilon_0\varepsilon_r}{\gamma_z}E_x. \tag{18}$$

These field components can also be obtained from Eqs. (7.2.6)–(7.2.9).

7.4. Square-Law Media

Miller [5] investigated the behavior of wave propagation in a continuous medium with square-law refractive index variation. It was shown that such a medium has a strong focusing effect relative to non-square-law media, where the beam will spread.

Consider a dielectric-sheet guide characterized by permittivity

$$\varepsilon(x) = \epsilon_0 \epsilon_r - x^2\,\delta\varepsilon = \epsilon_0 \epsilon_r (1 - x^2\,\delta\varepsilon_r), \tag{1}$$

where $\delta\varepsilon$ is a small constant and $\delta\varepsilon_r \equiv \delta\varepsilon/\epsilon_0\epsilon_r$. Equation (1) is valid when $\varepsilon(x) > 0$. The infinite-medium approximation is implied here; the signal is confined within a region where $x^2\,\delta\varepsilon \ll \epsilon_0$. It should be pointed out that the permittivity described by Eq. (1) is not practical; however, it provides an analytical solution to a dielectric-sheet guide and gives valuable insight into more complicated situations.

The equations to be solved are [Eqs. (7.1.4a) and (7.1.3a)]

TE modes:
$$\nabla^2 H_x - \gamma_0^2 \varepsilon_r(x) H_x = 0; \tag{2}$$

TM modes:
$$\nabla^2 E_x + \frac{\partial}{\partial x}\left[\frac{1}{\varepsilon_r(x)}\frac{d\varepsilon_r(x)}{dx}E_x\right] - \gamma_0^2 \varepsilon_r(x) E_x = 0. \tag{3}$$

The fields are assumed to have the form

$$\tilde{E}_x(\mathbf{r}, t) = E_x(\mathbf{r})e^{j\omega t} = e_x(x)e^{j\omega t - \gamma_z z},$$
$$\tilde{H}_x(\mathbf{r}, t) = H_x(\mathbf{r})e^{j\omega t} = h_x(x)e^{j\omega t - \gamma_z z}. \tag{4}$$

Equations (2) and (3) thus become

$$\text{TE modes:}\qquad \frac{d^2h_x}{dx^2}+\left[\gamma_z{}^2-\gamma_0^2\epsilon_r(1-\delta\varepsilon_r x^2)\right]h_x=0; \tag{5}$$

$$\text{TM modes:}\qquad \frac{d^2\bar{e}_x}{dx^2}+\left[\gamma_z{}^2-\gamma_0^2\varepsilon_r-\frac{3}{4}\left(\frac{1}{\varepsilon_r}\frac{d\varepsilon_r}{dx}\right)^2+\frac{1}{2}\frac{1}{\varepsilon_r}\frac{d^2\varepsilon_r}{dx^2}\right]\bar{e}_x=0,$$

$$\bar{e}_x\equiv\sqrt{\varepsilon_r(x)}\,e_x \qquad\text{and}\qquad \varepsilon_r\equiv 1-x^2\,\delta\varepsilon_r. \tag{6}$$

Equations (5) and (6) have the same general form, and their solution will be treated in the next section.

7.5. Hermite Differential Equation

The Schrodinger wave equation for certain potential energy has the form

$$\frac{d^2g}{dx^2}+(a-b^2x^2)g=0. \tag{1}$$

It is desired to find a finite, continuous, single-valued solution in the entire x space, $-\infty\le x\le\infty$. The polynominal method may be used to obtain the solution. This method consists of

a. determinating the asymptotic solution g_∞ of Eq. (1) for large values of x, and
b. determinating a correction factor $\hat{g}$ such that $g=\hat{g}g_\infty$ is the desired solution.

The asymptotic solution is determined from the asymptotic equation for Eq. (1) when $b^2x^2\gg a$,

$$\frac{d^2g_\infty}{dx^2}-b^2x^2g_\infty=0, \tag{2}$$

where $g_\infty\equiv g_{x=\infty}$ is the solution when the inequality $b^2x^2\gg a$ holds.

Let the trial solution for Eq. (2) be

$$g_\infty(x)=Ae^{px^2},\qquad \frac{dg_\infty}{dx}=2Apxe^{px^2},$$

$$\frac{d^2g_\infty}{dx^2}=(2px^2+1)2pAe^{px^2}\simeq(2px)^2Ae^{px^2},$$

$$A(2px)^2e^{px^2}-b^2x^2Ae^{px^2}=0,$$

where $p=\pm b/2$. Therefore,

$$g_\infty(x)=Ae^{-bx^2/2}. \tag{3}$$

The negative sign in the argument is chosen so that the function will remain finite at large values of x.

Next the solution for the entire x space is assumed to be the product of the asymptotic solution and a correction factor.

$$g(x) \equiv \hat{g}(x) e^{-bx^2/2}, \tag{4}$$

where $\hat{g}(x)$ is the correction factor to be determined. Then

$$\frac{dg}{dx} = \left(\frac{d\hat{g}}{dx} - bx\hat{g} \right) e^{-bx^2/2},$$

$$\frac{d^2g}{dx^2} = \left[\frac{d^2\hat{g}}{dx^2} - 2bx\frac{d\hat{g}}{dx} + (b^2x^2 - b)\hat{g} \right] e^{-bx^2/2}. \tag{5}$$

Substituting Eqs. (4) and (5) into (1) yields

$$\left[\frac{d^2\hat{g}}{dx^2} - 2bx\frac{d\hat{g}}{dx} + (b^2x^2 - b)\hat{g} \right] + (a - b^2x^2)\hat{g} = 0,$$

$$\frac{d^2\hat{g}}{dx^2} - 2bx\frac{d\hat{g}}{dx} + (a-b)\hat{g} = 0. \tag{6}$$

This equation may be transformed into normal form by a change of variable. Let

$$x = cq, \qquad q = \frac{x}{c}, \qquad \frac{dq}{dx} = \frac{1}{c},$$

$$\frac{d}{dx} = \frac{d}{dq}\frac{dq}{dx} = \frac{1}{c}\frac{d}{dq}, \qquad \frac{d^2}{dx^2} = \frac{1}{c^2}\frac{d^2}{dq^2},$$

$$\frac{1}{c^2}\frac{d^2\hat{g}}{dq^2} - 2bcq\frac{1}{c}\frac{d\hat{g}}{dq} + (a-b)\hat{g} = 0,$$

$$\frac{d^2\hat{g}}{dq^2} - 2bc^2q\frac{d\hat{g}}{dq} + c^2(a-b)\hat{g} = 0.$$

For the coefficient of $d\hat{g}/dq$ term to be $2q$, one must have

$$bc^2 = 1, \qquad c = 1/\sqrt{b}.$$

Then

$$\frac{d^2\hat{g}(q)}{dq^2} - 2q\frac{d\hat{g}(q)}{dq} + \left(\frac{a}{b} - 1 \right)\hat{g}(q) = 0. \tag{7}$$

This is known as the *Hermite differential equation* [1, 2, 6, 7]. The *power series method* may be used to obtain its solution. Let

$$\begin{aligned}\hat{g}(q) &= d_0 + d_1 q + d_2 q^2 + \cdots + d_n q^n + d_{n+1} q^{n+1} \\ &\quad + d_{n+2} q^{n+2} + \cdots, \\ \frac{d\hat{g}(q)}{dq} &= d_1 + 2d_2 q + \cdots + n d_n q^{n-1} + (n+1) d_{n+1} q^n \\ &\quad + (n+2) d_{n+2} q^{n+1} + \cdots, \\ \frac{d\hat{g}^2(q)}{dq^2} &= 2d_2 + \cdots + n(n-1) d_n q^{n-2} \\ &\quad + n(n+1) d_{n+1} q^{n-1} + (n+1)(n+2) d_{n+2} q^n + \cdots.\end{aligned} \tag{8}$$

Substituting Eq. (8) into Eq. (7) yields

$$\begin{aligned}&\left[2d_2 + \cdots + n(n-1) d_n q^{n-2} + n(n+1) d_{n+1} q^{n-1}\right. \\ &\quad \left. + (n+1)(n+2) d_{n+2} q^n + \cdots\right] \\ &\quad - 2q\left[d_1 + 2d_2 q + \cdots + n d_n q^{n-1} + (n+1) d_{n+1} q^n + \cdots\right] \\ &\quad + (a/b-1)\left[d_0 + d_1 q + \cdots + d_n q^n + d_{n+1} q^{n+1} + \cdots\right] = 0.\end{aligned} \tag{9}$$

The series given by Eq. (8) will be a solution of Eq. (7) if Eq. (9) can be satisfied. Equation (9) can be fulfilled for all values of q if the coefficient of each power of q individually equals zero. Thus

$$\begin{aligned}&\left[(n+1)(n+2) d_{n+2} - 2n d_n + (a/b-1) d_n\right] q^n = 0, \\ &d_{n+2} = \frac{2n-(a/b-1)}{(n+1)(n+2)} d_n, \qquad n = 0, 1, 2, \ldots .\end{aligned} \tag{10}$$

This is the recursion formula for the coefficients. The lowest two coefficients, d_0 and d_1, are the arbitrary constants of integration of the second-order ordinary differential equation. Coefficients with higher indexes $(d_2, d_3, \ldots)$ can be determined in terms of d_0 or d_1.

The series $\hat{g}(q)$ consists in general of an infinite number of terms and cannot serve as a desired solution because the series diverges rapidly with increasing q. To see this, consider the series expansion of e^{q^2},

$$\begin{aligned}e^{q^2} &= 1 + q^2 + \frac{(q^2)^2}{2!} + \cdots + \frac{(q^2)^m}{m!} + \frac{(q^2)^{m+1}}{(m+1)!} + \cdots \\ &= 1 + q^2 + \frac{q^4}{2!} + \cdots + \frac{q^{2m}}{m!} + \frac{q^{2m+2}}{(m+1)!} + \cdots \\ &= 1 + q^2 + \frac{q^4}{2!} + \cdots + \frac{q^n}{(n/2)!} + \frac{q^{n+2}}{(n/2+1)!} + \cdots \qquad (n \equiv 2m) \\ &= \delta_0 + \delta_1 q^2 + \delta_2 q^4 + \cdots + \delta_n q^n + \delta_{n+2} q^{n+2} + \cdots .\end{aligned} \tag{11}$$

The ratio of the coefficients of two successive terms of Eq. (11) is

$$\frac{\delta_{n+2}}{\delta_n}=\frac{(n/2)!}{(n/2+1)!}=\frac{1}{n/2+1}\simeq 2/n, \qquad n\gg 2. \tag{12}$$

The corresponding ratio for $\hat{g}(q)$ is

$$\frac{d_{n+2}}{d_n}=\frac{2n-(a/b-1)}{(n+1)(n+2)}\simeq\frac{2n}{n^2}=\frac{2}{n}, \qquad n\gg\tfrac{1}{2}(a/b-1) \quad \text{and} \quad n\gg 2. \tag{13}$$

Thus, for large values of n, the successive terms in each series [$\hat{g}(q)$ or e^{q^2}] are related by the same factor. If for some $n\gg 2$ and $\gg\frac{1}{2}(a/b-1)$ the ratio (δ_n/d_n) is given, its value will thus be preserved for all large n. Consequently, except for a constant factor, the series $\hat{g}(q)$ behaves very much like e^{q^2} for large values of q.

$$\lim_{q\to\infty}\hat{g}(q)=\lim_{q\to\infty}e^{q^2}. \tag{14}$$

Equation (4) now becomes

$$\lim_{q\to\infty}g(q)=\hat{g}(q)e^{-q^2/2}=e^{q^2}e^{-q^2/2}=\lim_{q\to\infty}e^{q^2/2}\to\infty. \tag{15}$$

Thus, if $\hat{g}(q)$ is an infinite series, the solution approaches infinity as q increases indefinitely. The series $\hat{g}(q)$ may be made to break off after a finite number of terms by choosing the values of the parameter a to make the numerator of Eq. (10) vanish at some desired value of n,

$$a=(1+2n)b. \tag{16}$$

It is also necessary to set $d_0=0$ or $d_1=0$ for n odd or even, respectively. This is because the choice of a can only cause either the odd or even series, but not both, to break off.

Summary

$$\begin{aligned} &g(x)=\hat{g}_n(x)e^{-bx^2/2},\\ &\hat{g}_n(q)=d_0+d_1q+\cdots+d_nq^n+\cdots, \qquad q=x/c=\sqrt{b}\,x,\\ &d_{n+2}=\frac{2n-(a/b-1)}{(n+1)(n+2)}d_n, \qquad n=0,1,2,\ldots . \end{aligned} \tag{17}$$

For the series to break off, $a=(1+2n)b$, and so

$$g(x)=e^{-bx^2/2}\sum_{k=0}^{n}d_k(\sqrt{b}\,x)^k.$$

The finite series $\hat{g}(q)$ are known as *Hermite polynomials* and can be determined by power series methods. (An easier method, by means of the generating function, will be taken up in the next section.) The first few Hermite polynomials are listed here for reference; the conventional notation for the Hermite polynomial of degree n (where n is an integer) is $H_n(q)$.

$$\begin{aligned} H_0(q) &\equiv \hat{g}_0(q) = 1, \\ H_1(q) &\equiv \hat{g}_1(q) = 2q, \\ H_2(q) &\equiv \hat{g}_2(q) = 4q^2 - 2, \\ H_3(q) &\equiv \hat{g}_3(q) = 8q^3 - 12q, \\ H_4(q) &\equiv \hat{g}(q) = 16q^4 - 48q^2 + 12. \end{aligned} \tag{18}$$

Example 8

Show that the normalized wave solution of the differential equation

$$\frac{d^2g}{dx^2} + (a - b^2x^2)g = 0 \tag{1}$$

can be expressed

$$\hat{g}_n(x) = M_n H_n(q) e^{-q^2/2}, \qquad q^2 \equiv bx^2, \tag{2}$$

$$M_n = \left(\sqrt{\frac{b}{\pi}}\ \frac{1}{2^n n!}\right)^{1/2}, \qquad n = 0, 1, 2, \ldots, \tag{3}$$

and where $H_n(q)$ is the Hermite polynomial [Eq. (7.5.8)].

SOLUTION

The general solution of Eq. (1) is given by Eq. (7.5.17),

$$g_n(x) = \hat{g}_n(x) e^{-q^2/2}, \tag{4}$$

$$\hat{g}_n(x) = d_0 + d_1 q + d_2 q^2 + \cdots + d_n q^n + \cdots, \qquad q^2 = bx^2, \tag{5}$$

$$d_{n+2} = \frac{2n - (a/b - 1)}{(n+1)(n+2)} d_n, \qquad n = 0, 1, 2, \ldots . \tag{6}$$

In this example, only the case of even n will be considered. One therefore sets d_1 equal to zero.

a. Evaluate $\hat{g}_0(x)$. In this case, the polynomial is terminated at $n=0$; that is, the numerator of Eq. (6) should be zero at $n=0$.

$$[2n - (a/b - 1)]_{n=0} = 0, \qquad a/b = 1. \tag{7}$$

With the fraction a/b as given by Eq. (7), Eq. (6) yields

$$d_{n+2} = 0 \qquad \text{for} \quad n = 0, 2, 4, \ldots . \tag{8}$$

The solution is then given by Eqs. (5) and (4),

$$\hat{g}_0(x) = d_0 \qquad \text{and} \qquad g_0(x) = d_0 e^{-q^2/2}. \tag{9}$$

The value of d_0 is determined from the normalization relation.

$$1=\int_{-\infty}^{\infty} g(x)g^*(x)\,dx=2\int_0^{\infty} d_0{}^2 e^{-q^2}\,dx$$

$$=2d_0{}^2\int_0^{\infty} e^{-bx^2}\,dx$$

$$=d_0{}^2\sqrt{\pi/b} \qquad \text{(Dwight [3, 860–11])},$$

$$d_0=(b/\pi)^{1/4}. \tag{10}$$

Thus

$$g_0(x)=\left(\sqrt{b/4}\,\right)^{1/2}e^{-q^2/2}=M_0e^{-q^2/2}H_0(q), \tag{11}$$

where $H_0(q)=1$ in Eq. (7.5.18).

b. Evaluate $\hat{g}_2(x)$. The numerator of Eq. (6) should be zero at $n=2$, so

$$[2n-(a/b-1)]_{n=2}=0, \qquad a/b=5. \tag{12}$$

Substituting Eq. (12) into Eq. (6) gives

$$n=0, \qquad d_2=-2d_0,$$
$$n=2, \qquad d_4=0. \tag{13}$$

Then

$$\hat{g}_2(x)=d_0+d_2q^2=-(2q^2-1)d_0,$$

$$g_2(x)=-(2q^2-1)d_0e^{-q^2/2}. \tag{14}$$

For normalized function, one has

$$1=\int_{-\infty}^{\infty}|g_2(x)|^2\,dx=2(-d_0)^2\int_0^{\infty}(2q^2-1)^2e^{-q^2}\,dx$$

$$=2(-d_0)^2\int_0^{\infty}(4b^2x^4-4bx^2+1)e^{-bx^2}\,dx$$

$$=2(-d_0)^2\sqrt{(\pi/b)}\left(\tfrac{3}{2}-1+\tfrac{1}{2}\right)=2(-d_0)^2\sqrt{\pi/b}\,,$$

$$-d_0=\left(\tfrac{1}{2}\sqrt{b/\pi}\,\right)^{1/2}; \tag{15}$$

$$g_2(x)=\left(\tfrac{1}{2}\sqrt{b/\pi}\,\right)^{1/2}(2q^2-1)e^{-q^2/2}$$

$$=\left(\sqrt{\frac{b}{\pi}}\,\frac{1}{2^2 2!}\right)^{1/2}2(2q^2-1)e^{-q^2/2}$$

$$=M_2e^{-q^2/2}H_2(q), \qquad H_2(q)=4q^2-2. \tag{16}$$

c. Evaluate $\hat{g}_4(x)$.

$$[2n-(a/b-1)]_{n=4}=0, \qquad a/b=9. \tag{17}$$

Equations (6) and (5) then give

$$d_2 = -4d_0, \qquad d_4 = \tfrac{4}{3}d_0, \qquad d_6 = 0, \tag{18}$$

and

$$\hat{g}_4(x) = d_0 + d_2q^2 + d_4q^4 = d_0\left(1 - 4q^2 + \tfrac{4}{3}q^4\right),$$

$$g_4(x) = d_0\left(1 - 4q^2 + \tfrac{4}{3}q^4\right)e^{-q^2/2}. \tag{19}$$

The normalization relation is

$$1 = \int_{-\infty}^{\infty} |g_4(x)|^2\,dx = 2d_0{}^2\int_0^{\infty}\left(1 - 4q^2 + \tfrac{4}{3}q^4\right)e^{-q^2/2}\,dx$$

$$= 2d_0{}^2\int_0^{\infty}\left(1 - 8bx^2 + \tfrac{56}{3}b^2x^4 - \tfrac{32}{3}b^3x^6 + \tfrac{16}{9}b^4x^4\right)e^{-bx^2}\,dx$$

$$= \tfrac{8}{3}d_0{}^2\sqrt{\pi/b}\,,$$

$$d_0 = \left(\tfrac{3}{8}\sqrt{b/\pi}\,\right)^{1/2}. \tag{20}$$

$$g_4(x) = \left(\tfrac{3}{8}\sqrt{b/\pi}\,\right)^{1/2}\left(1 - 4q^2 + \tfrac{4}{3}q^4\right)e^{-q^2/2}$$

$$= \left(\frac{1}{2^4 4!}\sqrt{\frac{b}{\pi}}\,\right)^{1/2} 12\left(1 - 4q^2 + \tfrac{4}{3}q^4\right)e^{-q^2/2}$$

$$= M_4e^{-q^2/2}H_4(q), \qquad H_4(q) = 12 - 48q^2 + 16q^4. \tag{21}$$

Higher-order terms can be verified similarly.

7.6. Generating Function of Hermite Polynomials

The Hermite polynomials are the finite polynomial solutions $H_n(q)$ of a differential equation. It is convenient, however, to study their properties through the *generating function* $G(q,s)$, defined as

$$G(q,s) \equiv e^{q^2-(s-q)^2} = e^{2qs-s^2} \tag{1}$$

$$= 1 + (2qs - s^2) + \frac{1}{2!}(2qs - s^2)^2 + \cdots + \frac{1}{n!}(2qs - s^2)^n + \cdots$$

$$= 1 + s(2q)/1! + s^2(4q - 2)/2! + s^3(8q^3 - 12q)/3! + \cdots$$

$$= 1 + \frac{s}{1!}H_1(q) + \frac{s^2}{2!}H_2(q) + \frac{s^3}{3!}H_3(q) + \cdots$$

$$= \sum_{k=0}^{\infty}\frac{s^k}{k!}H_k(q). \tag{2}$$

Hermite polynomials may be obtained from the generating function by Taylor's expansion.

$$G(q,s)=G(q,0)+\frac{s}{1!}G'(q,s)+\frac{s^2}{2!}G''(q,0)+\frac{s^3}{3!}G'''(q,0)+\cdots$$
$$=\sum_{k=0}^{\infty}\frac{s^k}{k!}\left[\frac{\partial^k G(q,s)}{\partial s^k}\right]_{s=0}. \tag{3}$$

Both expressions (2) and (3) represent identical generating functions, and therefore the coefficients of s^k must be equal. Thus,

$$H_k(q)=\left.\frac{\partial^k G(q,s)}{\partial s^k}\right|_{s=0}=\left[\frac{\partial^k}{\partial s^k}\left(e^{q^2-(s-q)^2}\right)\right]_{s=0}$$
$$=e^{q^2}\left.\frac{\partial^k}{\partial(s-q)^k}e^{-(s-q)^2}\right|_{s=0}=(-1)^k e^{q^2}\left.\frac{\partial^k}{\partial q^k}e^{-(s-q)^2}\right|_{s=0}$$
$$=(-1)^k e^{q^2}\frac{\partial^k(e^{-q^2})}{\partial q^k}. \tag{4}$$

This is a useful relation for obtaining individual polynomials. The relation between successive Hermite polynomials is obtained by differentiating (with respect to s) the two forms of $G(q,s)$, Eqs. (1) and (2).

$$\frac{\partial G}{\partial s}=\frac{\partial}{\partial s}\left[e^{q^2-(s-q)^2}\right]=-2(s-q)e^{q^2-(s-q)^2}=-2(s-q)G(q,s)$$
$$=-2(s-q)\left(1+\frac{s}{1!}H_1+\frac{s^2}{2!}H_2+\cdots+\frac{s^k}{k!}H_k+\cdots\right) \tag{5}$$

and

$$=\frac{\partial}{\partial s}\sum_{k=0}^{\infty}\frac{s^k}{k!}H_k(q)=H_1+sH_2+\frac{s^2}{2}H_3+\cdots+\frac{s^{k-1}}{(k-1)!}H_k+\cdots. \tag{6}$$

Equating the results of (5) and (6) yields

$$-\left(2s+\frac{2s^2}{1!}H_1+\frac{2s^3}{2!}H_2+\cdots+\frac{2s^{k+1}}{k!}H_k+\cdots\right)$$
$$+\left(2q+2q\frac{s}{1!}H_1+2q\frac{s^2}{2!}H_2+\cdots+2q\frac{s^k}{k!}H_k+\cdots\right)$$
$$=H_1+sH_2+\frac{s^2}{2}H_3+\cdots+\frac{s^{k-1}}{(k-1)!}H_k+\cdots,$$

$$(2q-H_1)+(2qH_1-2-H_2)s+\cdots$$
$$+\left[2qH_k/k!-2H_{k-1}/(k-1)!-H_{k+1}/k!\right]s^k+\cdots=0. \tag{7}$$

In order for this expression to be true for all values of s, the coefficients of each power of s must vanish, so that

$$\begin{aligned} 2q-H_1=0 &\quad \text{or} \quad H_1=2q, \\ 2qH_1-2-H_2=0 &\quad \text{or} \quad H_2=4q^2-2, \\ 2q\frac{H_k}{k!}-2\frac{H_{k-1}}{(k-1)!}-\frac{H_{k+1}}{k!}=0 &\quad \text{or} \quad H_{k+1}-2qH_k+2kH_{k-1}=0. \end{aligned} \tag{8}$$

Equation (8) is a recursion formula for the Hermite polynomials of different orders.

The relation for the derivatives of successive Hermite polynomials may be similarly obtained by differentiating the generating function with respect to q.

$$\frac{\partial G}{\partial q}=\frac{\partial}{\partial q}e^{2qs-s^2}=2sG(q,s)=2s\sum_{k=0}^{\infty}\frac{H_k}{k!}s^k$$

$$=\sum_{k=0}^{\infty}\frac{2H_k}{k!}s^{k+1} \tag{9}$$

$$=\frac{\partial}{\partial q}\sum_{k=0}^{\infty}\frac{H_k(q)}{k!}s^k$$

$$=\sum_{k=0}^{\infty}\frac{H_k'(q)}{k!}s^k. \tag{10}$$

Equating coefficients of s^k in Eqs. (9) and (10) yields

$$\frac{H_k'(q)}{k!}=2\frac{H_{k-1}(q)}{(k-1)!}$$

or

$$H_k'(q)=2kH_{k-1}(q). \tag{11}$$

Then

$$H_k''(q)=\frac{d}{dq}H_k'(q)=\frac{d}{dq}\left[2kH_{k-1}(q)\right]$$

$$=4k(k-1)H_{k-2}(q). \tag{12}$$

Equation (8) may be rearranged as

$$H_k(q)-2qH_{k-1}(q)+2(k-1)H_{k-2}(q)=0. \tag{13}$$

Application of Eqs. (11) and (12) yields

$$H_k-2q\frac{1}{2k}H_k'+2(k-1)\frac{1}{4k(k-1)}H_k''=0$$

or

$$H_k''(q)-2qH_k'(q)+2kH_k(q)=0. \tag{14}$$

From Eq. (7.5.16) $a=(1+2k)b$, or

$$k=\tfrac{1}{2}(a/b-1). \tag{15}$$

Then

$$H_k''(q)-2qH_k'(q)+(a/b-1)H_k(q)=0. \tag{16}$$

This is the same expression satisfied by the correction factor $\hat{g}(q)$ from Eq. (7.5.7). The general solution, Eq. (7.5.17), may be rearranged as

$$g_n(q)=M_nH_n(q)e^{-q^2/2}, \qquad q\equiv\sqrt{b}\,x, \tag{17}$$

where M_n is the magnitude of the nth solution, and the $g_n(x)$ are known as the *Hermite orthogonal functions* (see Figure 10). The function is normalized if

$$\int_{-\infty}^{\infty} g_n{}^2(x)\,dx=1. \tag{18}$$

Figure 10. Normalized wave functions: $g_n=M_nH_n(q)e^{-q^2/2}$. (a) $g_0(q)$, $n=0$; (b) $g_1(q)$, $n=1$; (c) $g_2(q)$, $n=2$; (d) $g_3(q)$, $n=3$.

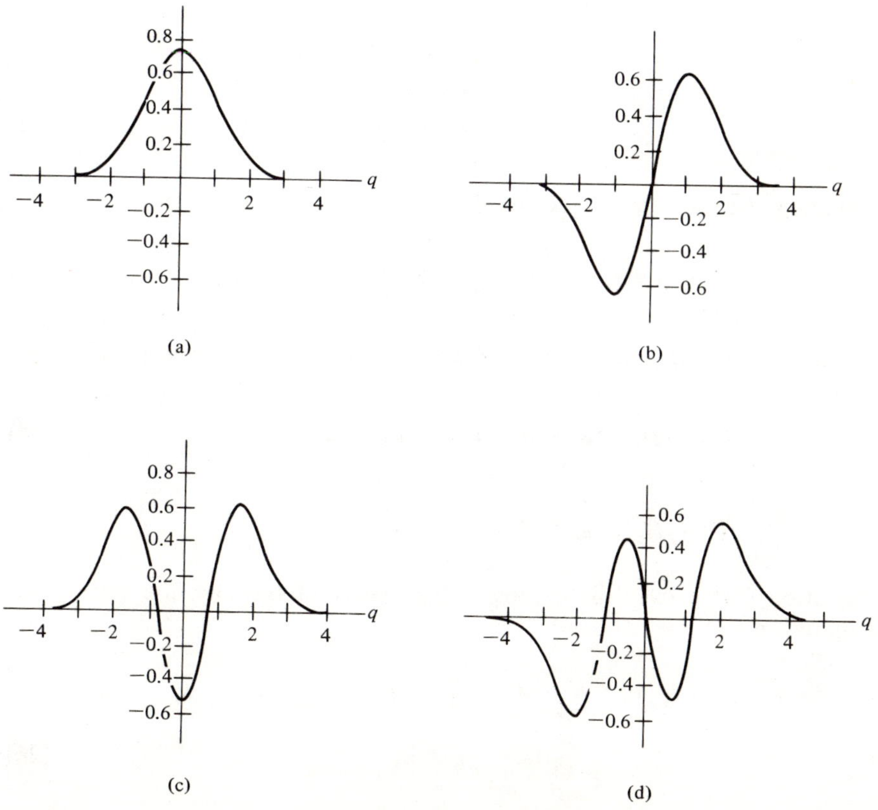

To show that these functions are orthogonal, consider two generating functions,

$$G(q,s)=e^{q^2-(s-q)^2}=\sum_{k=0}^{\infty}\frac{s^k}{k!}H_k(q), \tag{19}$$

$$F(q,t)=e^{q^2-(t-q)^2}=\sum_{l=0}^{\infty}\frac{t^l}{l!}H_l(q). \tag{20}$$

Then

$$\int_{-\infty}^{\infty}G(q,s)F(q,t)e^{-q^2}\,dq=\sum_{k=0}^{\infty}\sum_{l=0}^{\infty}\frac{s^k t^l}{k!\,l!}\times\int_{-\infty}^{\infty}H_k(q)H_l(q)e^{-q^2}\,dq. \tag{21}$$

But

$$\begin{aligned}\int_{-\infty}^{\infty}G(q,s)F(q,t)e^{-q^2}\,dq&=\int_{-\infty}^{\infty}e^{q^2-(s-q)^2}e^{q^2-(t-q)^2}e^{-q^2}\,dq\\ &=\int_{-\infty}^{\infty}e^{-[q+(s+t)]^2+2st}\,d(q+s+t)\\ &=\sqrt{\pi}\,e^{2st}\qquad\text{(Dwight [3, 860.11])}\\ &=\sqrt{\pi}\sum_{k=0}^{\infty}(2st)^k/k!.\end{aligned} \tag{22}$$

Equating the right-hand sides of Eqs. (21) and (22) yields

$$\sqrt{\pi}\sum_{k=0}^{\infty}(2st)^k/k!=\sum_{k=0}^{\infty}\sum_{l=0}^{\infty}\frac{s^k t^l}{k!\,l!}\int_{-\infty}^{\infty}H_k(q)H_l(q)e^{-q^2}\,dq. \tag{23}$$

Comparing the coefficients of $s^k t^l$ on both sides of Eq. (23), one concludes

$$\int_{-\infty}^{\infty}H_k(q)H_l(q)e^{-q^2}\,dq=0\qquad\text{for}\quad k\neq l, \tag{24}$$

$$\int_{-\infty}^{\infty}H_k(q)H_k(q)e^{-q^2}\,dq=\sqrt{\pi}\,2^k k!. \tag{25}$$

Equation (24) proves the orthogonal property of the Hermite orthogonal function $g_n(x)$.

$$\begin{aligned}\int_{-\infty}^{\infty}g_n(x)g_m(x)\,dx&=M_nM_m\int_{-\infty}^{\infty}H_n(q)H_m(q)e^{-q^2}\,dq\\ &=0,\qquad n\neq m.\end{aligned} \tag{26}$$

The magnitude M_n for the normalized function may now be evaluated by using the normalization relation (18),

$$1=\int_{-\infty}^{\infty} g_n(q)g_n(q)\,dq=M_n{}^2\int_{-\infty}^{\infty} H_n{}^2(q)e^{-q^2}\,dq$$
$$=M_n{}^2\sqrt{\pi}\,2^k k!,$$
$$M_n=\left(1/\sqrt{\pi}\,2^k n!\right)^{1/2}. \tag{27}$$

Equation (25) is used to obtain the above result.

7.7. TE Modes: $E_z=0$, $E_x=0$, $H_y=0$

The equation to be solved is Eq. (7.4.5), which can be rearranged as

$$\frac{d^2h_x}{dx^2}+(\Gamma^2-\Delta^2x^2)h_x=0, \tag{1}$$

where

$$\Gamma^2\equiv\gamma_z{}^2-\gamma_0^2\epsilon_r,\qquad \Delta^2=-\gamma_0^2\epsilon_r\delta\varepsilon_r,\qquad \gamma_0^2\equiv-\omega^2\mu\epsilon_0.$$

Equation (1) has the same form as Eq. (7.5.1), and its solution is given by Eq. (7.5.17). With the replacements

$$a\to\Gamma^2\qquad\text{and}\qquad b\to\Delta,$$

the solution of Eq. (1) is

$$h_x(x)=\hat{h}_n(x)e^{-\Delta x^2/2}, \tag{2}$$
$$\hat{h}_n(q)=d_0+d_1q+d_2q^2+\cdots+d_nq^n+\cdots\qquad\left(q\equiv\sqrt{\Delta}\,x\right), \tag{3}$$
$$d_{n+2}=\frac{2n-(\Gamma^2/\Delta-1)}{(n+1)(n+2)}d_n,\qquad n=0,1,2,\ldots. \tag{4}$$

Breakoff at

$$\Gamma^2=(1+2n)\Delta \tag{5}$$

is given by

$$\tilde{H}_x(\mathbf{r},t)=h_x(x)e^{j\omega t-\gamma_z z}. \tag{6}$$

The field components transverse to $\tilde{H}_x$ are given by Eq. (7.3.17).

$$\tilde{E}_y=\frac{-j\omega\mu}{\gamma_z}\tilde{H}_x\qquad\text{and}\qquad \tilde{H}_z=\frac{1}{\gamma_z}\frac{\partial\tilde{H}_x}{\partial x}. \tag{7}$$

The field $\tilde{H}_z$ may be evaluated as follows:

$$\frac{\partial\tilde{H}_x}{\partial x}=\frac{dh_x}{dx}e^{j\omega t-\gamma_z z},$$
$$\frac{dh_x}{dx}=\frac{d}{dx}\left[\hat{h}_n(x)e^{-\Delta x^2/2}\right]=\left(\frac{d\hat{h}_n}{dx}-x\Delta\hat{h}_n\right)e^{-\Delta x^2/2}$$
$$=\left(2n\hat{h}_{n-1}-x\Delta\hat{h}_n\right)e^{-\Delta x^2/2}, \tag{8}$$

where Eq. (7.6.11) has been used. Substituting Eq. (8) into the second relation in Eq. (7) yields

$$\tilde{H}_z = \frac{1}{\gamma_z}\left[2n\hat{h}_{n-1}(x) - x\hat{h}_n(x)\Delta\right]e^{-\Delta x^2/2}e^{j\omega t-\gamma_z z}. \tag{9}$$

The propagation constant can be evaluated from Eq. (1) and (5).

$$\gamma_{zn}{}^2 = \Gamma^2 - \omega^2\mu\epsilon_0\epsilon_r = (1+2n)\Delta - |\gamma_0^2|\epsilon_r$$

$$= (1+2n)|\gamma_0|\epsilon_r\sqrt{\delta\varepsilon_r} - |\gamma_0|^2\epsilon_r;$$

$$\gamma_{zn} = j|\gamma_0|\epsilon_r\left[1 - \frac{(1+2n)}{|\gamma_0|}\sqrt{\delta\varepsilon_r}\right]^{1/2}. \tag{10}$$

The additional subscript n has been added to the propagation constant for the obvious reason that it varies as n. For the lossless guide, $\gamma_{zn} = j\beta_{zn}$ and $\delta\varepsilon_r \to \delta\epsilon_r$,

$$\beta_{zn} = |\gamma_0|\epsilon_r\left[1 - \frac{(1+2n)}{|\gamma_0|}\sqrt{\gamma_0\epsilon_r}\right]^{1/2}. \tag{11}$$

Summary

For the TE modes, $E_z = 0$, $E_x = 0$, $H_y = 0$,

$$\tilde{H}_x(\mathbf{r},t) = \hat{h}_n(x)e^{-\Delta x^2/2}e^{j\omega t-\gamma_{zn}z}, \tag{12}$$

$$\tilde{E}_y(\mathbf{r},t) = -\frac{j\omega\mu}{\gamma_{zn}}\tilde{H}_x(\mathbf{r},t), \tag{13}$$

$$\tilde{H}_z(\mathbf{r},t) = \frac{1}{\gamma_{zn}}\left[2n\hat{h}_{n-1}(x) - x\hat{h}_n(x)\Delta\right]e^{-\Delta x^2/2}e^{j\omega t-\gamma_{zn}z}, \tag{14}$$

$$\hat{h}_n(q) = d_0 + d_1 q + \cdots + d_n q^n + \cdots \qquad \left(q = \sqrt{\Delta}\,x\right), \tag{15}$$

$$d_{n+2} = \frac{2n - (\Gamma^2/\Delta - 1)}{(n+1)(n+2)}d_n, \qquad n = 0,1,2,\ldots. \tag{16}$$

Breakoff at $\Gamma^2 = (1+2n)\Delta$ is given by

$$\Gamma^2 \equiv \gamma_{zn}{}^2 - \gamma_0^2\epsilon_r, \tag{17}$$

$$\Delta^2 \equiv -\gamma_0^2\varepsilon_r\delta\varepsilon_r, \qquad \gamma_0^2 \equiv -\omega^2\mu\epsilon_0, \tag{18}$$

$$\gamma_{zn} = j|\gamma_0|\epsilon_r\left[1 - \frac{1+2n}{|\gamma_0|}\sqrt{\delta\varepsilon_r}\right]^{1/2}. \tag{19}$$

The TE modes in a square-law dielectric slab are found to vary as Hermite orthogonal functions [Eq. (7.6.17)] in x. The plots of these wave functions (Figure 10) show that the fields are concentrated or oscillate near the region where x is small and diminish to zero at some larger value of x. This may be interpreted as the fields are being "focused" into the region where x is very small.

Example 9

A nonhomogeneous dielectric sheet guide has a thickness of 0.4 mm and $\epsilon\,\delta\epsilon_r=10^{-4}$ f/m². The guide is operating at a frequency of 10^{14} Hz. Determine the field distribution in the x direction of two TE modes, $n=0,2$.

SOLUTION

The TE field of nonhomogeneous dielectric sheet guide is given by Eq. (7.7.6),

$$H_x(\mathbf{r},t)=h_x(x)e^{j\omega t-\gamma_z z}, \tag{1}$$

$$h_x(x)=\hat{h}_n(x)e^{-\Delta x^2/2},$$

$$\hat{h}_n(x)=d_0+d_1q+d_2q^2+\cdots+d_nq^n+\cdots, \qquad q^2=\Delta x^2; \tag{2}$$

$$\Delta^2\equiv\omega^2\mu_0\epsilon\,\delta\epsilon_r$$

$$=(2\pi\times10^{14})^2 4\pi\times10^{-7}\times10^{-4}=4.96\times10^{19},$$

$$\Delta=7.04\times10^9.$$

From Example 8,

$$h_n(x)=M_nH_n(q)e^{-q^2/2}, \qquad q^2=\Delta x^2, \tag{3}$$

$$M_n=\left(\sqrt{\frac{\Delta}{\pi}}\,\frac{1}{2^n n!}\right)^{1/2}.$$

a. $h_0(x)=M_0H_0(q)e^{-q^2/2}$

$$=(\Delta/\pi)^{1/4}e^{-\Delta x^2/2}=(7.04\times10^9/\pi)^{1/4}e^{-\Delta x^2/2}$$

$$=217.6e^{-\Delta x^2/2}$$

x	$e^{-\Delta x^2/2}$	$h_0(x)$
0	1	217.6
10^{-5}	0.703	153.0
2×10^{-5}	0.245	53.2
3×10^{-5}	0.042	9.1
4×10^{-5}	3.57×10^{-3}	0.78

The function $h_0(x)$ is given in the accompanying table and plotted in Figure 11(a). It is an even function of x.

b. $h_2(x) = M_2 H_2(q) e^{-q^2/2}$

$$M_2 = \left(\sqrt{\frac{\Delta}{\pi}}\frac{1}{2^2 2!}\right)^{1/2} = \left(\frac{1}{8}\sqrt{\frac{7.043\times10^9}{\pi}}\right)^{1/2}$$

$$= 76.93,$$

$$H_2(q) = 4q^2 - 2 = 4\Delta x^2 - 2 = 4\times7.043\times10^9 x^2 - 2$$

$$= 2.817\times10^{10}x^2 - 2.$$

Figure 11. Example 9. (a) $h_0(x)$; (b) $h_2(x)$.

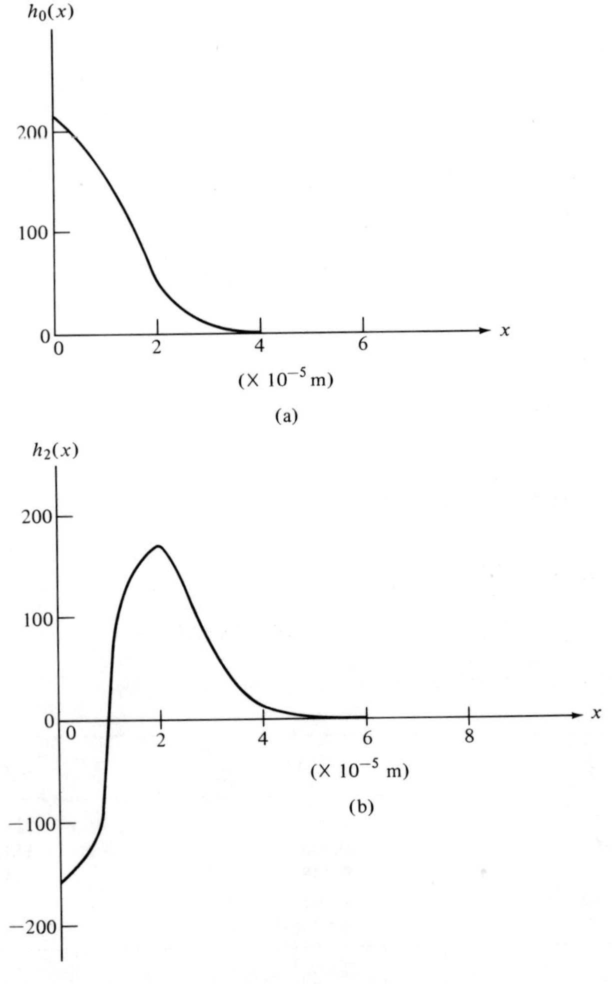

x	$H_2(q)$	$e^{-\Delta x^2/2}$	$h_2(x)$
0	-2	1	-153.86
4×10^{-6}	-1.955	0.9452	-142.16
6×10^{-6}	-1.898	0.8809	-128.62
8×10^{-6}	-1.81	0.798	-111.14
9×10^{-6}	0.772	0.752	-44.65
10^{-5}	0.817	0.703	44.22
1.5×10^{-5}	4.34	0.453	151.14
2×10^{-5}	9.27	0.245	174.36
2.5×10^{-5}	15.61	0.109	131.38
3×10^{-5}	23.36	0.042	75.46
4×10^{-5}	43.08	3.5×10^{-3}	11.60
6×10^{-5}	99.4	3.12×10^{-6}	0.024
10^{-4}	279.7	5.07×10^{-16}	1.09×10^{-11}

The function $h_2(x)$ is tabulated as shown and plotted in Figure 11(b). It is symmetrical with respect to $x=0$.

7.8. TM Modes: $H_z=0$, $H_x=0$, $E_y=0$

The equation to be solved is Eq. (7.4.6), which is repeated here.

$$\frac{\partial^2\bar{e}_x}{\partial x^2}+\left\{\gamma_z{}^2-\gamma_0^2\varepsilon_r(x)-\frac{3}{4}\left[\frac{1}{\varepsilon_r(x)}\frac{d\varepsilon_r(x)}{dx}\right]^2+\frac{1}{2}\frac{1}{\varepsilon_r(x)}\frac{d^2\varepsilon_r(x)}{dx^2}\right\}\bar{e}_x=0 \quad (1)$$

with

$$\varepsilon_r(x)=(1-x^2\,\delta\varepsilon_r)\epsilon_r \quad (2)$$

and

$$\frac{d\varepsilon_r}{dx}=-2x\varepsilon_r\,\delta\epsilon_r, \qquad \frac{d^2\varepsilon_r}{dx^2}=-2\varepsilon_r\,\delta\epsilon_r. \quad (3)$$

Substituting Eqs. (2) and (3) into Eq. (1) yields

$$\frac{\partial^2\bar{e}_x}{\partial x^2}+\left\{\gamma_z{}^2-\gamma_0^2\epsilon_r+\gamma_0^2x^2\epsilon_r\,\delta\varepsilon_r-\frac{3}{4}\left(\frac{-2x\,\delta\varepsilon_r}{1-x^2\,\delta\varepsilon_r}\right)^2+\frac{1}{2}\left(\frac{-2\,\delta\varepsilon_r}{1-x^2\,\delta\varepsilon_r}\right)\right\}\bar{e}_x=0,$$

$$\frac{\partial^2\bar{e}_x}{\partial x^2}+\left\{\gamma_z{}^2-\gamma_0^2\epsilon_r+\gamma_0^2x^2\epsilon_r\,\delta\varepsilon_r-3x^2\,\delta\varepsilon_r{}^2(1+x^2\,\delta\varepsilon_r+\cdots)^2\right.$$
$$\left.-\delta\varepsilon_r(1+x^2\,\delta\varepsilon_r+\cdots)\right\}\bar{e}_x=0,$$

$$\frac{\partial^2\bar{e}_x}{\partial x^2}+\{(\gamma_z{}^2-\gamma_0^2\epsilon_r-\epsilon_r\,\delta\varepsilon_r)+(\gamma_0^2\epsilon_r\,\delta\varepsilon_r-4\,\delta\varepsilon_r{}^2)x^2+\text{h.o.t.}\}\bar{e}_x=0, \quad (4)$$

where higher-order terms (h.o.t.) are of order $x^4 \delta\varepsilon_r{}^3$ or higher. If the higher-order terms are negligible, then Eq. (2) is simplified to

$$\frac{\partial^2 \bar{e}_x}{\partial x^2} + (\Gamma^2 - \Delta^2 x^2)\bar{e}_x = 0, \tag{5}$$

where

$$\begin{aligned} \Gamma^2 &\equiv \gamma_z{}^2 - \gamma_0^2 \epsilon_r - \epsilon_r \delta\varepsilon_r, \\ \Delta^2 &\equiv \left(\gamma_0^2 \epsilon_r \delta\varepsilon_r - 4\,\delta\epsilon_r{}^2\right) = -\gamma_0^2 \epsilon_r \delta\varepsilon_r \left(1 - 4\delta\varepsilon_r/\gamma_0^2\right). \end{aligned} \tag{6}$$

Equation (5) has the same form as that statisfied by h_x in the case of the TE mode, that is, Eq. (7.7.1); consequently the solution for $\bar{e}_x$ is the same as h_x, Eq. (7.7.2), except that here Γ and Δ are defined by Eq. (6).

The propagation constant can be evaluated from Eq. (6) and the breakoff relation, $\Gamma^2 = (1+2n)\Delta$.

$$\gamma_z{}^2 = \Gamma^2 + \gamma_0^2 \epsilon_r + \epsilon_r \delta\varepsilon_r = \gamma_0^2 \epsilon_r + \epsilon_r \delta\varepsilon_r (1+2n)\Delta,$$

$$\gamma_{zn}{}^2 = \gamma_0^2 \epsilon_r + \epsilon_r \delta\varepsilon_r (1+2n)\left[-\gamma_0^2\, \delta\varepsilon_r \left(1 - 4\,\delta\varepsilon_r/\gamma_0^2\right)\right]. \tag{7}$$

The other field components are given by Eqs. (7.3.15), (7.3.16), and (7.3.21).

$$E_z = \frac{1}{\gamma_{zn} \varepsilon_r(x)} \frac{\partial}{\partial x}\left[\varepsilon_r(x) E_x\right], \tag{8}$$

$$H_y = \frac{j\omega\epsilon_0 \varepsilon_r(x)}{\gamma_{zn}} E_x, \tag{9}$$

where

$$\tilde{E}_x(\mathbf{r}, t) = e_x(x) e^{j\omega t - \gamma_z z}, \tag{10}$$

$$\bar{e}_x = \sqrt{\varepsilon_r(x)}\; e_x. \tag{11}$$

References

1. G. Arfken, *Mathematical Methods for Physicists*, Academic, New York, 1966.
2. P. Dennery and A. Krzywicki, *Mathematics for Physicists*, Harper & Row, New York, 1967.
3. H. B. Dwight, *Tables of Integrals and Other Mathematical Data*, Macmillan, New York, 1947.
4. D. Marcuse, *Light Transmission Optics*, Van Nostrand Reinhold, Princeton, N. J., 1972.
5. S. E. Miller, Light Propagation in Generalized Lense-like Media, *Bell Syst. Tech. J.*, 44, 2017–2064 (1965).
6. L. Pauling and E. B. Wilson, *Introduction to Quantum Mechanics*, McGraw-Hill, New York, 1935.
7. L. I. Schiff, *Quantum Mechanics*, McGraw-Hill, New York, 1955.
8. M. S. Sodha and A. K. Ghatak, *Inhomogeneous Optical Waveguides*, Plenum, New York, 1977.
9. H. G. Unger, *Planar Optical Waveguides and Fibres*, Clarendon, Oxford, 1977.

8

Cladded Cylindrical Waveguides

A *cladded cylindrical waveguide* [6, 7, 9, 14, 15] consists of a dielectric cylinder surrounded by a concentric layer of a second dielectric. A cross-sectional view is shown in Figure 12.

Little error is introduced in the analysis if the thickness of the dielectric cladding is assumed to be infinite, because the fields decay exponentially in the cladding and are practically reduced to zero at the edge.

The dielectric shield serves the following purposes.

a. The fields of a dielectric guide are not fully contained within the core, $r \leq a_1$, decaying exponentially in the region $r > a_1$. Any object, such as a support, in contact with a bare waveguide alters the boundary conditions and consequently causes distortion in the fields. Dielectric cladding can minimize such an effect.
b. The number of propagating modes existing within a dielectric guide depends on its radius a_1 and the ratio of the dielectric constants of the media inside and outside the guide. With a proper choice of these parameters, it is possible to have only one propagating mode.

8.1. Numerical Aperture

The light gathering ability at the end of a dielectric guide is measured by a quality factor known as the numerical aperture (NA).

Consider a ray incident on the end of a dielectric guide, as shown in Figure 13. The ray is meridional; that is, its path is confined to a single plane. The application of Snell's law to the end surface yields

$$\sqrt{\epsilon_0} \sin\theta_0 = \sqrt{\epsilon_1} \sin\theta_1$$

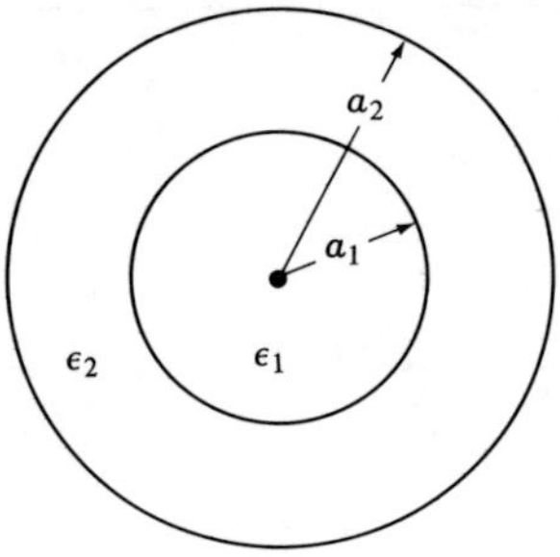

Figure 12. Cross section of a cladded cylindrical waveguide.

or

$$\sin\theta_0 = \sqrt{\epsilon_1/\epsilon_0}\,\sin\theta_1. \tag{1}$$

For an efficient guiding system, total reflection is expected at the core–cladding interface. One thus has

$$\sin(\pi/2-\theta_1) \geq \sqrt{\epsilon_2/\epsilon_1}$$

or

$$\cos\theta_1 \geq \sqrt{\epsilon_2/\epsilon_1}\,, \qquad \sin\theta_1 \leq \sqrt{1-\epsilon_2/\epsilon_1}\,. \tag{2}$$

Figure 13. Cross-sectional view of a cladded cylindrical guide.

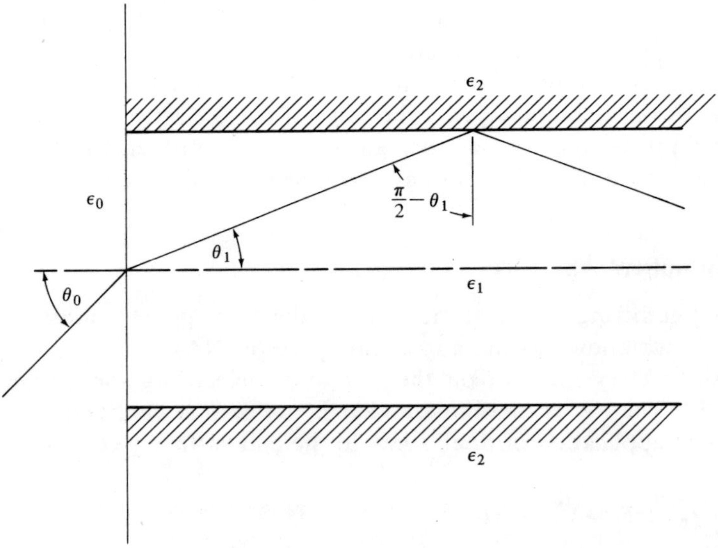

Substitution of Eq. (2) into Eq. (1) yields

$$\sin\theta_0 \leq \sqrt{(\epsilon_1-\epsilon_2)/\epsilon_0} \tag{3}$$

or

$$\sin\theta_0 \leq \sqrt{\epsilon_{1r}-\epsilon_{2r}}\,, \tag{4}$$

where $\epsilon_1 \equiv \epsilon_0\epsilon_{1r}$ and $\epsilon_2 \equiv \epsilon_0\epsilon_{2r}$. The maximum value θ_0 of the incident angle is then given by

$$\begin{aligned}\theta_{0\,\max} &= \sin^{-1}\sqrt{\epsilon_{1r}-\epsilon_{2r}} \qquad \text{for} \quad \sqrt{\epsilon_{1r}-\epsilon_{2r}}<1,\\ &= \pi/2 \qquad \text{for} \quad \sqrt{\epsilon_{1r}-\epsilon_{2r}}>1.\end{aligned} \tag{5}$$

The numerical aperture is defined as

$$\mathrm{NA} \equiv (\epsilon_{1r}-\epsilon_{2r})^{1/2} = (n_1{}^2-n_2{}^2)^{1/2}/n_0, \tag{6}$$

where $n_i \equiv \sqrt{\epsilon_i}$ is the index of refraction. The second form is generally used with the assumption that $n_0=1$.

8.2. Guided Modes in Circular Waveguides

In cladded waveguides, the permittivities in the core and in the cladding remains constant within each region. Therefore, Maxwell's equations for homogeneous media will be used:

$$\nabla\times\tilde{\mathbf{E}} = -\mu\frac{\partial\tilde{\mathbf{H}}}{\partial t}, \qquad \nabla\times\tilde{\mathbf{H}} = \varepsilon\frac{\partial\tilde{\mathbf{E}}}{\partial t}. \tag{1}$$

Let the fields be

$$\tilde{\mathbf{E}}(\mathbf{r},t) = \mathbf{E}(\mathbf{r})e^{j\omega t}, \qquad \tilde{\mathbf{H}}(\mathbf{r},t) = \mathbf{H}(\mathbf{r})e^{j\omega t}. \tag{2}$$

Then

$$\nabla\times\mathbf{E} = -j\omega\mu\mathbf{H}, \qquad \nabla\times\mathbf{H} = j\omega\varepsilon\mathbf{E}. \tag{3}$$

The equations to be solved are

$$\nabla^2\mathbf{E}-\gamma^2\mathbf{E}=0, \quad \nabla^2\mathbf{H}-\gamma^2\mathbf{H}=0, \qquad \gamma^2\equiv-\omega^2\mu\varepsilon. \tag{4}$$

In cylindrical coordinates, Eq. (4) has the form

$$\frac{1}{r}\frac{\partial}{\partial r}\left(r\frac{\partial\mathbf{f}}{\partial r}\right)+\frac{1}{r^2}\frac{\partial^2\mathbf{f}}{\partial\phi^2}+\frac{\partial^2\mathbf{f}}{\partial z^2}-\gamma^2\mathbf{f}=0, \tag{5}$$

where $\mathbf{f}$ stands for either $\mathbf{E}$ or $\mathbf{H}$. Equation (5) may be rearranged to give

$$\frac{\partial^2\mathbf{f}}{\partial r^2}+\frac{1}{r}\frac{\partial\mathbf{f}}{\partial r}+\frac{1}{r^2}\frac{\partial^2\mathbf{f}}{\partial\phi^2}+\frac{\partial^2\mathbf{f}}{\partial z^2}-\gamma^2\mathbf{f}=0. \tag{6}$$

This is a vector equation; the scalar equation

$$\frac{\partial^2 f}{\partial r^2}+\frac{1}{r}\frac{\partial f}{\partial r}+\frac{1}{r^2}\frac{\partial^2 f}{\partial\phi^2}+\frac{\partial^2 f}{\partial z^2}-\gamma^2 f=0 \tag{7}$$

is valid for the component of **f** whose unit vector is independent of the coordinates.

The method of separation of variables may be used to find the solution of Eq. (7). Let

$$f(r,\phi,z)=R(r)\Phi(\phi)Z(z). \tag{8}$$

Substitution of Eq. (8) into Eq. (7) yields

$$\Phi Z\left(\frac{d^2R}{dr^2}+\frac{1}{r}\frac{dR}{dr}\right)+\frac{1}{r^2}RZ\frac{d^2\Phi}{d\phi^2}+R\Phi\frac{d^2Z}{dz^2}-\gamma^2R\Phi Z=0,$$

or

$$\frac{1}{R}\left(\frac{d^2R}{dr^2}+\frac{1}{r}\frac{dR}{dr}\right)+\frac{1}{r^2}\frac{1}{\Phi}\frac{d^2\Phi}{d\phi^2}-\gamma^2=-\frac{1}{Z}\frac{d^2Z}{dz^2}. \tag{9}$$

For Eq. (9) to be valid for all values of r, ϕ, and z, each side must equal a constant.

$$\frac{1}{Z}\frac{d^2Z}{dz^2}\equiv\gamma_z{}^2, \qquad Z(z)=Z_1e^{\gamma_z z}+Z_2c^{-\gamma_z z}. \tag{10}$$

Therefore,

$$r^2\frac{1}{R}\left(\frac{d^2R}{dr^2}+\frac{1}{r}\frac{dR}{dr}\right)+(\gamma_z{}^2-\gamma^2)r^2=\frac{-1}{\Phi}\frac{d^2\Phi}{d\phi^2}, \tag{11}$$

$$\frac{1}{\Phi}\frac{d^2\Phi}{d\phi^2}\equiv\gamma_\phi{}^2, \qquad \Phi(\phi)=\Phi_1e^{\gamma_\phi\phi}+\Phi_2e^{-\gamma_\phi\phi}.$$

If $\Phi(\phi)$ is to be single valued, $\Phi(\phi)=\Phi(\phi+2m\pi)$, and one must choose

$$\frac{1}{\Phi}\frac{d^2\Phi}{d\phi^2}=-n^2, \qquad \Phi(\phi)=\Phi_1e^{jn\phi}+\Phi_2e^{-jn\phi} \tag{12}$$

with $n=1,2,\ldots$. The notation has been chosen for later convenience. Substituting Eq. (12) into Eq. (11) yields

$$\frac{d^2R}{dr^2}+\frac{1}{r}\frac{dR}{dr}+\left[(\gamma_z{}^2-\gamma^2)-\frac{n^2}{r^2}\right]R=0. \tag{13}$$

This is *Bessel's differential equation* and can be put into normal form by a change of variable. Let $u=kr$, where $k^2\equiv\gamma_z{}^2-\gamma^2>0$. Then

$$\frac{d}{dr}=k\frac{d}{du} \qquad \text{and} \qquad \frac{d^2}{dr^2}=k^2\frac{d^2}{du^2},$$

$$\frac{d^2R}{du^2}+\frac{1}{u}\frac{dR}{du}+\left(1-\frac{n^2}{u^2}\right)R=0. \tag{14}$$

The solution is known to be

$$R(u) = R_1 J_n(u) + R_2 Y_n(u)$$
$$= R_1 J_n(kr) + R_2 Y_n(kr), \qquad k^2 \equiv \gamma_z{}^2 - \gamma^2, \tag{15}$$

where J_n and Y_n are *Bessel functions of the first and second kind*, respectively. The general solution is then given by

$$f(r,\phi,z) = [R_1 J_n(kr) + R_2 Y_n(kr)](\Phi_1 e^{jn\phi} + \Phi_2{}^{-jn\phi})(Z_1 e^{\gamma_z z} + Z_2 e^{-\gamma_z z}). \tag{16}$$

Equation (16) is the general solution of the scalar wave equation, Eq. (7), in cylindrical coordinates. It will be shown in the following section that the transverse field components can be expressed in terms of the axial field components. The general problem of wave propagation in circular waveguide then becomes determining the solution of the scalar wave equation for the axial field components.

8.3. Transverse Fields in Terms of Axial Fields

It is convenient to express the transverse field components in terms of the axial field components, the z components. It has been shown [Eqs. (A4.7) and (A4.10)] that

$$-\nabla_z \times (\nabla_z \times \mathbf{E}_\perp) - \gamma^2 \mathbf{E}_\perp = \nabla_z \times (\nabla_\perp \times \mathbf{E}_z) - j\omega\mu \nabla_\perp \times \mathbf{H}_z, \tag{1}$$

$$-\nabla_z \times (\nabla_z \times \mathbf{H}_\perp) - \gamma^2 \mathbf{H}_\perp = \nabla_z \times (\nabla_\perp \times \mathbf{H}_z) + j\omega\varepsilon \nabla_\perp \times \mathbf{E}_z, \tag{2}$$

where $\gamma^2 \equiv -\omega^2\mu\varepsilon$ and ε is uniform within the region.

The curl of a vector $\mathbf{A}$ in cylindrical coordinates is given by Eq. (A5.4.11),

$$\nabla \times \mathbf{A} = \frac{1}{r} \begin{vmatrix} \hat{\mathbf{r}} & r\hat{\boldsymbol{\phi}} & \hat{\mathbf{z}} \\ \dfrac{\partial}{\partial r} & \dfrac{\partial}{\partial \phi} & \dfrac{\partial}{\partial z} \\ A_r & rA_\phi & A_z \end{vmatrix}, \tag{3}$$

$$\nabla_z \times \mathbf{E}_\perp = \frac{1}{r} \begin{vmatrix} \hat{\mathbf{r}} & r\hat{\boldsymbol{\phi}} & \hat{\mathbf{z}} \\ 0 & 0 & \dfrac{\partial}{\partial z} \\ E_r & rE_\phi & 0 \end{vmatrix} = -\hat{\mathbf{r}}\frac{\partial E_\phi}{\partial z} + \hat{\boldsymbol{\phi}}\frac{\partial E_r}{\partial z}, \tag{4}$$

$$\nabla_z \times (\nabla_z \times \mathbf{E}_\perp) = -\hat{\mathbf{r}}\frac{\partial^2 E_r}{\partial z^2} - \hat{\boldsymbol{\phi}}\frac{\partial^2 E_\phi}{\partial z^2}, \tag{5}$$

$$\nabla_\perp \times \mathbf{E}_z = \hat{\mathbf{r}}\frac{1}{r}\frac{\partial E_z}{\partial \phi} - \hat{\boldsymbol{\phi}}\frac{\partial E_z}{\partial r}, \tag{6}$$

$$\nabla_z \times (\nabla_\perp \times \mathbf{E}_z) = \hat{\mathbf{r}}\frac{\partial^2 E_z}{\partial r\,\partial z} + \hat{\boldsymbol{\phi}}\frac{1}{r}\frac{\partial^2 E_z}{\partial \phi\,\partial z}. \tag{7}$$

Similarly,

$$\nabla_z \times (\nabla_z \times \mathbf{H}_\perp) = -\hat{\mathbf{r}}\frac{\partial^2 H_r}{\partial z^2} - \hat{\boldsymbol{\phi}}\frac{\partial^2 H_\phi}{\partial z^2}, \tag{8}$$

$$\nabla_z \times (\nabla_\perp \times \mathbf{H}_z) = \hat{\mathbf{r}}\frac{\partial^2 H_z}{\partial r\,\partial z} + \hat{\boldsymbol{\phi}}\frac{1}{r}\frac{\partial^2 H_z}{\partial \phi\,\partial z}. \tag{9}$$

Substituting Eqs. (4)–(7) into Eq. (1) yields

$$\hat{\mathbf{r}}\frac{\partial^2 E_r}{\partial z^2} + \hat{\boldsymbol{\phi}}\frac{\partial^2 E_\phi}{\partial z^2} - \gamma^2(\hat{\mathbf{r}}E_r + \hat{\boldsymbol{\phi}}E_\phi) = \left(\hat{\mathbf{r}}\frac{\partial^2 E_z}{\partial r\,\partial z} + \hat{\boldsymbol{\phi}}\frac{1}{r}\frac{\partial^2 E_z}{\partial \phi\,\partial z}\right) - j\omega\mu\left(\hat{\mathbf{r}}\frac{1}{r}\frac{\partial H_z}{\partial \phi} - \hat{\boldsymbol{\phi}}\frac{\partial H_z}{\partial r}\right). \tag{10}$$

Substituting Eqs. (6), (8), and (9) into Eq. (2) yields

$$\hat{\mathbf{r}}\frac{\partial^2 H_r}{\partial z^2} + \hat{\boldsymbol{\phi}}\frac{\partial^2 H_\phi}{\partial z^2} - \gamma^2(\hat{\mathbf{r}}H_r + \hat{\boldsymbol{\phi}}H_\phi) = \left(\hat{\mathbf{r}}\frac{\partial^2 H_z}{\partial r\,\partial z} + \hat{\boldsymbol{\phi}}\frac{1}{r}\frac{\partial^2 H_z}{\partial \phi\,\partial z}\right) + j\omega\varepsilon\left(\hat{\mathbf{r}}\frac{1}{r}\frac{\partial E_z}{\partial \phi} - \hat{\boldsymbol{\phi}}\frac{\partial E_z}{\partial r}\right). \tag{11}$$

For a guided wave with variation in ϕ and z of the exponential form,

$$e^{+jn\phi - \gamma_z z}, \tag{12}$$

Eqs. (10) and (11) become

$$(\gamma_z{}^2 - \gamma^2)(\hat{\mathbf{r}}E_r + \hat{\boldsymbol{\phi}}E_\phi) = \left(-\hat{\mathbf{r}}\gamma_z\frac{\partial E_z}{\partial r} + \hat{\boldsymbol{\phi}}\frac{-jn\gamma_z}{r}E_z\right) - j\omega\mu\left(\hat{\mathbf{r}}\frac{jn}{r}H_z - \hat{\boldsymbol{\phi}}\frac{\partial H_z}{\partial r}\right)$$

or

$$E_r = \frac{1}{\gamma_z{}^2 - \gamma^2}\left(-\gamma_z\frac{\partial E_z}{\partial r} + \frac{\omega\mu n}{r}H_z\right), \tag{13}$$

$$E_\phi = \frac{1}{\gamma_z{}^2 - \gamma^2}\left(\frac{-jn\gamma_z}{r}E_z + j\omega\mu\frac{\partial H_z}{\partial r}\right), \tag{14}$$

$$(\gamma_z{}^2 - \gamma^2)(\hat{\mathbf{r}}H_r + \hat{\boldsymbol{\phi}}H_\phi) = \left(-\hat{\mathbf{r}}\gamma_z\frac{\partial H_z}{\partial r} - \hat{\boldsymbol{\phi}}\frac{jn\gamma_z}{r}H_z\right) + j\omega\varepsilon\left(\hat{\mathbf{r}}\frac{jn}{r}E_z - \hat{\boldsymbol{\phi}}\frac{\partial E_z}{\partial r}\right),$$

or

$$H_r = \frac{1}{\gamma_z{}^2 - \gamma^2}\left(\frac{-\omega\varepsilon n}{r}E_z - \gamma_z\frac{\partial H_z}{\partial r}\right), \tag{15}$$

$$H_\phi = \frac{1}{\gamma_z{}^2 - \gamma^2}\left(-j\omega\epsilon\frac{\partial E_z}{\partial r} - \frac{jn\gamma_z}{r}H_z\right). \tag{16}$$

Equations (13)–(16) express the transverse field components in terms of the axial components. Hence, the problem becomes that of finding the axial fields. These equations then supply the remaining components of the field.

8.4. Axial Field Components

It was shown in Section 8.2 that the axial field components satisfy the wave equation in cylindrical coordinates [Eq. (8.2.7)],

$$\frac{\partial^2 f}{\partial r^2}+\frac{1}{r}\frac{\partial f}{\partial r}+\frac{1}{r^2}\frac{\partial^2 f}{\partial \phi^2}+\frac{\partial^2 f}{\partial z^2}-\gamma^2 f=0, \tag{1}$$

where f stands for either E_z or H_z. By the method of separation of variables, f is found to be [Eq. (8.2.16)]

$$f(r,\phi,z)=R(r)e^{\pm\gamma_z z\pm jn\phi}, \tag{2}$$

and $R(r)$ satisfies Bessel's equation (8.2.13),

$$\frac{d^2R}{dr^2}+\frac{1}{r}\frac{dR}{dr}+\left[\left(\gamma_z{}^2-\gamma^2\right)-\frac{n^2}{r^2}\right]R=0, \tag{3}$$

where $\gamma^2\equiv-\omega^2\mu\varepsilon\equiv-|\gamma^2|$. For the case that $k^2\equiv|\gamma^2|+\gamma_z{}^2>0$, Eq. (3) may be transformed into normal form by the change of variable $u=kr$.

$$\frac{d^2R}{du^2}+\frac{1}{u}\frac{dR}{du}+\left(1-\frac{n^2}{u^2}\right)R=0. \tag{4}$$

The solution is

$$R(r)=R_1J_n(kr)+R_2Y_n(kr), \tag{5}$$

where J_n and Y_n are Bessel functions of the first and second kind, respectively. The small-argument asymptotic forms of these functions are [1, 2, 4, 5, 8, 12, 16]

$$\begin{aligned}
&\lim_{u\to 0} J_n(u)\simeq\frac{1}{\Gamma(n+1)}\left(\frac{u}{2}\right)^n \qquad \text{for all } n,\\
&\lim_{u\to 0} Y_0(u)\simeq\frac{2}{\pi}\ln u,\\
&\lim_{u\to 0} Y_n(u)\simeq\frac{-\Gamma(n)}{\pi}\left(\frac{2}{u}\right)^n \qquad \text{for } \quad n>0,
\end{aligned} \tag{6}$$

where $\Gamma(p+1)=p!$ for p an integer.

The problem to be considered is a cylindrical dielectric guide with dielectric constant ε_1 immersed in a second medium of infinite extent with dielectric constant ε_2 (Figure 14).

In the waveguide, $0\le r\le a$, a finite guided wave solution is desired. Therefore,

$$R_1(r)=R_0J_n(k_1r), \qquad 0\le r\le a, \tag{7}$$

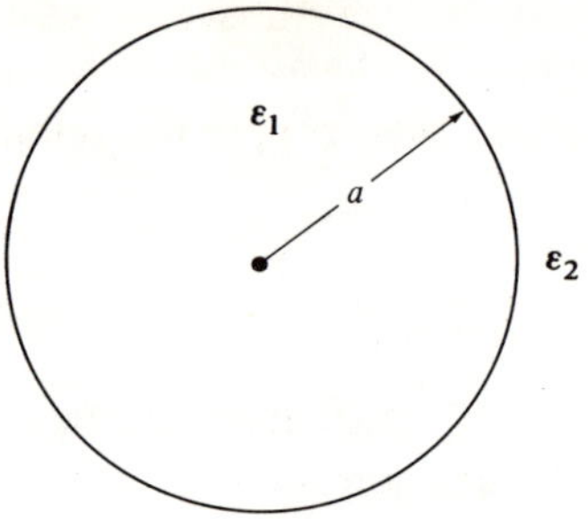

Figure 14. Approximate model of a cylindrical guide of radius a and dielectric constant ε_1 immersed in a second medium of infinite extent of constant ε_2.

where $k_1{}^2=\gamma_z{}^2-\gamma_1{}^2=\omega^2\mu_0\varepsilon_1+\gamma_z{}^2>0$; $Y_n(k_1r)$ is excluded as a solution because, since the waveguide region encloses the origin, $Y_n(0)=\infty$. The axial field components are

$$E_{z1}=E_1J_n(k_1r)e^{-\gamma_z z+jn\phi},\qquad H_{z1}=H_1J_n(k_1r)e^{-\gamma_z z+jn\phi}. \tag{8}$$

The corresponding transverse field components are given by Eqs. (8.3.13)–(8.3.16),

$$E_{r1}=\frac{1}{k_1{}^2}\left[-\gamma_z k_1E_1\dot{J}_n(k_1r)+\frac{\omega\mu_0 nH_1}{r}J_n(k_1r)\right]e^{-\gamma_z z+jn\phi}, \tag{9}$$

$$E_{\phi1}=\frac{1}{k_1{}^2}\left[\frac{-j\gamma_z nE_1}{r}J_n(k_1r)+j\omega\mu_0k_1H_1\dot{J}_n(k_1r)\right]e^{-\gamma_z z+jn\phi}, \tag{10}$$

$$H_{r1}=\frac{1}{k_1{}^2}\left[\frac{-\omega\varepsilon_1 nE_1}{r}J_n(k_1r)-\gamma_z k_1H_1\dot{J}_n(k_1r)\right]e^{-\gamma_z z+jn\phi}, \tag{11}$$

$$H_{\phi1}=\frac{1}{k_1{}^2}\left[-j\omega\varepsilon_1k_1E_1\dot{J}_n(k_1r)-\frac{jn\gamma_z H_1}{r}J_n(k_1r)\right]e^{-\gamma_z z+jn\phi}. \tag{12}$$

In a region exterior to the guide ($r>a$), the fields should be attenuated so that the structure forms an efficient guiding system. The solution to the modified Bessel equation

$$\frac{d^2R}{dr^2}+\frac{1}{r}\frac{dR}{dr}-\left(k^2+\frac{n^2}{r^2}\right)R=0 \tag{13}$$

has the desired property. Equation (3) will have this form if

$$\gamma_z{}^2-\gamma_2{}^2<0,\qquad \gamma_2{}^2\equiv-\omega^2\mu_0\varepsilon_2. \tag{14}$$

Let

$$|\gamma_2|^2+\gamma_z{}^2\equiv-k_2{}^2,\qquad\text{or}\qquad jk_2=\sqrt{|\gamma_2|^2+\gamma_z{}^2}\,. \tag{15}$$

Equation (3) becomes

$$\frac{d^2R}{dr^2}+\frac{1}{r}\frac{dR}{dr}+\left[(jk_2)^2-\frac{n^2}{r^2}\right]R=0. \tag{16}$$

The general solution is

$$R(jk_2r)=R_a j^{-n}J_n(jk_2r)+R_b j^{n+1}H_n^{(1)}(jk_2r). \tag{17}$$

The large-argument asymptotic forms of $J_n(jk_2r)$ and $H_n^{(1)}(jk_2r)$ are

$$\begin{aligned}\lim_{u\to\infty} H_n^{(1)}(ju)&\simeq\sqrt{\pi/2u}\;e^{-u},\\ \lim_{u\to\infty} J_n(ju)&\simeq\sqrt{1/2u\pi}\;e^{u}.\end{aligned} \tag{18}$$

since region II extends to infinity and $J_n(jk_2r)$ cannot be accepted as a solution. Therefore, $H_n^{(1)}(jk_2r)$ is the solution for $r>a$.

$$E_{z2}=E_2H_n^{(1)}(jk_2r)e^{-\gamma_z z+jn\phi}, \tag{19}$$

$$H_{z2}=H_2H_n^{(1)}(jk_2r)e^{-\gamma_z z+jn\phi}, \tag{20}$$

$$E_{r2}=\frac{-1}{k_2^{\,2}}\left[-\gamma_z E_2 jk_2\dot{H}_n^{(1)}(jk_2r)+\frac{\omega\mu_0 nH_2}{r}H_n^{(1)}(jk_2r)\right]e^{-\gamma_z z+jn\phi}, \tag{21}$$

$$E_{\phi 2}=\frac{-1}{k_2^{\,2}}\left[\frac{-j\gamma_z nE_2}{r}H_n^{(1)}(jk_2r)-k_2\omega\mu_0H_2\dot{H}_n^{(1)}(jk_2r)\right]e^{-\gamma_z z+jn\phi}, \tag{22}$$

$$H_{r2}=\frac{-1}{k_2^{\,2}}\left[\frac{-\omega\varepsilon_2 nE_2}{r}H_n^{(1)}(jk_2r)-jk_2\gamma_z H_2\dot{H}_n^{(1)}(jk_2r)\right]e^{-\gamma_z z+jn\phi}, \tag{23}$$

$$H_{\phi 2}=\frac{-1}{k_2^{\,2}}\left[k_2\omega\varepsilon_2E_2\dot{H}_n^{(1)}(jk_2r)-\frac{j\gamma_z nH_2}{r}H_n^{(1)}(jk_2r)\right]e^{-\gamma_z z+jn\phi}. \tag{24}$$

The boundary conditions at $r=a$ require that the tangential components of the fields must be continuous.

$$E_{z1}(a)=E_{z2}(a),\qquad E_1J_n(k_1a)=E_2H_n^{(1)}(jk_2a), \tag{25}$$

$$H_{z1}(a)=H_{z2}(a),\qquad H_1J_n(k_1a)=H_2H_n^{(1)}(jk_2a); \tag{26}$$

$$E_{\phi 1}(a)=E_{\phi 2}(a),$$

$$\begin{aligned}&\frac{1}{k_1^{\,2}}\left[\frac{-j\gamma_z nE_1}{a}J_n(k_1a)+j\omega\mu_0k_1H_1\dot{J}_n(k_1a)\right]\\ &\quad=\frac{-1}{k_2^{\,2}}\left[\frac{-j\gamma_z nE_2}{a}H_n^{(1)}(jk_2a)-k_2\omega\mu_0H_2\dot{H}_n^{(1)}(jk_2a)\right],\end{aligned} \tag{27}$$

$$H_{\phi 1}(a)=H_{\phi 2}(a),$$

$$\frac{1}{k_1^{\,2}}\left[-j\omega\varepsilon_1 k_1 E_1 \dot{J}_n(k_1 a)-\frac{jn\gamma_z H_1}{a}J_n(k_1 a)\right]$$

$$=\frac{-1}{k_2^{\,2}}\left[k_2\omega\varepsilon_2 E_2 \dot{H}_n^{(1)}(jk_2 a)-\frac{j\gamma_z n H_2}{a}H_n^{(1)}(jk_2 a)\right]. \tag{28}$$

The coefficients E_1, E_2, H_1, and H_2 can now be evaluated. Equations (25) and (26) yield

$$E_2=\frac{J_n(k_1 a)}{H_n^{(1)}(jk_2 a)}E_1, \tag{29}$$

$$H_2=\frac{J_n(k_1 a)}{H_n^{(1)}(jk_2 a)}H_1. \tag{30}$$

The following notation is introduced to simplify mathematical manipulation.

$$\begin{aligned} &K\equiv k_2^{\,2}/k_1^{\,2}, \qquad J\equiv J_n(k_1 a), \qquad H\equiv H_n^{(1)}(jk_2 a),\\ &M\equiv J/H, \qquad \hat{J}\equiv \dot{J}/J, \qquad \hat{H}\equiv \dot{H}/H. \end{aligned} \tag{31}$$

Now Eqs. (29) and (30) become

$$E_2=ME_1 \qquad \text{and} \qquad H_2=MH_1. \tag{32}$$

From Eq. (28),

$$K\left(-j\omega\varepsilon_1 k_1 E_1 \dot{J}-\frac{jn\gamma_z H_1}{a}J\right)=-\left(\omega\varepsilon_2 k_2 M E_1 \dot{H}-\frac{j\gamma_z n}{a}MH_1 H\right),$$

$$E_1\left(-j\omega\varepsilon_1 k_1 \dot{J}K+\omega\varepsilon_2 k_2 \frac{J}{H}\dot{H}\right)=H_1\left(\frac{j\gamma_z n}{a}\frac{J}{H}H+\frac{jn\gamma_z}{a}KJ\right),$$

$$E_1 J\left(-j\omega\varepsilon_1 k_1 \frac{k_2^{\,2}}{k_1^{\,2}}\frac{\dot{J}}{J}+\omega\varepsilon_2 k_2 \frac{\dot{H}}{H}\right)=H_1 J\frac{j\gamma_z n}{a}(1+K),$$

$$\frac{H_1}{E_1}=\frac{1}{k_1^{\,2}}\,\frac{-j\omega\varepsilon_1 k_1 k_2^{\,2}\hat{J}+\omega\varepsilon_2 k_2 k_1^{\,2}\hat{H}}{(j\gamma_z n/a)(k_1^{\,2}+k_2^{\,2})/k_1^{\,2}}$$

$$=\frac{a}{\gamma_z n}\,\frac{-j\omega\varepsilon_2\left[(\varepsilon_1/\varepsilon_2)k_1 k_2^{\,2}\hat{J}+jk_1^{\,2}k_2\hat{H}\right]}{j\omega^2\mu_0(\varepsilon_1-\varepsilon_2)}$$

$$=\frac{-a}{\gamma_z n}\,\frac{(\varepsilon_1/\varepsilon_2)k_1 k_2^{\,2}\hat{J}+jk_1^{\,2}k_2\hat{H}}{\omega\mu_0(\varepsilon_1/\varepsilon_2-1)}. \tag{33}$$

The ratio H_1/E_1 can also be obtained from Eq. (27).

$$K\left(\frac{-j\gamma_z n}{a}JE_1+j\omega\mu_0 k_1\dot{J}H_1\right)=\frac{j\gamma_z n}{a}HME_1+k_2\omega\mu_0\dot{H}MH_1,$$

$$\frac{-j\gamma_z n}{a}(KJ+MH)E_1=\omega\mu_0\left(k_2M\dot{H}-jk_1K\dot{J}\right)H_1,$$

$$\begin{aligned}\frac{H_1}{E_1}&=\frac{(-j\gamma_z n/a)\left(KJ+J\dot{H}/H\right)}{\omega\mu_0\left(k_2J\dot{H}/H-jk_1K\dot{J}\right)}\\&=\frac{-j\gamma_z n}{\omega\mu_0 a}\,\frac{1+K}{k_2\hat{H}-jk_1K\hat{J}}\\&=\frac{\gamma_z n}{\omega\mu_0 a}\,\frac{k_1{}^2+k_2{}^2}{k_1k_2\left(k_2\hat{J}+jk_1\hat{H}\right)}.\end{aligned}\tag{33a}$$

Equations (25)–(28) may be expressed in matrix form:

$$\begin{bmatrix} J & -H & 0 & 0\\ 0 & 0 & J & -H\\ \dfrac{-1}{a}j\gamma_z nKJ & \dfrac{-j}{a}\gamma_z nH & j\omega\mu_0Kk_1\dot{J} & -k_2\omega\mu_0\dot{H}\\ -j\omega\varepsilon_1Kk_1\dot{J} & \omega\varepsilon_2k_2\dot{H} & \dfrac{-j}{a}\gamma_z nKJ & \dfrac{-j}{a}\gamma_z nH\end{bmatrix}\begin{bmatrix}E_1\\E_2\\H_1\\H_2\end{bmatrix}=0.\tag{34}$$

This is a homogeneous equation, and there will be a nontrivial solution only if the determinant vanishes:

$$J\begin{vmatrix}0 & J & -H\\ \dfrac{-j}{a}\gamma_z nH & j\omega\mu_0Kk_1\dot{J} & -\omega\mu_0k_2\dot{H}\\ \omega\varepsilon_2k_2\dot{H} & \dfrac{-j}{a}\gamma_z nKJ & \dfrac{-j}{a}\gamma_z nH\end{vmatrix}$$

$$+H\begin{vmatrix}0 & J & -H\\ \dfrac{-j}{a}\gamma_z nKJ & j\omega\mu_0Kk_1\dot{J} & -\omega\mu_0k_2\dot{H}\\ -j\omega\varepsilon_1Kk_1\dot{J} & \dfrac{-j}{a}\gamma_z nKJ & \dfrac{-j}{a}\gamma_z nH\end{vmatrix}=0,$$

$$J\Bigg\{-J\left[\left(\frac{-j}{a}\gamma_z nH\right)^2+\omega^2\mu_0\varepsilon_2k_2{}^2\dot{H}^2\right]$$

$$-H\left[\left(\frac{-j}{a}\gamma_z n\right)^2KJH-j\omega^2\mu_0\varepsilon_2Kk_1k_2\dot{H}\dot{J}\right]\Bigg\}$$

$$+H\Bigg\{-J\left[\left(\frac{-j}{a}\gamma_z n\right)^2KJH-j\omega^2\mu_0\varepsilon_1k_1k_2K\dot{J}\dot{H}\right]$$

$$-H\left[\left(\frac{-j}{a}\gamma_z n\right)^2 K^2 J^2 - \omega^2\mu_0\varepsilon_1 K^2 k_1{}^2 \hat{J}^2\right]\bigg\} = 0,$$

$$(1+K)^2\left(\frac{-j}{a}\gamma_z n\right)^2 = \omega^2\mu_0\left[\varepsilon_1\frac{k_2{}^4}{k_1{}^2}\hat{J}^2 + j(\varepsilon_2+\varepsilon_1)\frac{k_2{}^3}{k_1}\hat{H}\hat{J} + (jk_2)^2\varepsilon_2\hat{H}^2\right],$$

$$\left(\frac{k_1{}^2+k_2{}^2}{k_1{}^2}\right)^2\left(-j\frac{\gamma_z n}{a}\right)^2 = \omega^2\mu_0\varepsilon_2\left[\frac{\varepsilon_1}{\varepsilon_2}\frac{k_2{}^4}{k_1{}^2}\hat{J}^2 + j\left(1+\frac{\varepsilon_1}{\varepsilon_2}\right)\frac{k_2{}^3}{k_1}\hat{H}\hat{J} + (jk_2)^2\hat{H}^2\right],$$

$$\frac{\left[\omega^2\mu_0\varepsilon_2(\varepsilon_1/\varepsilon_2-1)\right]^2}{k_1{}^4}\left(-j\frac{\gamma_z n}{a}\right)^2 = \omega^2\mu_0\varepsilon_2\left(\frac{\varepsilon_1}{\varepsilon_2}\frac{k_2{}^2}{k_1}\hat{J}+jk_2\hat{H}\right) \times\left(\frac{k_2{}^2}{k_1}\hat{J}+jk_2\hat{H}\right),$$

$$\left[\frac{-j\gamma_z n\gamma_2}{k_1{}^2}\left(\frac{\varepsilon_1}{\varepsilon_2}-1\right)\right]^2 = \left(\frac{\varepsilon_1}{\varepsilon_2}\frac{k_2{}^2 a}{k_1}\hat{J}+jk_2 a\hat{H}\right)\left(\frac{k_2{}^2 a}{k_1}\hat{J}+jk_2 a\hat{H}\right). \tag{35}$$

This is the eigenvalue equation for determining the values of the propagation constant γ_z. The index n can be either positive or negative.

The fields in a dielectric guide have all six components, and it is impossible, in general, to subdivide them into TE and TM modes as in the hollow metallic guides. The only exception is for $n=0$, when Eq. (35) becomes

$$\left(\frac{\varepsilon_1}{\varepsilon_2}\frac{k_2{}^2 a}{k_1}\hat{J}+jk_2 a\hat{H}\right)\left(\frac{k_2{}^2 a}{k_1}\hat{J}+jk_2 a\hat{H}\right) = 0 \quad \text{at} \quad n=0. \tag{36}$$

This implies one of the following conditions:

$$\frac{\varepsilon_1 k_2}{\varepsilon_2 k_1}\frac{\dot{J}_0(k_1 a)}{J_0(k_1 a)} + j\frac{\dot{H}_0{}^{(1)}(jk_2 a)}{H_0{}^{(1)}(jk_2 a)} = 0, \tag{37}$$

$$\frac{k_2}{k_1}\frac{\dot{J}_0(k_1 a)}{J_0(k_1 a)} + j\frac{\dot{H}_0{}^{(1)}(jk_2 a)}{H_0{}^{(1)}(jk_2 a)} = 0. \tag{38}$$

With $\dot{J}_0(u)=J_1(u)$ and $\dot{H}_0{}^{(1)}(u)=H_1{}^{(1)}(u)$, Eqs. (37) and (38) become

$$\text{TM modes:}\quad \frac{\varepsilon_1 k_2}{\varepsilon_2 k_1}\frac{J_1(k_1 a)}{J_0(k_1 a)}+j\frac{H_1{}^{(1)}(jk_2 a)}{H_0{}^{(1)}(jk_2 a)}=0,\quad n=0; \tag{39}$$

$$\text{TE modes:}\quad \frac{k_2}{k_1}\frac{J_1(k_1 a)}{J_0(k_1 a)}+j\frac{H_1{}^{(1)}(jk_2 a)}{H_0{}^{(1)}(jk_2 a)}=0,\quad n=0. \tag{40}$$

The index n of the Bessel function can be obtained from Eq. (35):

$$n^2=\left[\frac{-j\gamma_z\gamma_2}{k_1{}^2 a}\left(\frac{\varepsilon_1}{\varepsilon_2}-1\right)\right]^{-2}\left(\frac{\varepsilon_1}{\varepsilon_2}\frac{k_2{}^2}{k_1}\hat{J}+jk_2\hat{H}\right)\left(\frac{k_2{}^2}{k_1}\hat{J}+jk_2\hat{H}\right). \tag{41}$$

Substituting Eq. (41) into Eq. (33) yields

$$\frac{H_1}{E_1}=\frac{j\gamma_2}{\omega\mu_0}\left[\frac{(\varepsilon_1/\varepsilon_2)(k_2{}^2/k_1)\hat{J}+jk_2\hat{H}}{(k_2{}^2/k_1)\hat{J}+jk_2\hat{H}}\right]^{1/2}. \tag{42}$$

TM modes. Equation (42) can be rearranged as

$$H_1=\frac{j\gamma_2}{\omega\mu_0}\left[\frac{(\varepsilon_1/\varepsilon_2)(k_2{}^2/k_1)\hat{J}+jk_2\hat{H}}{(k_2{}^2/k_1)\hat{J}+jk_2\hat{H}}\right]^{1/2} E_1. \tag{43}$$

When Eq. (39) is imposed at $n=0$, Eq. (43) vanishes. This implies that $H_1=0$, and so the **H** field is transverse to the axial direction.

TE modes. Equation (42) can be rearranged as

$$E_1=\frac{\omega\mu_0}{j\gamma_2}\left[\frac{(k_2{}^2/k_1)\hat{J}+jk_2\hat{H}}{(\varepsilon_1/\varepsilon_2)(k_2{}^2/k_1)\hat{J}+jk_2\hat{H}}\right]^{1/2} H_1. \tag{44}$$

When the condition Eq. (40) is imposed at $n=0$, one obtains $E_1=0$, and thus the **E** field is transverse to the axial direction.

8.5. Cutoff Frequency

One of the most important properties of a dielectric waveguide is its *cutoff frequency*. The field of a guided wave is defined to be *cut off* when it does not decay outside the core.

The fields exterior to the core are given by Eqs. (8.4.19)–(8.4.24). Each field component varies as $H_n^{(1)}(jk_2r)$ since

$$\dot{H}_v^{(1)}(u)=\tfrac{1}{2}\left[H_{v-1}^{(1)}(u)-H_{v+1}^{(1)}(u)\right] \qquad \text{for} \quad v\neq 0 \tag{1}$$

and

$$\dot{H}_0^{(1)}(u)=-H_1^{(1)}(u). \tag{2}$$

The asymptotic expression of $H_v^{(1)}(u)$ for large arguments is

$$\lim_{u\to\infty} H_v^{(1)}(u)\simeq\sqrt{2/\pi u}\; e^{j(u-\pi/4-\pi v/2)}. \tag{3}$$

In the present case, $u=jk_2r$, one has

$$\lim_{jk_2r\to\infty} H_v^{(1)}(jk_2r)\simeq\sqrt{2/j\pi k_2 r}\; e^{-k_2r-j\pi(1+2v)/4}. \tag{4}$$

It is this property of $H_v^{(1)}$ (that it decays exponentially for large values of k_2r) that makes it an acceptable solution for the guided modes. In other words, the guided fields adhere close to the guide. This decaying behavior will disappear at $k_2=0$, when the fields are no longer attached to the guide, and this is the condition for cutoff; that is,

$$jk_2\equiv\sqrt{|\gamma_2|^2+\gamma_z^{\,2}}=0, \qquad |\gamma^2|^2\equiv\omega^2\mu_0\varepsilon_2. \tag{5}$$

The eigenvalue γ_z satisfies Eq. (8.4.35), which is repeated here in a slightly altered form.

$$\left[\frac{-j\gamma_z n\gamma_2}{k_1^{\,2}k_2^{\,2}a^2}(\epsilon-1)\right]^2=\left(\frac{\epsilon}{k_1a}\hat{J}-\frac{1}{jk_2a}\hat{H}\right)\left(\frac{1}{k_1a}\hat{J}-\frac{1}{jk_2a}\hat{H}\right), \tag{6}$$

where

$$\epsilon\equiv\varepsilon_1/\varepsilon_2, \qquad \hat{J}(u)\equiv\dot{J}_n(u)/J_n(u), \qquad \hat{H}\equiv\dot{H}_n^{(1)}(u)/H_n^{(1)}(u). \tag{7}$$

Equation (6) is not convenient to use at cutoff. However, with the identity

$$\dot{G}_v(u)=\tfrac{1}{2}\left[G_{v-1}(u)-G_{v+1}(u)\right], \qquad v\neq 0, \tag{8}$$

where G_v stands for either J_v or $H_v^{(1)}$,

$$\frac{1}{k_1a}\hat{J}=\frac{1}{k_1a}\frac{\dot{J}_n}{J_n}=\frac{1}{2k_1a}\left[\frac{J_{n-1}}{J_n}-\frac{J_{n+1}}{J_n}\right]$$

$$\equiv J^- - J^+ \qquad \text{(Schlesinger et al. [10, 11])}, \tag{9}$$

$$\frac{1}{jk_2a}\hat{H}=\frac{1}{jk_2a}\frac{\dot{H}_n^{(1)}}{H_n}=\frac{1}{2jk_2a}\left[\frac{H_{n-1}^{(1)}}{H_n^{(1)}}-\frac{H_{n+1}^{(1)}}{H_n^{(1)}}\right]$$

$$\equiv H^- - H^+. \tag{10}$$

Substituting Eqs. (9) and (10) into Eq. (6) yields

$$
\begin{aligned}
&\left[\frac{-j\gamma_z n\gamma_2}{k_1{}^2k_2{}^2a^2}(\epsilon-1)\right]^2\\
&\quad=[\epsilon(J^- - J^+)-(H^- - H^+)][(J^- - J^+)-(H^- - H^+)]\\
&\quad=[(\epsilon J^- - H^-)-(\epsilon J^+ - H^+)][(J^- - H^-)-(J^+ - H^+)]\\
&\quad=[(\epsilon J^- - H^-)(J^- - H^-)+(\epsilon J^+ - H^+)(J^+ - H^+)]\\
&\qquad-(\epsilon J^+ - H^+)(J^- - H^-)-(\epsilon J^- - H^-)(J^+ - H^+)\\
&\quad=[(\epsilon J^- - H^-)(J^- - H^-)+(\epsilon J^+ - H^+)(J^+ - H^+)]\\
&\qquad+[(\epsilon J^+ - H^+)(J^- - H^-)+(\epsilon J^- - H^-)(J^+ - H^+)]\\
&\qquad-2[(\epsilon J^+ - H^+)(J^- - H^-)+(\epsilon J^- - H^-)(J^+ - H^+)]\\
&\quad=-2[(\epsilon J^+ - H^+)(J^- - H^-)+(\epsilon J^- - H^-)(J^+ - H^+)]\\
&\qquad+(\epsilon J^+ - H^+)[(J^+ - H^+)+(J^- - H^-)]\\
&\qquad+(\epsilon J^- - H^-)[(J^+ - H^+)+(J^- - H^-)]\\
&\quad=-2[(\epsilon J^+ - H^+)(J^- - H^-)+(\epsilon J^- - H^-)(J^+ - H^+)]\\
&\qquad+[\epsilon(J^+ + J^-)-(H^+ + H^-)][(J^+ + J^-)-(H^+ - H^-)].
\end{aligned}
\tag{11}
$$

Using the relation

$$\frac{2v}{u}G_v(u)=G_{v+1}(u)+G_{v-1}(u), \qquad G_v=J_v \quad \text{or} \quad H_v{}^{(1)} \tag{12}$$

or

$$\frac{2v}{u}=\frac{G_{v+1}(u)}{G_v(u)}+\frac{G_{v-1}(u)}{G_v(u)}, \tag{13}$$

one has

$$J^+(k_1a)+J^-(k_1a)=\frac{n}{(k_1a)^2}, \qquad J^\pm\equiv\frac{1}{2k_1a}\frac{J_{v\pm1}}{J_v}, \tag{14}$$

$$H^+(jk_2a)+H^-(jk_2a)=\frac{-n}{(k_2a)^2}. \tag{15}$$

The second term on the right-hand side of Eq. (11) then becomes

$$[\epsilon(J^+ + J^-) - (H^+ + H^-)][(J^+ + J^-) - (H^+ + H^-)]$$

$$= \left[\epsilon\frac{n}{(k_1 a)^2} - \frac{-n}{(k_2 a)^2}\right]\left[\frac{n}{(k_1 a)^2} - \frac{-n}{(k_2 a)^2}\right]$$

$$= \left(\frac{n}{a^2}\right)^2\left(\frac{\epsilon}{k_1{}^2} + \frac{1}{k_2{}^2}\right)\left(\frac{1}{k_1{}^2} + \frac{1}{k_2{}^2}\right)$$

$$= \left(\frac{n}{k_1{}^2 k_2{}^2 a^2}\right)^2 (\epsilon k_2{}^2 + k_1{}^2)(k_1{}^2 + k_2{}^2)$$

$$= \left(\frac{n}{k_1{}^2 k_2{}^2 a^2}\right)^2 \left[-\epsilon(|\gamma_2|^2 + \gamma_z{}^2) + (|\gamma_1|^2 + \gamma_z{}^2)\right]\omega^2\mu_0(\varepsilon_1 - \varepsilon_2)$$

$$= \left(\frac{n}{k_1{}^2 k_2{}^2 a^2}\right)^2 \omega^2\mu_0\varepsilon_2(\epsilon - 1)\left[(1-\epsilon)\gamma_z{}^2\right] \qquad (\text{because } \epsilon\gamma_2{}^2 = \gamma_1{}^2)$$

$$= \left[-j\frac{n\gamma_2\gamma_z}{k_1{}^2 k_2{}^2 a^2}(\epsilon - 1)\right]^2 \qquad (\text{because } \gamma_2{}^2 = \omega^2\mu_0\varepsilon_2). \tag{16}$$

Note that the right-hand side of Eq. (16) is identical to the left-hand side of Eq. (11). Substituting Eq. (16) into Eq. (11) yields

$$(\epsilon J^+ - H^+)(J^- - H^-) + (\epsilon J^- - H^-)(J^+ - H^+) = 0. \tag{17}$$

Equation (17) does not contain k_2 as a coefficient and is therefore more suitable at cutoff. The asymptotic expressions for the Hankel function of the first kind are

$$\lim_{k_2 \to 0} H_0{}^{(1)}(jk_2 a) \simeq \frac{2j}{\pi}\ln\frac{\Gamma k_2 a}{2} \qquad (\Gamma \equiv 1.78167), \tag{18}$$

$$\lim_{k_2 \to 0} H_v{}^{(1)}(jk_2 a) \simeq -\frac{j(v-1)!}{\pi}\left(\frac{2}{jk_2 a}\right)^v \qquad \text{for} \quad v = 1,2,3,\ldots. \tag{19}$$

Accordingly,

$$\lim_{k_2 \to 0} H^+ = \lim_{k_2 \to 0} \frac{1}{2jk_2 a}\frac{H_{n+1}{}^{(1)}}{H_n{}^{(1)}} = \frac{-n}{(k_2 a)^2}, \qquad n = 1,2,\ldots, \tag{20}$$

$$\lim_{k_2 \to 0} H^+ = \lim_{k_2 \to 0} \frac{1}{2jk_2 a}\frac{H_1{}^{(1)}}{H_0{}^{(1)}} = \frac{1}{2(k_2 a)^2 \ln(\Gamma k_2 a/2)}, \qquad n = 0, \tag{21}$$

$$\lim_{k_2\to 0} H^- = \lim_{k_2\to 0} \frac{1}{2jk_2a}\frac{H_{n-1}{}^{(1)}}{H_n{}^{(1)}} = -\frac{1}{2}\ln\frac{\Gamma k_2 a}{2}, \qquad n=1, \tag{22}$$

$$\lim_{k_2\to 0} H^- = \lim_{k_2\to 0} \frac{1}{2jk_2a}\frac{H_{n-1}{}^{(1)}}{H_n{}^{(1)}} = \frac{1}{4(n-1)}, \qquad n=2,3,\ldots. \tag{23}$$

Equation (17) can be expressed

$$\left(\frac{\epsilon}{2k_1a}\frac{J_{n+1}}{J_n} - H^+\right)\left(\frac{1}{2k_1a}\frac{J_{n-1}}{J_n} - H^-\right) + \left(\frac{\epsilon}{2k_1a}\frac{J_{n-1}}{J_n} - H^-\right)\left(\frac{J_{n+1}}{2k_1aJ_n} - H^+\right) = 0,$$

$$(\epsilon J_{n+1} - 2k_1aJ_nH^+)(J_{n-1} - 2k_1aJ_nH^-) + (\epsilon J_{n-1} - 2k_1aJ_nH^-)(J_{n+1} - 2k_1aJ_nH^+) = 0. \tag{24}$$

For the case where $n\neq 0$, substituting Eq. (20) into Eq. (24) yields

$$\left[\epsilon J_{n+1} + 2nk_1aJ_n/(k_2a)^2\right](J_{n-1} - 2k_1aJ_nH^-) + (\epsilon J_{n-1} - 2k_1aJ_nH^-)\left[J_{n+1} + 2nk_1aJ_n/(k_2a)^2\right] = 0,$$

$$\left[(k_2a)^2\epsilon J_{n+1} + 2nk_1aJ_n\right](J_{n-1} - 2k_1aJ_nH^-) + (\epsilon J_{n-1} - 2k_1aJ_nH^-)\left[(k_2a)^2J_{n+1} + 2nk_1aJ_n\right] = 0. \tag{25}$$

In the limit $k_2\to 0$, one has

$$2nk_1aJ_n\left[(\epsilon+1)J_{n-1} - 4k_1aJ_nH^-\right]_{k_2\to 0} = 0, \qquad n\neq 0. \tag{26}$$

a. For $n=1$, Eq. (26) becomes

$$2k_1aJ_1(k_1a)\left[(\epsilon+1)J_0(k_1a) + 2k_1aJ_1\ln(\Gamma k_2a/2)\right]_{k_2\to 0} = 0$$

or

$$\left[k_1aJ_1(k_1a)\right]^2\ln(\Gamma k_2a/2)\big|_{k_2\to 0} = 0, \tag{27}$$

since $J_0(k_1a)$ is finite and may be neglected in comparison with $\ln(\Gamma k_2a/2)|_{k_2\to 0}$.

This implies

$$k_1a = 0 \tag{28}$$

or

$$J_1(k_1a) = 0. \tag{29}$$

b. For $n>1$, then Eq. (26) becomes

$$\lim_{k_2\to 0}\left\{2nk_1aJ_n(k_1a)\left[(\epsilon+1)J_{n-1}(k_1a) - k_1aJ_n(k_1a)\frac{1}{(n-1)}\right]\right\} = 0. \tag{30}$$

The possible solutions are

$$k_1a=0, \tag{31}$$

$$J_n(k_1a)=0 \quad \text{for} \quad k_1a\neq 0, \quad n=2,3,\ldots, \tag{32}$$

$$(\epsilon+1)J_{n-1}(k_1a)=\frac{k_1a}{n-1}J_n(k_1a), \qquad n=2,3,\ldots. \tag{33}$$

For both the $n=1$ and $n>1$ case, $k_1a=0$ is a possible solution. To investigate whether $k_1a=0$ is an acceptable solution, it is necessary to find the asymptotic expression for Eq. (17) in the limit as $k_2\to 0$ and $k_1a\to 0$. This will then be analyzed for three separate cases: $n=0$, $n=1$, and $n>1$.

The small-argument approximation for the Bessel function $J_n(k_1a)$ is

$$\lim_{k_1a\to 0} J_0(k_1a)\simeq 1,$$

$$\lim_{k_1a\to 0} J_\nu(k_1a)\simeq\frac{1}{\nu!}\left(\frac{k_1a}{2}\right)^\nu \quad \text{for} \quad \nu=1,2,3,\ldots.$$

Then

$$\lim_{k_1a\to 0} J^+ = \lim_{k_1a\to 0}\frac{1}{2k_1a}\frac{J_1(k_1a)}{J_0(k_1a)}\simeq\frac{1}{4}, \qquad n=0, \tag{34}$$

$$\lim_{k_1a\to 0} J^+ = \lim_{k_1a\to 0}\frac{1}{2k_1a}\frac{J_{n+1}(k_1a)}{J_n(k_1a)}\simeq\frac{1}{4(n+1)}, \qquad n=1,2,3,\ldots, \tag{35}$$

$$\lim_{k_1a\to 0} J^- = \lim_{k_1a\to 0}\frac{1}{2k_1a}\frac{J_{n-1}(k_1a)}{J_n(k_1a)}\simeq\frac{n}{(k_1a)^2}, \qquad n=1,2,3,\ldots. \tag{36}$$

Case a: $n=0$. For the $n=0$ case, the eigenvalue equations simplify to Eqs. (8.4.39) and (8.4.40), which are repeated here.

$$\text{TM modes:} \qquad \epsilon\frac{k_2}{k_1}\frac{J_1(k_1a)}{J_0(k_1a)}+j\frac{H_1^{(1)}(jk_2a)}{H_0^{(1)}(jk_2a)}=0; \tag{37}$$

$$\text{TE modes:} \qquad \frac{k_2}{k_1}\frac{J_1(k_1a)}{J_0(k_1a)}+j\frac{H_1^{(1)}(jk_2a)}{H_0^{(1)}(jk_2a)}=0. \tag{38}$$

For the present purpose, it is only necessary to investigate one of these relations, since they differ merely by the constant ϵ in the first term.

$$\epsilon\frac{k_2}{k_1}\frac{J_1(k_1a)}{J_0(k_1a)}=-j\frac{H_1^{(1)}(jk_2a)}{H_0^{(1)}(jk_2a)}=\left(k_2a\ln\frac{\Gamma k_2a}{2}\right)^{-1},$$

$$\frac{k_1a}{\epsilon}\frac{J_0(k_1a)}{J_1(k_1a)}=(k_2a)^2\ln\frac{\Gamma k_2a}{2}=(k_2a)^2\left[\left(\tfrac{1}{2}\Gamma k_2a-1\right)-\tfrac{1}{2}\left(\tfrac{1}{2}\Gamma k_2a-1\right)^2\right.$$
$$\left.+\tfrac{1}{3}\left(\tfrac{1}{2}\Gamma k_2a-1\right)^3-\cdots\right].$$

Therefore,

$$\lim_{\substack{k_1a\to 0\\ k_2a\to 0}}\left[\frac{k_1a}{\epsilon}\frac{J_0(k_1a)}{J_1(k_1a)}\right]=0. \tag{39}$$

Equation (39) can be satisfied if

$$k_1a=0 \tag{40}$$

or

$$J_0(k_1a)=0. \tag{41}$$

Since

$$\lim_{k_1a\to 0} J_1(k_1a)\simeq\tfrac{1}{2}k_1a,$$

it follows that

$$\begin{aligned}\lim_{k_1a\to 0}\frac{k_1a}{\epsilon}\frac{J_0(k_1a)}{J_1(k_1a)} &= \lim_{k_1a\to 0}\frac{k_1a}{\epsilon}\frac{J_0(k_1a)}{\frac{1}{2}k_1a}\\ &= \lim_{k_1a\to 0}\frac{2}{\epsilon}J_0(k_1a)\neq 0.\end{aligned} \tag{42}$$

Thus Eq. (40) is not a solution of Eq. (39). The acceptable cutoff solution is $k_1=k_{1c}$ such that

$$J_0(k_{1c}a)=0. \tag{43}$$

Case b: $n=1$. The small-argument approximation of Eq. (17) is

$$\lim_{\substack{k_1a\to 0\\ k_2a\to 0}}\left\{\left[\frac{\epsilon}{4(n+1)}+\frac{n}{(k_2a)^2}\right]\left[\frac{n}{(k_1a)^2}+\frac{1}{2}\ln\frac{\Gamma k_2a}{2}\right]+\left[\frac{n\epsilon}{(k_1a)^2}+\frac{1}{2}\ln\frac{\Gamma k_2a}{2}\right]\left[\frac{1}{4(n+1)}+\frac{n}{(k_2a)^2}\right]\right\}=0,$$

$$\lim_{\substack{k_1a\to 0\\ k_2a\to 0}}\left\{\frac{1}{(k_1k_2a^2)^2}\left\{\left[\frac{\epsilon(k_2a)^2}{4(n+1)}+n\right]\left[n+\frac{(k_1a)^2}{2}\ln\frac{\Gamma k_2a}{2}\right]+\left[n\epsilon+\frac{(k_1a)^2}{2}\ln\frac{\Gamma k_2a}{2}\right]\left[\frac{(k_2a)^2}{4(n+1)}+n\right]\right\}\right\}=0,$$

$$\lim_{\substack{k_1a\to 0\\ k_2a\to 0}}\frac{n}{(k_1k_2a^2)^2}\left\{\left[n+\frac{(k_1a)^2}{2}\ln\frac{\Gamma k_2a}{2}\right]+\left[n\epsilon+\frac{(k_1a)^2}{2}\ln\frac{\Gamma k_2a}{2}\right]\right\}=0,$$

$$\lim_{\substack{k_1a\to 0\\ k_2a\to 0}}\frac{n}{(k_2a)^2}\left[\frac{n(\epsilon+1)}{(k_1a)^2}+\ln\frac{\Gamma k_2a}{2}\right]=0. \tag{44}$$

Equation (44) can be satisfied if

$$\lim_{k_1a\to 0}\frac{n(\epsilon+1)}{(k_1a)^2}=\lim_{k_2a\to 0}\left(-\ln\frac{\Gamma k_2a}{2}\right)$$

$$=\lim_{k_2a\to 0}\left(\ln\frac{2}{\Gamma k_2a}\right),\qquad n=1. \tag{45}$$

Thus $k_1a=0$ is an acceptable solution for $n=1$.

Case c: $n>1$. Substituting Eqs. (20), (23), (35), and (36) into Eq. (17) yields

$$\lim_{\substack{k_1a\to 0\\ k_2a\to 0}}\left\{\left[\frac{\epsilon}{4(n+1)}+\frac{n}{(k_2a)^2}\right]\left[\frac{n}{(k_1a)^2}-\frac{1}{4(n-1)}\right]\right.$$

$$\left.+\left[\epsilon\frac{n}{(k_1a)^2}-\frac{1}{4(n-1)}\right]\left[\frac{1}{4(n+1)}+\frac{n}{(k_2a)^2}\right]\right\}=0,$$

$$\lim_{\substack{k_1a\to 0\\ k_2a\to 0}}\frac{1}{(k_1k_2a^2)^2}\left\{\left[\frac{\epsilon(k_2a)^2}{4(n+1)}+n\right]\left[n-\frac{(k_1a)^2}{4(n-1)}\right]\right.$$

$$\left.+\left[n\epsilon-\frac{(k_1a)^2}{4(n-1)}\right]\left[\frac{(k_2a)^2}{4(n+1)}+n\right]\right\}=0,$$

$$\lim_{\substack{k_1a\to 0\\ k_2a\to 0}}\frac{1}{(k_1k_2a^2)^2}n^2(1+\epsilon)=0. \tag{46}$$

Since Eq. (46) is impossible in the limit, both k_1a and k_2a are vanishingly small. Therefore $k_1a=0$ is not an acceptable solution to Eq. (30).

Summary

If $n=0$,

$$k_1=k_{1c}\qquad\text{such that}\quad J_0(k_{1c}a)=J_0(u_{0m})=0, \tag{47}$$

where u_{0m} is the mth root of the equation.

If $n=1$,

i. $k_1a=0$; (48)

ii. $k_1a=u_{1m}$, the mth root of

$$J_1(u_{1m})=0. \tag{49}$$

If $n>1$, there are two cases.

i. The EH_{nm} modes are given by

$$J_n(k_1a)=0, \tag{50}$$

where $k_1a=u_{nm}$, the mth root of the equation, which does not equal zero.

ii. The HE_{nm} modes are

$$(\epsilon+1)J_{n-1}(k_1a)=\frac{k_1a}{n-1}J_n(k_1a). \tag{51}$$

The cutoff frequency f_c may be determined as follows:

$$k_1{}^2=\gamma_z{}^2-\gamma_1{}^2 \quad \text{and} \quad -k_2{}^2=\gamma_z{}^2-\gamma_2{}^2, \tag{52}$$

$$k_1{}^2=(\gamma_2{}^2-k_2{}^2)-\gamma_1{}^2 \quad \text{or} \quad k_1=\sqrt{-\gamma_1{}^2+\gamma_2{}^2-k_2{}^2}. \tag{53}$$

At cutoff frequency, $k_2=0$, and

$$k_{10}\equiv k_1|_{k_2=0}=\sqrt{-\gamma_1{}^2+\gamma_2{}^2}=\sqrt{\omega^2\mu_0(\varepsilon_1-\varepsilon_2)},$$

$$f_c=\frac{k_{10}}{2\pi\sqrt{\mu_0(\varepsilon_1-\varepsilon_2)}}. \tag{54}$$

In the $n=1$ case, $k_1a=0$ is a solution, and therefore, this particular mode has a cutoff frequency $f_c=0$; this mode does not cut off. It is possible to choose the radius a of the core such that all other modes are cut off, leaving only a single mode. This is known as single-mode operation.

8.6. Designation of Modes

For cylindrical metallic waveguides, the guided wave solutions may be subdivided into transverse electric waves and transverse magnetic waves.

In dielectric waveguides, only the $n=0$ modes can be divided into transverse electric modes, TE_{0m} or H_{0m}, and transverse magnetic modes, TM_0 or E_0. The higher-order solutions ($n>0$) are hybrid modes; both H_z and E_z have nonzero values.

The scheme for designating these modes suggested by Beam et al. [3, 13] is based upon that for metallic guides. Instead of classifying the solutions entirely by one axial field component (say E_z), they are classified according to the axial field component which contributes more. The E-like modes, labeled EH_{nm}, are the modes which depend more heavily on E_z than H_z. Similarly, the H-like modes are designated by HE_{nm}.

The relative contribution from E_z and H_z in a hybrid mode can be determined from the boundary relations (8.4.25)–(8.4.28). Equation (8.4.27) is repeated here.

$$\frac{1}{k_1{}^2}\left(\frac{-j\gamma_z n}{a}J_n E_1 + j\omega\mu_0 k_1 \dot{J}_n H_1\right) = \frac{-1}{k_2{}^2}\left(\frac{-j\gamma_z n}{a}H_n{}^{(1)}E_2 - k_2\omega\mu_0 \dot{H}_n{}^{(1)}H_2\right), \tag{1}$$

$$K\left[\frac{n}{a} + \left(\frac{\omega\mu_0}{-j\gamma_z}\frac{H_1}{E_1}\right)jk_1\hat{J}\right] = -\left[\frac{n}{a} - \left(\frac{\omega\mu_0}{-j\gamma_z}\frac{H_1}{E_1}\right)k_2\hat{H}\right], \tag{2}$$

where

$$K \equiv \frac{k_2{}^2}{k_1{}^2}, \qquad E_2 = \frac{J_n}{H_n{}^{(1)}}E_1, \qquad H_2 = \frac{J_n}{H_n{}^{(1)}}H_1.$$

Let

$$P \equiv \frac{\omega\mu}{-j\gamma_z}\frac{H_1}{E_1}. \tag{3}$$

Then

$$K\left(n/a + jPk_1\hat{J}\right) = -n/a + k_2 P\hat{H},$$

$$\left(jKk_1\hat{J} - k_2\hat{H}\right)P = -\frac{n}{a}(1+K),$$

$$P = \frac{-(n/a)\left(k_1{}^2 + k_2{}^2\right)}{jk_1k_2\left(k_2\hat{J} + jk_1\hat{H}\right)}. \tag{4}$$

The ratio P can also be obtained from Eq. (8.4.28):

$$K\left(-j\omega\varepsilon_1 k_1 \dot{J}_n E_1 - \frac{jn\gamma_z}{a}J_n H_1\right) = -\left(k_2\omega\varepsilon_2 \dot{H}_n{}^{(1)}E_2 - \frac{j\gamma_z n}{a}H_n{}^{(1)}H_2\right),$$

$$K\left[\frac{-j\omega^2\mu_0\varepsilon_1 k_1}{-j\gamma_z}\dot{J}_n - \left(\frac{\omega\mu_0}{-j\gamma_z}\frac{H_1}{E_1}\right)j\frac{n\gamma_z}{a}J_n\right] = -\left[\frac{k_2\omega^2\mu_0\varepsilon_2}{-j\gamma_z}\hat{H}J_n - \left(\frac{\omega\mu_0}{-j\gamma_z}\frac{H_1}{E_1}\right)j\frac{\gamma_z n}{a}J_n\right],$$

$$K\left(\frac{-\gamma_1{}^2k_1}{\gamma_z}\hat{J} - j\frac{\gamma_z n}{a}P\right) = \frac{-\gamma_2{}^2k_2}{j\gamma_z}\hat{H} + j\frac{\gamma_z n}{a}P,$$

$$-\frac{j\gamma_z n}{a}(K+1)P = \frac{k_2{}^2}{k_1{}^2}\frac{\gamma_1{}^2k_1}{\gamma_z}\hat{J} + j\frac{\gamma_2{}^2k_2}{\gamma_z}\hat{H},$$

$$P = \frac{(1/k_1{}^2\gamma_z)k_1k_2\left(\gamma_1{}^2k_2\hat{J} + j\gamma_2{}^2k_1\hat{H}\right)}{-(j\gamma_z n/a)\left[\left(k_1{}^2 + k_2{}^2\right)/k_1{}^2\right]} \tag{5}$$

$$= \frac{\omega^2\mu_0}{j\gamma_z{}^2 n/a} \frac{k_1k_2\left(\varepsilon_1 k_2\hat{J}+j\varepsilon_2 k_1\hat{H}\right)}{\left(k_1{}^2+k_2{}^2\right)}$$

$$= \frac{\omega^2\mu_0\varepsilon_2 k_1k_2\left[(\varepsilon_1/\varepsilon_2)k_2\hat{J}+jk_1\hat{H}\right]}{\left(j\gamma_z{}^2 n/a\right)\left(k_1{}^2+k_2{}^2\right)}. \tag{6}$$

The two expressions for P, Eqs. (4) and (6), can be shown to be equivalent by using the identity of Eqs. (8.4.33) and (8.4.33a):

$$\frac{1}{-\gamma_z n/a} \frac{k_1k_2\left(\frac{\varepsilon_1}{\varepsilon_2}k_2\hat{J}+jk_1\hat{H}\right)}{(\omega\mu_0/\varepsilon_2)(\varepsilon_1-\varepsilon_2)} = \frac{1}{\omega\mu_0} \frac{k_1{}^2+k_2{}^2}{k_1k_2\left(k_2\hat{J}+jk_1\hat{H}\right)} \frac{\gamma_z n}{a} \tag{7}$$

since

$$k_1{}^2+k_2{}^2 = \omega^2\mu_0(\varepsilon_1-\varepsilon_2) \qquad \text{or} \qquad \varepsilon_1-\varepsilon_2 = \frac{k_1{}^2+k_2{}^2}{\omega^2\mu_0}. \tag{8}$$

Therefore,

$$\left(\frac{\varepsilon_1}{\varepsilon_2}k_2\hat{J}+jk_1\hat{H}\right) = \frac{-(\gamma_z n/a)^2\left(k_1{}^2+k_2{}^2\right)^2}{\omega^2\mu_0\varepsilon_2 k_1{}^2k_2{}^2\left(k_2\hat{J}+jk_1\hat{H}\right)}. \tag{9}$$

Subsituting Eq. (9) into Eq. (6) yields

$$P = \frac{\omega^2\mu_0\varepsilon_2 k_1k_2}{j\gamma_z{}^2(n/a)\left(k_1{}^2+k_2{}^2\right)}\left[\frac{-(\gamma_z n/a)^2}{\omega^2\mu_0\varepsilon_2} \frac{\left(k_1{}^2+k_2{}^2\right)^2}{k_1{}^2k_2{}^2\left(k_2\hat{J}+jk_1\hat{H}\right)}\right]$$

$$= \frac{-(n/a)\left(k_1{}^2+k_2{}^2\right)}{jk_1k_2\left(k_2\hat{J}+jk_1\hat{H}\right)}, \tag{10}$$

which is identical to Eq. (4).

The field components, given by Eqs. (8.4.8)–(8.4.12), can now be expressed in terms of ratio P:

$$E_{z1} = E_1 J_n(k_1 r)e^{-\gamma_z z+jn\phi},$$

$$H_{z1} = \frac{-j\gamma_z}{\omega\mu_0} PE_1 J_n(k_1 r)e^{-\gamma_z z+jn\phi},$$

$$E_{r1} = \frac{1}{k_1{}^2}\left[-\gamma_z k_1 E_1 \dot{J}_n(k_1 r) - \frac{j\gamma_z n}{r}PE_1 J_n(k_1 r)\right]e^{-\gamma_z z+jn\phi}$$

$$= \frac{-\gamma_z}{k_1{}^2}\left[k_1\dot{J}_n(k_1 r)+\frac{jn}{r}PJ_n(k_1 r)\right]E_1 e^{-\gamma_z z+jn\phi},$$

$$E_{\phi 1} = \frac{-\gamma_z}{k_1{}^2}\left[\frac{jn}{r}J_n(k_1 r)+Pk_1\dot{J}_n(k_1 r)\right]E_1 e^{-\gamma_z z+jn\phi},$$

$$H_{r1} = \frac{1}{k_1{}^2}\left[\frac{-\omega\varepsilon_1 n}{r}J_n(k_1 r) + \frac{j\gamma_z{}^2 k_1}{\omega\mu_0}P\dot{J}_n(k_1 r)\right]E_1 e^{-\gamma_z z + jn\phi},$$

$$H_{\phi 1} = \frac{1}{k_1{}^2}\left[-j\omega\varepsilon_1 k_1 \dot{J}_n(k_1 r) - \frac{n\gamma_z{}^2}{\omega\mu_0 r}PJ_n(k_1 r)\right]E_1 e^{-\gamma_z z + jn\phi}$$

$$= \frac{1}{k_1\omega\mu_0}\left[j\gamma_1{}^2 \dot{J}_n(k_1 r) - \frac{n\gamma_z{}^2}{k_1 r}PJ_n(k_1 r)\right]E_1 e^{-\gamma_z z + jn\phi}$$

$$= \frac{j\gamma_1{}^2}{k_1\omega\mu_0}\left[\dot{J}_n(k_1 r) + j\frac{\gamma_z{}^2 n}{k_1 r\gamma_1{}^2}PJ_n(k_1 r)\right]E_1 e^{-\gamma_z z + jn\phi}.$$

The weight of E_{z1} and H_{z1} can be evaluated once P is determined.

References

1. M. Abramowitz and I. A. Stegun, *Handbook of Mathematical Functions*, Dover, New York, 1965.
2. G. Arfken, *Mathematical Methods for Physicists*, Academic, New York, 1966.
3. R. E. Beam, M. M. Astrahan, W. C. Jakes, H. M. Wachowski, and W. L. Firestone, Northwestern University Report ATI 94929, 1949.
4. P. Dennery and A. Krzywicki, *Mathematics for Physicists*, Harper & Row, New York, 1967.
5. E. Jahnke and F. Emde, *Tables of Functions with Formulae and Curves*, Dover, New York, 1945.
6. D. Marcuse, *Light Transmission Optics*, Van Nostrand Reinhold, Princeton, N. J., 1972.
7. D. Marcuse, *Theory of Dielectric Optical Waveguides*, Academic, New York, 1974.
8. N. W. McLachlan, *Bessel Functions for Engineers*, Oxford University Press, London, 1955.
9. J. E. Midwinter, *Optical Fibers for Transmission*, Wiley, New York, 1979.
10. S. P. Schlesinger and D. D. King, Dielectric Image Lines, *IRE Trans. Microwave Theory Tech*. MTT-6, 291–299 (1958).
11. S. P. Schlesinger, P. Diament, and A. Vigants, On Higher Order Hybrid Modes of Dielectric Cylinders, *IRE Trans. Microwave Theory Tech*. MTT-8, 252–253 (1960).
12. S. A. Schelkunoff, *Applied Mathematics for Engineers and Scientists*, Van Nostrand, Princeton, N. J., 1965.
13. E. Snitzer, Cylindrical Dielectric Waveguide Modes, *J. Opt. Soc. Am*. 51, 491–498 (1961).
14 M. S. Sodha and A. K. Ghatak, *Inhomogeneous Optical Waveguides*, Plenum, New York, 1977.
15. H. G. Unger, *Planar Optical Waveguides and Fibres*, Clarendon, Oxford, 1977.
16. G. N. Watson, *A Treatise on the Theory of Bessel Functions*, Cambridge University Press, New York, 1958.

9

Methods of Approximation

Exact solutions for inhomogeneous dielectric waveguides can be obtained for only a few types of inhomogeneity. In more complex cases methods of approximation are employed. Some basic techniques of perturbation as well as the WKB method are introduced in this chapter.

9.1. Perturbation Method

Consider a differential equation of the form

$$b_0 D^n r(t) + [b_1 + \Upsilon p_1(t)] D^{n-1} r(t) + \cdots + [b_n + \Upsilon p_n(t)] r(t) = f(t), \quad (1)$$

where $b_0, b_1, \ldots, b_n$ are constants; $f(t), p_1(t), \ldots, p_n(t)$ are continuous functions for $t \geq 0$; and Υ is a constant *parameter of perturbation* [1, 5, 6, 11], which can in general be complex. The initial conditions are

$$D^k r(t)\Big|_{t=t_0} = \frac{d^k r}{dt^k}\Big|_{t=t_0} = r^{(k)}(t_0), \qquad k = 0, 1, 2, \ldots, n-1. \quad (2)$$

For each given value of Υ, the solution of Eq. (1), $r(t)$, depends on Υ. From the theory of differential equations, the solution $r(t)$ is continuous in t and Υ. For every value of t, $r(t)$ is an analytic function of Υ. Thus, $r(t)$ can be expressed in terms of a power series,

$$r(t) = \sum_{i=0}^{\infty} q_i(t) \Upsilon^i. \quad (3)$$

It is permissible to differentiate the series termwise,

$$r^{(k)}(t) = \sum_{i=0}^{\infty} q_i^{(k)}(t) \Upsilon^i, \quad (4)$$

and at initial time $t=t_0$ one has

$$r^{(k)}(t_0)=\sum_{i=0}^{\infty} q_i{}^{(k)}(t_0)\Upsilon^i. \tag{5}$$

It is a property of power series that two series $\sum_{i=0}^{\infty} a_i x^i$ and $\sum_{i=0}^{\infty} b_i x^i$ are identical that have nonzero radii of convergence and have equal sums whenever both converge:

$$a_i=b_i, \qquad i=0,1,2\ldots.$$

Expanding Eq. (5),

$$r^{(k)}(t_0)=q_0{}^{(k)}(t_0)+q_i{}^{(k)}(t_0)\Upsilon+\cdots+q_n{}^{(k)}(t_0)\Upsilon^n+\cdots, \tag{6}$$

and equating the coefficients of identical powers of Υ^i on each side of Eq. (6) yields

$$r^{(k)}(t_0)=q_0{}^{(k)}(t_0), \qquad q_i{}^{(k)}(t_0)=0, \quad i=1,2,3,\ldots. \tag{7}$$

The differential equation (1) may be written in the more compact form

$$L_0[r(t)]+\Upsilon L_1[r(t)]=f(t), \tag{8}$$

where

$$\begin{aligned} L_0 &\equiv b_0 D^n+b_1 D^{n-1}+\cdots+b_n, \\ L_1 &\equiv p_1(t)D^{n-1}+p_2(t)D^{n-2}+\cdots+p_n(t). \end{aligned} \tag{9}$$

Substituting Eq. (3) into Eq. (8) yields

$$L_0\left[\sum_{i=0}^{\infty} q_i\Upsilon^i\right]+\Upsilon L_1\left[\sum_{i=0}^{\infty} q_i\Upsilon^i\right]=f(t),$$

$$L_0[q_0+q_1\Upsilon+\cdots+q_k\Upsilon^k+\cdots]$$
$$+\Upsilon L_1[q_0+q_1\Upsilon+\cdots+q_k\Upsilon^k+\cdots]=f(t),$$

$$L_0[q_0]+\sum_{i=1}^{\infty}(L_0[q_i]+L_1[q_{i-1}])\Upsilon^i=f(t). \tag{10}$$

Since it is valid independent of the value of Υ, this equation can be satisfied if and only if the coefficient of each power of Υ^i vanishes, that is, if and only if

$$L_0[q_i(t)]+L_1[q_{i-1}(t)]=0, \qquad i=1,2,3,\ldots. \tag{11}$$

Then

$$L_0[q_0(t)]=f(t). \tag{12}$$

Thus $q_0(t)$ is the solution of the unperturbed differential equation

$$b_0 D^n q_0(t)+b_1 D^{n-1}q_0(t)+\cdots+b_n q_0(t)=f(t) \tag{13}$$

with the original initial conditions, Eqs. (2) and (7):

$$q_0{}^{(k)}(t_0)=r^{(k)}(t_0). \tag{14}$$

The correction term $q_i(t)$ is the solution of Eq. (11), which may be rearranged as

$$L_0[q_i(t)] = -L_1[q_{i-1}(t)]$$

or

$$b_0 D^n q_i(t) + b_1 D^{n-1} q_i(t) + \cdots + b_n q_i(t) = f_i(t), \qquad i = 1,2,\ldots, \tag{15}$$

and

$$\begin{aligned} f_i(t) &\equiv -L_1[q_{i-1}(t)] \\ &= -[p_1(t)D^{n-1} + p_2(t)D^{n-2} + \cdots + p_n(t)]q_{i-1}(t), \\ & i = 1,2,\ldots, \end{aligned} \tag{16}$$

with the initial conditions given by Eq. (6). Once the solution of Eq. (13) is obtained, that is, once $q_0(t)$ is known, $f_1(t)$ can be evaluated from Eq. (16) in terms of $q_0(t)$. Consequently $q_1(t)$ can be determined, in principle, from Eq. (15). All higher-order correction terms $q_i(t)$ can be obtained successively, and the complete solution is given by the sum in Eq. (3).

9.2. Schrödinger First-Order Perturbation Theory

Consider a generalized wave equation of the form

$$L[g(\mathbf{r}_i)] - Wg(\mathbf{r}_i) = 0, \tag{1}$$

where L is the spatial operator

$$L \equiv \sum_{i=1}^{n} a_i \nabla_i^{\,2} + a_0, \qquad \nabla_i^{\,2} \equiv \frac{\partial^2}{\partial x_i^{\,2}} + \frac{\partial^2}{\partial y_i^{\,2}} + \frac{\partial^2}{\partial z_i^{\,2}}, \tag{2}$$

$\mathbf{r}_i = \hat{\mathbf{x}} x_i + \hat{\mathbf{y}} y_i + \hat{\mathbf{z}} z_i$ is the position vector of the ith particle, W is the energy of the system, and $a_0, a_1, \ldots$ are constant coefficients. Let the operator L be arranged into the following perturbed form [2, 7, 12, 13]:

$$L \equiv L_0 + \Upsilon L_1 + \Upsilon^2 L_2 + \cdots, \tag{3}$$

where Υ is the constant of perturbation and L_0 is the operator of the unperturbed system,

$$L_0[g_0(\mathbf{r})] - W_0 g_0(\mathbf{r}) = 0. \tag{4}$$

The subscript 0 is used to identify the quantity related to the unperturbed state. The unperturbed equation, Eq. (4), is assumed to have solution $g_{0k}(\mathbf{r})$, $k = 0, 1, 2, \ldots$ and this is the eigenfunction corresponding to the kth eigenvalue W_{0k}. The functions g_{0k} form a complete orthogonal set, and for convenience it is assumed that they are normalized:

$$\begin{aligned} \int g_{0k} g_{0j}\, d\tau &= 0 \quad \text{if} \quad k \neq j, \\ &= 1 \quad \text{if} \quad k = j. \end{aligned} \tag{5}$$

For a small perturbation Υ, the wave function $g(\mathbf{r})$ and the energy W for the perturbed system should not deviate very far from the unperturbed system. In other words, the introduction of a small perturbation should not create large variations. With this in mind, the wave function of the perturbed system can be expanded as the wave function of the unperturbed system plus some correction terms:

$$g_m(\mathbf{r}) = g_{0m} + \Upsilon\bar{g}_{m1} + \Upsilon^2\bar{g}_{m2} + \cdots, \tag{6}$$

where the $\bar{g}_{mi}$, $i = 1, 2, \ldots$, are the correction terms to be determined. The energy function W_m can be expressed similarly,

$$W_m = W_{0m} + \Upsilon\overline{W}_{m1} + \Upsilon^2\overline{W}_{m2} + \cdots. \tag{7}$$

These series [(6) and (7)] will converge for a reasonably small perturbation Υ.

Substituting Eqs. (3), (6), and (7) into Eq. (1) yields for the mth mode

$$(L_0 + \Upsilon L_1 + \Upsilon^2 L_2 + \cdots)[g_{0m} + \Upsilon\bar{g}_{m1} + \Upsilon^2\bar{g}_{m2} + \cdots]$$
$$-(W_{0m} + \Upsilon\overline{W}_{m1} + \Upsilon^2\overline{W}_{m2} + \cdots)[g_{0m} + \Upsilon\bar{g}_{m1} + \Upsilon^2\bar{g}_{m2} + \cdots] = 0,$$

$$(L_0[g_{0m}] - W_{om}g_{0m}) + (L_0[\bar{g}_{m1}] + L_1[g_{0m}] - W_{0m}\bar{g}_{m1} - \overline{W}_{m1}g_{0m})\Upsilon$$
$$+ (L_0[\bar{g}_{m2}] + L_1[\bar{g}_{m1}] + L_2[g_{0m}]$$
$$- W_{0m}\bar{g}_{m2} - W_{m1}\bar{g}_{m1} - \overline{W}_{m2}\bar{g}_{0m})\Upsilon^2 + \cdots = 0. \tag{8}$$

This equation can be satisfied for all values of Υ if the coefficient of each power of Υ vanishes individually. Thus

$$L_0[g_{0m}] - W_{0m}g_{0m} = 0, \tag{9}$$

$$L_0[\bar{g}_{m1}] - W_{0m}\bar{g}_{m1} = -(L_1[g_{0m}] - \overline{W}_{m1}g_{0m}), \tag{10}$$

$$\vdots$$

Equation (9) is identical to Eq. (4). The zero-order term of the solution Eq. (6) for the perturbed system is the solution of the unperturbed system.

To obtain a solution of Eq. (10), $\bar{g}_{m1}$ is expressed in terms of the eigenfunctions for the unperturbed system:

$$\bar{g}_{m1} = a_{m0}g_{00} + a_{m1}g_{01} + \cdots + a_{mk}g_{0k} + \cdots = \sum_{i=0}^{\infty} a_{mi}g_{0i}. \tag{11}$$

Substituting Eq. (11) into Eq. (10) yields

$$L_0\left[\sum_{i=1}^{\infty} a_{mi}g_{0i}\right] - \sum_{i=0}^{\infty} a_{mi}W_{0m}[g_{0i}] = \overline{W}_{mi}g_{0m} - L_1[g_{0m}], \tag{12}$$

$$L_0\left[\sum_{i=0}^{\infty} a_{mi}g_{0i}\right] = \sum_{i=0}^{\infty} a_{mi}(W_{0i}g_{0i}) \tag{13}$$

by virtue of Eq. (9). Substituting Eq. (13) into Eq. (12) yields

$$\sum_{i=0}^{\infty} a_{mi}(W_{0i} - W_{0m})g_{0i} = \overline{W}_{mi}g_{0m} - L_1[g_{0m}]. \tag{14}$$

Multiplying Eq. (14) by g_{0m}^* and integrating over the entire region V produces

$$\int_V g_{0m}^* \sum_{i=0}^{\infty} a_{mi}(W_{0i} - W_{0m})g_{0i}\, d\tau = \int_V g_{0m}^*(\overline{W}_{mi}g_{0m} - L_1[g_{0m}])\, d\tau. \tag{15}$$

The left-hand side of Eq. (15) vanishes for all values of i, since

$$\int_V g_{0m}^* \sum_{i=0}^{\infty} a_{mi}(W_{0i} - W_{0m})g_{0i}\, d\tau = \sum_{i=0}^{\infty} a_{mi}(W_{0i} - W_{0m})\int_V g_{0m}^* g_{0i}\, d\tau;$$

$$\begin{aligned} \int_V g_{0m}^* g_{0i}\, d\tau &= 0 \quad &&\text{for} \quad m \neq i,\\ &= 1 \quad &&\text{for} \quad m = i;\\ W_{0i} - W_{0m} &= 0 \quad &&\text{for} \quad m = i,\\ &\neq 0 \quad &&\text{for} \quad m \neq i. \end{aligned}$$

Therefore,

$$\int_V g_{0m}^* \overline{W}_{mi} g_{0m}\, d\tau = \int_V g_{0m}^* L_1[g_{0m}]\, d\tau$$

or

$$\Upsilon \overline{W}_{mi} = \Upsilon \int_V g_{0m}^* L_1[g_{0m}]\, d\tau. \tag{16}$$

The integral on the left-hand side can be carried out directly since $\overline{W}_{mi}$ is a constant. The parameter of perturbation is included in Eq. (16) for convenience so that all first-order corrections have the common factor Υ.

Summary

In first-order perturbation theory

$$L = L_0 + \Upsilon L_1,$$

$$g_m(\mathbf{r}) = g_{0m} + \Upsilon \bar{g}_{m1}, \qquad W_m = W_{0m} + \Upsilon \overline{W}_{m1},$$

$$\overline{W}_{m1} = \int_V g_{0m}^* L_1[g_{0m}]\, d\tau, \qquad \bar{g}_{m1} = \sum_{i=0}^{\infty} a_{mi} g_{0i}.$$

The coefficient a_{mn} can be evaluated from Eq. (14) by multiplying it by $g_{0n}{}^*$, $n \neq m$, and integrating over all space:

$$\int_V g_{0n}{}^* \sum_{i=0}^{\infty} a_{mi}(W_{0i} - W_{0m}) g_{0i}\, d\tau = \int_V g_{0n}{}^* \left(\overline{W}_{m1} g_{0m} - L[g_{0m}] \right) d\tau,$$

$$\sum_{i=0}^{\infty} a_{mi}(W_{0i} - W_{0m}) \int_V g_{0n}{}^* g_{0i}\, d\tau = \overline{W}_{m1} \int_V g_{0n}{}^* g_{0m}\, d\tau - \int_V g_{0n}{}^* L[g_{0m}]\, d\tau,$$

$$a_{mn}(W_{0n} - W_{0m}) = -\int_V g_{0n}{}^* L[g_{0m}]\, d\tau,$$

$$a_{mn} = -\int_V g_{0n}{}^* L[g_{0m}]\, d\tau / (W_{0n} - W_{0m}), \qquad n \neq m. \tag{17}$$

The value of a_{nn} is not given by this process and is chosen so as to normalize the function $g(\mathbf{r})$. If only first-order terms are considered, then $a_{nn} = 0$.

9.3. WKB Method

A number of physical problems are described by second-order differential equations with varying coefficients. If these coefficients change slowly about a large mean value, the *WKB method* [3, 4, 8–10, 14] may be used to obtain an approximate solution. (WKB abbreviates Wentzel, Kramers, and Brillouin.)

Consider a system described by a second-order equation of the form

$$\frac{d^2 r(s)}{ds^2} + P^2(s) r(s) = 0, \tag{1}$$

where $P^2(s)$ is the varying coefficient. If $P^2(s)$ is a constant, the solution is known to have the form $e^{\pm jPs}$. If $P^2(s)$ is not constant but nearly so, the solution should not be greatly different. Let the trial solution be

$$r(s) = R(s) e^{jQ(s)}. \tag{2}$$

Then

$$\frac{dr}{ds} = \dot{R} e^{jQ} + R j \dot{Q} e^{jQ} = (\dot{R} + jR\dot{Q}) e^{jQ}, \tag{3}$$

$$\frac{d^2 r}{ds^2} = \left[(\ddot{R} - R\dot{Q}^2) + j(2\dot{R}\dot{Q} + R\ddot{Q}) \right] e^{jQ}. \tag{4}$$

Substituting Eqs. (2) and (4) into Eq. (1) yields

$$\left[(\ddot{R} - R\dot{Q}^2) + j(2\dot{R}\dot{Q} + R\ddot{Q}) \right] e^{jQ} + P^2 R e^{jQ} = 0,$$

$$(\ddot{R} - R\dot{Q}^2 + P^2 R) + j(2\dot{R}\dot{Q} + R\ddot{Q}) = 0. \tag{5}$$

This is a complex equation, and its real and imaginary parts should vanish independently; that is,

$$\ddot{R}-R\dot{Q}^2+P^2R=0, \tag{6}$$

$$2\dot{R}\dot{Q}+R\ddot{Q}=0. \tag{7}$$

Since $R(s)$ is almost constant if $P^2(s)$ is almost constant, the rate of variation of $R(s)$ should be very small. Hence if $\ddot{R}(s)$ is negligible, Eq. (6) becomes

$$P=\dot{Q}, \qquad \text{or} \qquad Q(s)=\int_0^s P(s_1)\,ds_1. \tag{8}$$

This implies that the phase coefficient is given by the mean value of $P(s)$, instead of by $P(s)$ as in the constant-coefficient case.

The magnitude function $R(s)$ can be determined from Eq. (7):

$$2\dot{R}/R+\ddot{Q}/\dot{Q}=0,$$

$$2\ln R+\ln\dot{Q}=\text{const}\equiv K_1,$$

$$\ln R^2\dot{Q}=K_1,$$

$$R^2\dot{Q}=e^{K_1}\equiv K,$$

$$R=\sqrt{K/\dot{Q}}=\sqrt{K/P}\equiv R_0P^{-1/2}, \tag{9}$$

where R_0 is an arbitrary constant. Thus the complete solution is

$$r(s)=R_0P^{-1/2}e^{\pm jQ}, \qquad Q(s)=\int_0^s P(s_1)\,ds_1. \tag{10}$$

The restriction of this solution may be obtained by differentiating Eq. (10) twice with respect to s:

$$\frac{dr}{ds}=R_0\left(-\tfrac{1}{2}P^{-3/2}\dot{P}\pm jP^{-1/2}\dot{Q}\right)e^{\pm jQ}=r(s)\left(-\frac{1}{2}\frac{\dot{P}}{P}\pm jP\right),$$

$$P=\dot{Q};$$

$$\frac{d^2r}{ds^2}=\dot{r}\left(-\frac{1}{2}\frac{\dot{P}}{P}\pm jP\right)+r\left[\frac{1}{2}\left(\frac{\dot{P}}{P}\right)^2-\frac{\ddot{P}}{2P}\pm j\dot{P}\right]$$

$$=r\left[\left(-\frac{1}{2}\frac{\dot{P}}{P}\pm jP\right)^2+\frac{1}{2}\left(\frac{\dot{P}}{P}\right)^2-\frac{\ddot{P}}{2P}\pm j\dot{P}\right]$$

$$=-r(s)\left[P^2-\frac{3}{4}\left(\frac{\dot{P}}{P}\right)^2+\frac{\ddot{P}}{2P}\right]$$

or

$$\frac{d^2r}{ds^2}+\left[P^2+\frac{\ddot{P}}{2P}-\frac{3}{4}\left(\frac{\dot{P}}{P}\right)^2\right]r(s)=0. \tag{11}$$

Equation (11) will reduce to Eq. (1) if the following inequality is satisfied:

$$\frac{\ddot{P}}{2P}-\frac{3}{4}\left(\frac{\dot{P}}{P}\right)^2 \ll P^2$$

or

$$\frac{1}{P^2}\left[\frac{\ddot{P}}{2P}-\frac{3}{4}\left(\frac{\dot{P}}{P}\right)^2\right]\ll 1. \tag{12}$$

For a nonhomogeneous differential equation, the particular integral can be obtained by the method of variation of parameters.

References

1. R. Courant and D. Hilbert, *Methods of Mathematical Physics*, Interscience, New York, 1953.
2. R. H. Dicke and J. P. Wittke, *Introduction to Quantum Mechanics*, Addison-Wesley, Reading, Mass., 1960.
3. W. H. Furry, Two Notes on Phase-Integral Methods, *Phys. Rev.* 71, 360–371 (1947).
4. J. Heading, *An Introduction to Phase-Integral Methods*, Wiley, New York, 1962.
5. J. Irving and N. Mullineux, *Mathematics in Physics and Engineering*, Academic, New York, 1959.
6. K. Kaplan, *Operational Methods for Linear Systems*, Addison-Wesley, Reading, Mass., 1962.
7. H. A. Kramers, *Quantum Mechanics*, Dover, New York, 1964.
8. H. A. Kramers, Wellenmechanik und halbzahlige Quantisierung, *Z. Phys.* 39, 828 (1926).
9. R. E. Langer, On the Connection Formulas and the Solutions of the Wave Equation, *Phys. Rev.* 51, 669–676 (1937).
10. S. C. Miller, Jr., and R. H. Good, Jr., A WBK-Type Approximation to the Schrodinger Equation, *Phys. Rev.* 91, 174–179 (1953).
11. R. M. Morse and H. Feshbach, *Methods of Theoretical Physics*, McGraw-Hill, New York, 1953.
12. J. L. Powell, and B. Crasemann, *Quantum Mechanics*, Addison-Wesley, Reading, Mass., 1961.
13. L. I. Schiff, *Quantum Mechanics*, McGraw-Hill, New York, 1955.
14. G. Wentzel, Eine Verallgemeinerung der Quantenbedingungen fur die Zwecke der Wellenmechanik, *Z. Phys.* 38, 518 (1926).

10

Inhomogeneous Circular Waveguides

Inhomogeneous circular waveguides with radial variation in the dielectric constant will now be investigated. The classical case of square-law media is also examined. The chapter concludes with a discussion of one of the most important characteristics of a dielectric waveguide—the dispersion.

10.1. Radially Inhomogeneous Waveguides

Consider a circular waveguide [1, 3, 4] made of dielectric characterized by

$$\varepsilon(r)=\hat{\varepsilon}[1-\Delta\epsilon(r)], \tag{1}$$

where $\hat{\varepsilon}$ is the permittivity at the center of the guide and $\Delta\epsilon(r)$ is the *parameter of inhomogeneity*, which obeys $\Delta\epsilon(0)=0$, and suppose

$$0<\Delta\epsilon(r)\ll 1 \qquad \text{for} \quad 0<r<a, \tag{2}$$

where a is the radius of the core.

The electromagnetic fields in the guide satisfy the following generalized wave equation [Eqs. (1.3.10) and (1.3.11)]:

$$\nabla^2\mathbf{E}+\nabla\left(\frac{1}{\varepsilon}\nabla\,\varepsilon\cdot\mathbf{E}\right)+\frac{1}{\mu}\nabla\,\mu\times(\nabla\times\mathbf{E})-\gamma^2\mathbf{E}=0, \tag{3}$$

$$\nabla^2\mathbf{H}+\nabla\left(\frac{1}{\mu}\nabla\,\mu\cdot\mathbf{H}\right)+\frac{1}{\varepsilon}\nabla\,\varepsilon\times(\nabla\times\mathbf{H})-\gamma^2\mathbf{H}=0, \tag{4}$$

where $\gamma^2\equiv-\omega^2\mu\varepsilon$. In the present case, $\mu=\mu_0=\text{const}$, and Eqs. (3) and (4) become

$$\nabla^2\mathbf{E}+\nabla\left(\frac{1}{\varepsilon}\nabla\,\varepsilon\cdot\mathbf{E}\right)-\gamma^2\mathbf{E}=0, \tag{5}$$

$$\nabla^2\mathbf{H}+\frac{1}{\varepsilon}\nabla\,\varepsilon\times(\nabla\times\mathbf{H})-\gamma^2\mathbf{H}=0. \tag{6}$$

The z components of above equations are

$$\nabla^2 E_z + \frac{\partial}{\partial z}\left(\frac{1}{\varepsilon}\frac{d\varepsilon}{dr}E_r\right) - \gamma^2 E_z = 0, \tag{7}$$

$$\nabla^2 H_z + \frac{d\varepsilon}{\varepsilon dr}\left(\frac{\partial H_r}{\partial z} - \frac{\partial H_z}{\partial r}\right) - \gamma^2 H_z = 0. \tag{8}$$

Transverse field components can be expressed in terms of axial components. Let

$$\mathbf{E} \equiv \mathbf{E}_{\parallel} + \mathbf{E}_{\perp}, \qquad \mathbf{H} \equiv \mathbf{H}_{\parallel} + \mathbf{H}_{\perp}, \qquad \nabla \equiv \nabla_{\parallel} + \nabla_{\perp}, \tag{9}$$

where $\mathbf{E}_{\parallel}$ represents that portion of $\mathbf{E}$ which is parallel to and $\mathbf{E}_{\perp}$ that portion of $\mathbf{E}$ transverse to the axial direction. Maxwell's equations take the following form:

$$\nabla \times \mathbf{E} = -j\omega\mu_0 \mathbf{H},$$

$$(\nabla_{\parallel} + \nabla_{\perp}) \times (\mathbf{E}_{\parallel} + \mathbf{E}_{\perp}) = -j\omega\mu_0(\mathbf{H}_{\parallel} + \mathbf{H}_{\perp}),$$

$$\nabla_{\parallel} \times \mathbf{E}_{\perp} + \nabla_{\perp} \times \mathbf{E}_{\parallel} + \nabla_{\perp} \times \mathbf{E}_{\perp} = -j\omega\mu_0(\mathbf{H}_{\parallel} + \mathbf{H}_{\perp}),$$

$$\nabla_{\perp} \times \mathbf{E}_{\perp} = -j\omega\mu_0 \mathbf{H}_{\parallel}, \tag{10}$$

$$\nabla_{\parallel} \times \mathbf{E}_{\perp} + \nabla_{\perp} \times \mathbf{E}_{\parallel} = -j\omega\mu_0 \mathbf{H}_{\perp}. \tag{11}$$

Similarly,

$$\nabla \times \mathbf{H} = j\omega\epsilon_0\varepsilon_r \mathbf{E},$$

$$(\nabla_{\parallel} + \nabla_{\perp}) \times (\mathbf{H}_{\parallel} + \mathbf{H}_{\perp}) = j\omega\epsilon_0\varepsilon_r(\mathbf{E}_{\parallel} + \mathbf{E}_{\perp}),$$

$$\nabla_{\perp} \times \mathbf{H}_{\perp} = j\omega\epsilon_0\varepsilon_r \mathbf{E}_{\parallel}, \tag{12}$$

$$\nabla_{\parallel} \times \mathbf{H}_{\perp} + \nabla_{\perp} \times \mathbf{H}_{\parallel} = j\omega\epsilon_0\varepsilon_r \mathbf{E}_{\perp}. \tag{13}$$

$\mathbf{E}_{\perp}$ can be obtained from Eq. (13),

$$\mathbf{E}_{\perp} = \frac{1}{j\omega\epsilon_0\varepsilon_r}(\nabla_{\parallel} \times \mathbf{H}_{\perp} + \nabla_{\perp} \times \mathbf{H}_{\parallel}). \tag{14}$$

Substituting Eq. (14) into Eq. (11) yields

$$\nabla_{\parallel} \times \left[\frac{1}{j\omega\epsilon_0\varepsilon_r}(\nabla_{\parallel} \times \mathbf{H}_{\perp} + \nabla_{\perp} \times \mathbf{H}_{\parallel})\right] + \nabla_{\perp} \times \mathbf{E}_{\parallel} = -j\omega\mu_0 \mathbf{H}_{\perp},$$

$$\nabla_{\parallel}\left(\frac{1}{\varepsilon_r}\right) \times (\nabla_{\parallel} \times \mathbf{H}_{\perp} + \nabla_{\perp} \times \mathbf{H}_{\parallel})$$

$$+ \frac{1}{\varepsilon_r}\left[\nabla_{\parallel} \times (\nabla_{\parallel} \times \mathbf{H}_{\perp}) + \nabla_{\parallel} \times (\nabla_{\perp} \times \mathbf{H}_{\parallel})\right]$$

$$+ j\omega\epsilon_0 \nabla_{\perp} \times \mathbf{E}_{\parallel} = -\gamma_0^2 \mathbf{H}_{\perp} \tag{15}$$

($\gamma_0^2 \equiv -\omega^2\mu_0\epsilon_0$). Since $\varepsilon_r = \varepsilon_r(r)$, $\nabla_\parallel(1/\varepsilon_r)=0$. Therefore,

$$\nabla_\parallel \times(\nabla_\parallel \times \mathbf{H}_\perp) + \nabla_\parallel \times(\nabla_\perp \times \mathbf{H}_\parallel) + j\omega\epsilon_0\varepsilon_r \nabla_\perp \times \mathbf{E}_\parallel = -\gamma_0^2\epsilon_r \mathbf{H}_\perp . \tag{16}$$

$\mathbf{H}_\perp$ is obtained from Eq. (11),

$$\mathbf{H}_\perp = \frac{-1}{j\omega\mu_0}(\nabla_\parallel \times \mathbf{E}_\perp + \nabla_\perp \times \mathbf{E}_\parallel). \tag{17}$$

Eliminating $\mathbf{H}_\perp$ between Eqs. (17) and (13) yields

$$\nabla_\parallel \times \left[\frac{-1}{j\omega\mu_0}(\nabla_\parallel \times \mathbf{E}_\perp + \nabla_\perp \times \mathbf{E}_\parallel)\right] + \nabla_\perp \times \mathbf{H}_\parallel = j\omega\epsilon_0\varepsilon_r \mathbf{E}_\perp ,$$

$$\nabla_\parallel \times(\nabla_\parallel \times \mathbf{E}_\perp) + \nabla_\parallel \times(\nabla_\perp \times \mathbf{E}_\parallel) - j\omega\mu_0 \nabla_\perp \times \mathbf{H}_\parallel = -\gamma_0^2\varepsilon_r \mathbf{E}_\perp . \tag{18}$$

Equations (16) and (18) are identical to Eqs. (8.3.2) and (8.3.1) respectively. Therefore the transverse field components are given by Eqs. (8.3.13)–(8.3.16):

$$E_r = \frac{1}{\gamma_z{}^2 - \gamma^2}\left(-\gamma_z \frac{\partial E_z}{\partial r} + \frac{\omega\mu_0 n}{r} H_z\right), \tag{19}$$

$$E_\phi = \frac{1}{\gamma_z{}^2 - \gamma^2}\left(\frac{-jn\gamma_z}{r} E_z + j\omega\mu_0 \frac{\partial H_z}{\partial r}\right), \tag{20}$$

$$H_r = \frac{1}{\gamma_z{}^2 - \gamma^2}\left(\frac{-\omega\varepsilon n}{r} E_z - \gamma_z \frac{\partial H_z}{\partial r}\right), \tag{21}$$

$$H_\phi = \frac{1}{\gamma_z{}^2 - \gamma^2}\left(-j\omega\varepsilon \frac{\partial E_z}{\partial r} - \frac{jn\gamma_z}{r} H_z\right). \tag{22}$$

In cylindrical coordinates,

$$\nabla^2 U = \frac{\partial^2 U}{\partial r^2} + \frac{1}{r}\frac{\partial U}{\partial r} + \frac{1}{r^2}\frac{\partial^2 U}{\partial \phi^2} + \frac{\partial^2 U}{\partial z^2}. \tag{23}$$

Applying Eq. (23) to Eqs. (7) and (8) yields

$$\frac{d^2 e_z}{dr^2} + \frac{1}{r}\frac{de_z}{dr} - \frac{n^2}{r^2} e_z + \gamma_z{}^2 e_z - \frac{\gamma_z}{\varepsilon}\frac{d\varepsilon}{dr} e_r - \gamma^2 e_z = 0, \tag{24}$$

$$\frac{d^2 h_z}{dr^2} + \frac{1}{r}\frac{dh_z}{dr} - \frac{n^2}{r^2} h_z + \gamma_z{}^2 h_z - \frac{\gamma_z}{\varepsilon}\frac{d\varepsilon}{dr} h_r - \frac{1}{\varepsilon}\frac{d\varepsilon}{dr}\frac{dh_z}{dr} - \gamma^2 h_z = 0, \tag{25}$$

where the fields are assumed to have the form

$$E_z(\mathbf{r}) \equiv e_z(r)e^{jn\phi-\gamma_z z}, \qquad E_r(\mathbf{r}) \equiv e_r(r)e^{jn\phi-\gamma_z z},$$
$$H_z(\mathbf{r}) \equiv h_z(r)e^{jn\phi-\gamma_z z}, \qquad H_r(\mathbf{r}) \equiv h_r(r)e^{jn\phi-\gamma_z z}. \tag{26}$$

Applying Eq. (26) to Eqs. (19) and (21) yields

$$e_r = \frac{1}{\gamma_z^{\,2}-\gamma^2}\left(-\gamma_z\frac{de_z}{dr} + \frac{\omega\mu_0 n}{r}h_z\right), \tag{27}$$

$$h_r = \frac{1}{\gamma_z^{\,2}-\gamma^2}\left(\frac{-\omega\varepsilon n}{r}e_z - \gamma_z\frac{dh_z}{dr}\right). \tag{28}$$

Substituting Eq. (27) into Eq. (24) yields

$$\frac{d^2e_z}{dr^2} + \frac{1}{r}\frac{de_z}{dr} + \left(\gamma_z^{\,2}-\gamma^2-\frac{n^2}{r^2}\right)e_z$$
$$-\frac{\gamma_z}{\varepsilon}\frac{d\varepsilon}{dr}\frac{1}{\gamma_z^{\,2}-\gamma^2}\left(-\gamma_z\frac{de_z}{dr} + \frac{\omega\mu_0 n}{r}h_z\right) = 0,$$

$$\frac{d^2e_z}{dr^2} + \left[\frac{1}{r} + \frac{1}{\varepsilon}\frac{d\varepsilon}{dr}\frac{\gamma_z^{\,2}}{\gamma_z^{\,2}-\gamma^2}\right]\frac{de_z}{dr}$$
$$+\left(\gamma_z^{\,2}-\gamma^2-\frac{n^2}{r^2}\right)e_z = \frac{1}{\varepsilon}\frac{d\varepsilon}{dr}\frac{\gamma_z\omega\mu_0 n}{r(\gamma_z^{\,2}-\gamma^2)}h_z. \tag{29}$$

Substituting Eq. (28) into Eq. (25) yields

$$\frac{d^2h_z}{dr^2} + \frac{1}{r}\frac{dh_z}{dr} + \left(\gamma_z^{\,2}-\gamma^2-\frac{n^2}{r^2}\right)h_z$$
$$-\frac{\gamma_z}{\varepsilon}\frac{d\varepsilon}{dr}\frac{1}{\gamma_z^{\,2}-\gamma^2}\left(\frac{-\omega\varepsilon n}{r}e_z - \gamma_z\frac{dh_z}{dr}\right) - \frac{1}{\varepsilon}\frac{d\varepsilon}{dr}\frac{dh_z}{dr} = 0,$$

$$\frac{d^2h_z}{dr^2} + \left(\frac{1}{r} - \frac{1}{\varepsilon}\frac{d\varepsilon}{dr} + \frac{1}{\varepsilon}\frac{d\varepsilon}{dr}\frac{\gamma_z^{\,2}}{\gamma_z^{\,2}-\gamma^2}\right)\frac{dh_z}{dr}$$
$$+\left(\gamma_z^{\,2}-\gamma^2-\frac{n^2}{r^2}\right)h_z = -\frac{d\varepsilon}{dr}\frac{\gamma_z\omega n}{r(\gamma_z^{\,2}-\gamma^2)}e_z,$$

$$\frac{d^2h_z}{dr^2} + \left(\frac{1}{r} + \frac{1}{\varepsilon}\frac{d\varepsilon}{dr}\frac{\gamma^2}{\gamma_z^{\,2}-\gamma^2}\right)\frac{dh_z}{dr}$$
$$+\left(\gamma_z^{\,2}-\gamma^2-\frac{n^2}{r^2}\right)h_z = -\frac{d\varepsilon}{dr}\frac{\gamma_z\omega n}{r(\gamma_z^{\,2}-\gamma^2)}e_z. \tag{30}$$

The following analysis follows that of Kurtz and Streifer [1]. For a lossless medium

$$\varepsilon(r)=\hat{\epsilon}[1-\Delta(r)]=\epsilon_0\epsilon_r[1-\Delta(r)]; \tag{31}$$

$$\frac{1}{\varepsilon(r)}\frac{d\varepsilon(r)}{dr}=\frac{-\dot{\Delta}}{1-\Delta}; \tag{32}$$

$$\gamma_z=j\beta_z,$$

$$\gamma_z{}^2-\gamma^2=-\beta_z{}^2+\omega^2\mu_0\hat{\epsilon}(1-\Delta)=\omega^2\mu\hat{\epsilon}\left[(1-\beta_z{}^2/\omega^2\mu_0\hat{\epsilon})-\Delta\right]. \tag{33}$$

For a small variation $\Delta(\mathbf{r})\ll 1$, the propagation constant γ_z should not be very different from that for free space, so that

$$\gamma_z{}^2=-\beta_z{}^2\simeq-\omega^2\mu_0\hat{\epsilon}.$$

Let

$$m\equiv 1-\beta_z{}^2/\omega^2\mu_0\hat{\epsilon} \qquad \text{or} \qquad \beta_z=\omega\sqrt{\mu_0\hat{\epsilon}}\,(1-m)^{1/2}. \tag{34}$$

Substituting Eq. (34) into Eq. (33) yields

$$\gamma_z{}^2-\gamma^2=\omega^2\mu_0\hat{\epsilon}(m-\Delta). \tag{35}$$

On substituting Eqs. (32) and (35) into Eqs. (29) and (30), one obtains

$$\ddot{e}_z+\left[\frac{1}{r}+\frac{-\dot{\Delta}}{1-\Delta}\frac{-\beta_z{}^2}{\omega^2\mu\hat{\epsilon}(m-\Delta)}\right]\dot{e}_z$$

$$+\left[\omega^2\mu_0\hat{\epsilon}(m-\Delta)-\frac{n^2}{r^2}\right]e_z=\frac{-\dot{\Delta}}{1-\Delta}\frac{j\beta_z\omega\mu_0 n}{r\omega^2\mu_0\hat{\epsilon}(m-\Delta)}h_z,$$

$$\ddot{e}_z+\left[\frac{1}{r}+\frac{(1-m)\dot{\Delta}}{(1-\Delta)(m-\Delta)}\right]\dot{e}_z$$

$$+\left[\omega^2\mu_0\hat{\epsilon}(m-\Delta)-\frac{n^2}{r^2}\right]e_z=\frac{-j\dot{\Delta}}{1-\Delta}\frac{(1-m)^{1/2}n\mu_0}{r(\mu_0\hat{\epsilon})^{1/2}(m-\Delta)}h_z, \tag{36}$$

$$\ddot{h}_z+\left[\frac{1}{r}-\frac{-\dot{\Delta}}{1-\Delta}\frac{-\omega^2\mu_0\hat{\epsilon}(1-\Delta)}{\omega^2\mu_0\hat{\epsilon}(m-\Delta)}\right]\dot{h}_z$$

$$+\left[\omega^2\mu_0\hat{\epsilon}(m-\Delta)-\frac{n^2}{r^2}\right]h_z=\frac{\hat{\epsilon}\Delta\, j\beta_z\omega n}{r\omega^2\mu_0\hat{\epsilon}(m-\Delta)}e_z,$$

$$\ddot{h}_z+\left(\frac{1}{r}+\frac{\dot{\Delta}}{m-\Delta}\right)\dot{h}_z$$

$$+\left[\omega^2\mu_0\hat{\epsilon}(m-\Delta)-\frac{n^2}{r^2}\right]h_z=j\frac{\hat{\epsilon}\dot{\Delta}(1-m)^{1/2}n}{r(\mu_0\hat{\epsilon})^{1/2}(m-\Delta)}e_z. \tag{37}$$

These are coupled equations. For small variations in permittivity, that is, for $\Delta(r) \ll 1$,

$$1-\Delta \simeq 1$$

and

$$(1-m)/(1-\Delta) \simeq (1-m)(1+\Delta+\cdots)=1-(m-\Delta)-\cdots \simeq 1$$

to a first-order approximation. Then Eqs. (36) and (37) become

$$\ddot{e}_z+\left(\frac{1}{r}+\frac{\dot{\Delta}}{m-\Delta}\right)\dot{e}_z+\left[\omega^2\mu_0\hat{\epsilon}(m-\Delta)-\frac{n^2}{r^2}\right]e_z=M(-j\mu_0 h_z), \quad (38)$$

$$\ddot{h}_z+\left(\frac{1}{r}+\frac{\dot{\Delta}}{m-\Delta}\right)\dot{h}_z+\left[\omega^2\mu_0\hat{\epsilon}(m-\Delta)-\frac{n^2}{r^2}\right]h_z=M(j\hat{\epsilon}e_z), \quad (39)$$

where

$$M \equiv n\dot{\Delta}/r\sqrt{\mu_0\hat{\epsilon}}\,(m-\Delta). \quad (40)$$

Equations (38) and (39) are of the form

$$L_1[e_z]=M_1 h_z, \qquad M_1 \equiv -j\mu_0 M, \quad (41)$$

$$L_1[h_z]=M_2 e_z, \qquad M_2 \equiv j\hat{\epsilon} M, \quad (42)$$

where

$$L_1 \equiv \frac{d^2}{dr^2}+\left(\frac{1}{r}+\frac{\dot{\Delta}}{m-\Delta}\right)\frac{d}{dr}+\left[\omega^2\mu_0\hat{\epsilon}(m-\Delta)-\frac{n^2}{r^2}\right]. \quad (43)$$

It would be simpler to solve these coupled equations if M_1 and M_2 were made equal to each other. This can be accomplished by adjusting the magnitude of the functions e_z and h_z. Let

$$g_e \equiv a e_z \qquad \text{and} \qquad g_h \equiv b h_z, \quad (44)$$

where a and b are the magnitude factors to be determined. It is desired to have

$$L_1[g_e]=N g_h, \quad (45)$$

$$L_1[g_h]=N g_e. \quad (46)$$

Substituting Eq. (44) into Eq. (45) yields

$$L_1[a e_z]=N b h_z \qquad \text{or} \qquad L_1[e_z]=(Nb/a)h_z. \quad (47)$$

Comparison of Eq. (47) with Eq. (41) yields

$$Nb/a=M_1=-j\mu_0 M. \quad (48)$$

Similarly, from Eq. (46),

$$L_1[h_z]=(Na/b)e_z$$

and

$$Na/b=M_2=j\hat{\epsilon}M. \quad (49)$$

Dividing Eq. (48) by Eq. (49) yields

$$b/a = j\sqrt{\mu_0/\hat{\epsilon}}\,. \tag{50}$$

One may therefore set $a=1$ and $b=j\sqrt{\mu_0/\hat{\epsilon}} \equiv j\hat{\eta}$. With this choice

$$g_e = e_z \qquad \text{and} \qquad g_h = j\hat{\eta}h_z. \tag{51}$$

Then, from Eq. (41),

$$L_1[e_z] = L_1[g_e] = M_1\frac{g_h}{b} = \frac{-j\mu_0 M}{b}g_h$$

or

$$L_1[g_e] = Ng_h, \tag{52}$$

where

$$N = \frac{-j\mu_0}{b}M = \frac{-\dot{\Delta}n}{r(m-\Delta)}. \tag{53}$$

Similarly, from Eq. (42)

$$L_1[g_h] = bM_2g_e = Ng_e. \tag{54}$$

Equations (52) and (54) may be further simplified as follows:

$$L[g_e] = g_h, \tag{55}$$

$$L[g_h] = g_e, \tag{56}$$

where

$$L \equiv \frac{L_1}{N} = \frac{-r(m-\Delta)}{\dot{\Delta}n}\left\{\frac{d^2}{dr^2} + \left(\frac{1}{r} + \frac{\dot{\Delta}}{m-\Delta}\right)\frac{d}{dr} + \left[\omega^2\mu_0\hat{\epsilon}(m-\Delta) - \frac{n^2}{r^2}\right]\right\}. \tag{57}$$

Equations (55) and (56) can be expressed

$$L[L[g_e]] = L[g_h] = g_e, \tag{58}$$

$$L[L[g_h]] = L[g_e] = g_h. \tag{59}$$

Any function that satisfies the relation

$$L[g_k] = \pm g_k \tag{60}$$

also satisfies Eqs. (58) and (59); in other words, g_k is a solution of Eqs. (58) and (59).

The solution of Eq. (58) can be written

$$g_e = \sum_{k=1}^{4} A_k g_k, \tag{61}$$

and g_h is given by

$$g_h = L[g_e] = \sum_{k=1}^{4} A_k L[g_k] = \sum_{k=1}^{4} (-1)^{k+1} A_k g_k. \tag{62}$$

The problem now is to solve Eq. (60), which is

$$\frac{d^2 g_k}{dr^2} + \left(\frac{1}{r} + \frac{\dot{\Delta}}{m-\Delta}\right)\frac{dg_k}{dr} + \left[\omega^2 \mu_0 \hat{\epsilon}(m-\Delta) - \frac{n^2}{r^2} \pm \frac{n\dot{\Delta}}{r(m-\Delta)}\right] g_k = 0. \tag{63}$$

In the last term choice of the upper sign correspond to the case of odd k and the lower sign to even k.

Some properties of the solutions g_e and g_h can be easily visualized by rearranging Eqs. (61) and (62) as follows:

$$g_e = (A_1 g_1 + A_2 g_2) + (A_3 g_3 + A_4 g_4) \equiv g_{ea} + g_{eb}, \tag{64}$$

$$g_{ea} \equiv A_1 g_1 + A_2 g_2 \qquad \text{and} \qquad g_{eb} \equiv A_3 g_3 + A_4 g_4, \tag{65}$$

$$g_h = (A_1 g_1 - A_2 g_2) + (A_3 g_3 - A_4 g_4) \equiv g_{ha} + g_{hb}, \tag{66}$$

$$g_{ha} \equiv A_1 g_1 - A_2 g_2 \qquad \text{and} \qquad g_{hb} \equiv A_3 g_3 - A_4 g_4. \tag{67}$$

If the solutions singular at the origin (g_{eb} and g_{hb}) are discarded, then

$$g_e = g_{ea} = A_1 g_1 + A_2 g_2 \qquad \text{and} \qquad g_h = g_{ha} = A_1 g_1 - A_2 g_2$$

by virtue of Eq. (63), $g_1 = g_2$ when $n = 0$, and one may choose either $A_1 = A_2$ or $A_1 = -A_2$. For the choice $A_1 = A_2$, then $g_{ha} = 0$ and the TM modes result. On the other hand, if $A_1 = -A_2$, then $g_{ea} = 0$, yielding the TE modes. The other set of solutions may be treated similarly.

The transverse field components can be obtained by substituting Eq. (51) into Eqs. (19)–(22). To simplify the treatment, only one term of Eqs. (64) and (66) will be used.

$$g_e = A_k g_k \equiv Ag, \qquad g_h = A_k g_k \equiv \pm Ag. \tag{68}$$

Then

$$e_z = g_e = Ag \qquad \text{and} \qquad h_z = \frac{1}{j\hat{\eta}} g_h = \frac{\pm A}{j\hat{\eta}} g \qquad \left(\hat{\eta} = \sqrt{\mu_0/\hat{\epsilon}}\right). \tag{69}$$

From Eq. (19),

$$\begin{aligned} e_r &= \frac{1}{\gamma_z{}^2 - \gamma^2}\left(-\gamma_z \dot{e}_z + \frac{\omega\mu_0 n}{r} h_z\right) \\ &= \frac{A}{\omega^2 \mu \hat{\epsilon}(m-\Delta)}\left(-j\omega\sqrt{\mu_0 \hat{\epsilon}}\, \dot{g} + \frac{\omega\mu_0 n}{r}\,\frac{\pm g}{j(\mu_0/\hat{\epsilon})^{1/2}}\right) \\ &= \frac{-jA}{\omega(\mu_0\hat{\epsilon})^{1/2}(m-\Delta)}\left(\pm\frac{n}{r} g + \dot{g}\right) \\ &= \frac{-jA}{\beta_z(m-\Delta)}\hat{g}, \end{aligned} \tag{70}$$

$$\hat{g} \equiv \dot{g} \pm ng/r. \tag{71}$$

Similarly,

$$e_\phi = \frac{1}{\gamma_z^{\,2} - \gamma^2}\left(\frac{-jn\gamma_z}{r}e_z + j\omega\mu_0 \dot{h}_z\right) = \frac{\pm A}{\beta_z(m-\Delta)}\hat{g}, \tag{72}$$

$$\mathbf{e}_\perp = \hat{\mathbf{r}}e_r + \hat{\boldsymbol{\phi}}e_\phi = \frac{A}{\beta_z(m-\Delta)}(-j\hat{\mathbf{r}} \pm \hat{\boldsymbol{\phi}})\hat{g}. \tag{73}$$

The corresponding transverse components of the magnetic field are given by

$$\mathbf{h}_\perp = \frac{1}{\hat{\eta}}\hat{\mathbf{z}} \times \mathbf{e}_\perp = \frac{A}{\beta_z(m-\Delta)\hat{\eta}}(\mp\hat{\mathbf{r}} - j\hat{\boldsymbol{\phi}})\hat{g}. \tag{74}$$

These components may also be obtained from Eqs. (21) and (22):

$$h_r = \frac{1}{\gamma_z^{\,2} - \gamma^2}\left(\frac{-\omega\varepsilon n}{r}e_z - \gamma_z \dot{h}_z\right) = \frac{\mp A}{\omega\mu_0(m-\Delta)}\hat{g}, \tag{75}$$

$$h_\phi = \frac{1}{\gamma_z^{\,2} - \gamma^2}\left(-j\omega\varepsilon\dot{e}_z - \frac{jn\gamma_z}{r}h_z\right) = \frac{-jA}{\omega\mu_0(m-\Delta)}\hat{g}. \tag{76}$$

Note that each transverse field component is proportional to a scalar function G, where

$$G \equiv \frac{\hat{g}}{1-\Delta/m}, \qquad \hat{g} \equiv \dot{g} \pm ng/r \tag{77}$$

or

$$r\dot{g} \pm ng = (1-\Delta/m)rG. \tag{78}$$

Differentiating Eq. (78) with respect to r, one has

$$r\ddot{g} + (1 \pm n)\dot{g} = \frac{-\dot{\Delta}}{m}rG + \left(1 - \frac{\Delta}{m}\right)(r\dot{G} + G). \tag{79}$$

From Eq. (63),

$$r\ddot{g} = -\left(1 + \frac{r\dot{\Delta}}{m-\Delta}\right)\dot{g} - \left[r\omega^2\mu_0\hat{\epsilon}(m-\Delta) - \frac{n^2}{r} \pm \frac{n\dot{\Delta}}{m-\Delta}\right]g. \tag{80}$$

Then

$$r\ddot{g} + (1 \pm n)\dot{g} = \left(-\frac{r\dot{\Delta}}{m-\Delta} \pm n\right)\dot{g} - \left[r\omega^2\mu_0\hat{\epsilon}(m-\Delta) - \frac{n^2}{r} \pm \frac{n\dot{\Delta}}{m-\Delta}\right]g. \tag{81}$$

From Eq. (77) one has

$$\dot{g} = (1-\Delta/m)G \mp ng/r. \tag{82}$$

Substituting Eq. (82) into Eq. (81) yields

$$r\ddot{g}+(1\pm n)\dot{g}=\left(\frac{-r\dot{\Delta}}{m-\Delta}\pm n\right)\left(\frac{m-\Delta}{m}G\mp\frac{n}{r}g\right)$$

$$-\left[r\omega^2\mu_0\hat{\epsilon}(m-\Delta)-\frac{n^2}{r}\pm\frac{n\dot{\Delta}}{m-\Delta}\right]g$$

$$=\frac{-r\dot{\Delta}}{m}G\pm n\left(1-\frac{\Delta}{m}\right)G\pm\frac{n\dot{\Delta}}{m-\Delta}g-\frac{n^2}{r}g$$

$$-\left[r\omega^2\mu_0\hat{\epsilon}(m-\Delta)-\frac{n^2}{r}\pm\frac{n\dot{\Delta}}{m-\Delta}\right]g$$

$$=\frac{-r\dot{\Delta}}{m}G\pm n\left(1-\frac{\Delta}{m}\right)G-r\omega^2\mu_0\hat{\epsilon}(m-\Delta)g. \tag{83}$$

Substituting Eq. (83) into Eq. (79) gives

$$-r\omega^2\mu_0\hat{\epsilon}(m-\Delta)g=\frac{m-\Delta}{m}r\dot{G}+\frac{m-\Delta}{m}(1\mp n)G,$$

$$g=\frac{-1}{m\omega^2\mu_0\hat{\epsilon}}\left[\dot{G}\mp\frac{(n\mp 1)}{r}G\right]. \tag{84}$$

Differentiating Eq. (84) with respect to r yields

$$\dot{g}=\frac{-1}{m\omega^2\mu_0\hat{\epsilon}}\left(\ddot{G}\mp\frac{n\mp 1}{r}\dot{G}\pm\frac{n\mp 1}{r^2}G\right). \tag{85}$$

Substituting Eqs. (84) and (85) into Eq. (78) produces

$$\left(1-\frac{\Delta}{m}\right)rG=\frac{-r}{m\omega^2\mu_0\hat{\epsilon}}\left(\ddot{G}\mp\frac{n\mp 1}{r}\dot{G}\pm\frac{n\mp 1}{r^2}G\right)$$

$$\pm\frac{-n}{m\omega^2\mu_0\hat{\epsilon}}\left(\dot{G}\mp\frac{n\mp 1}{r}G\right),$$

$$\ddot{G}+\frac{1}{r}\dot{G}[\mp(n\mp 1)\pm n]+\left[\omega^2\mu_0\hat{\epsilon}m\left(1-\frac{\Delta}{m}\right)\pm(n\mp 1)(1\mp n)\frac{1}{r^2}\right]G=0,$$

$$\ddot{G}+\dot{G}/r+\left[\omega^2\mu_0\hat{\epsilon}m(1-\Delta/m)-(n\mp 1)^2/r^2\right]G=0. \tag{86}$$

It has just shown that the scalar function G satisfies a second-order differential equation. This is a simpler equation than Eq. (63) which is satisfied by the function g_k.

10.2. Square-Law Media

Consider a guide [1, 2, 4] made of lossless dielectric characterized by

$$\epsilon(r)=\hat{\epsilon}(1-kr^2), \tag{1}$$

where k is a small constant. The infinite-medium approximation will be used in the following; that is, the beam is assumed to be confined within a region where Eq. (1) is valid. It is then justified to use Eq. (1) for all values of r.

The equation to be solved is either (10.1.54) or (10.1.86). The latter is simpler and will be considered here.

$$\ddot{G}+\frac{1}{r}\dot{G}+\left[\omega^2\mu_0\hat{\epsilon}m\left(1-\frac{\Delta}{m}\right)-\frac{(n\mp1)^2}{r^2}\right]G=0. \tag{2}$$

For the present case of $\Delta\equiv kr^2$,

$$\ddot{G}+\frac{1}{r}\dot{G}+\left[\omega^2\mu_0\hat{\epsilon}m\left(1-\frac{kr^2}{m}\right)-\left(\frac{(n\mp1)}{r}\right)^2\right]G=0. \tag{3}$$

This is a second-order ordinary differential equation with variable coefficients. One of the most general methods for solving ordinary differential equations is to obtain a series solution in powers of the dependent variable. Another method is to transform the given equation into one of the well-known forms.

Equation (3) can be simplified by the transformation

$$G=r^s\mathcal{G}, \tag{4}$$

where s is a parameter to be determined.

$$\begin{aligned}\dot{G}&=r^s\dot{\mathcal{G}}+sr^{s-1}\mathcal{G},\\ \ddot{G}&=r^s\ddot{\mathcal{G}}+2sr^{s-1}\dot{\mathcal{G}}+s(s-1)r^{s-2}\mathcal{G}.\end{aligned} \tag{5}$$

Substituting Eqs. (4) and (5) into Eq. (3) yields

$$\ddot{\mathcal{G}}+\frac{2s+1}{r}\dot{\mathcal{G}}+\left[\omega^2\mu_0\hat{\epsilon}m\left(1-\frac{kr^2}{m}\right)+\frac{s^2-(n\mp1)^2}{r^2}\right]\mathcal{G}=0. \tag{6}$$

The r^{-2} term in the bracket can be eliminated by choosing

$$s^2=(n\mp1)^2 \qquad \text{or} \qquad s=n\mp1. \tag{7}$$

Then

$$\ddot{\mathcal{G}}+\frac{2s+1}{r}\dot{\mathcal{G}}+\omega^2\mu_0\hat{\epsilon}m\left(1-\frac{kr^2}{m}\right)\mathcal{G}=0, \qquad \mathcal{G}\equiv Gr^{-s}, \quad s=n\mp1. \tag{8}$$

This equation can be further simplified by a change of the independent variable. Let

$$\rho=a\left(\sqrt{k/m}\,r\right)^2, \tag{9}$$

$$\begin{aligned}\dot{\mathcal{G}}&=\frac{d\mathcal{G}}{d\rho}\frac{d\rho}{dr}=2a\frac{k}{m}r\mathcal{G}', \qquad \mathcal{G}'\equiv\frac{d\mathcal{G}}{d\rho},\\ \ddot{\mathcal{G}}&=\left(2a\frac{k}{m}r\right)^2\mathcal{G}''+\frac{2ak}{m}\mathcal{G}', \qquad \mathcal{G}''\equiv\frac{d^2\mathcal{G}}{d\rho^2}.\end{aligned} \tag{10}$$

Substituting Eqs. (9) and (10) into Eq. (8) yields

$$\rho\mathcal{G}''+(s+1)\mathcal{G}'+\frac{1}{4}\,\frac{\omega^2\mu_0\hat{\epsilon}m}{a^2(k/m)}(a-\rho)\mathcal{G}=0. \tag{11}$$

It will be convenient to set

$$\frac{\omega^2\mu_0\hat{\epsilon}}{a^2(k/m)}=1 \qquad \text{or} \qquad a^2=\omega^2\mu_0\hat{\epsilon}(m/k) \tag{12}$$

so that

$$\rho\mathcal{G}''+(s+1)\mathcal{G}'+\tfrac{1}{4}(a-\rho)\mathcal{G}=0. \tag{13}$$

The ρ in the last term can be removed by the transformation

$$\mathcal{G}=e^{c\rho}\hat{G}, \tag{14}$$

where c is a parameter to be determined.

$$\begin{aligned}\mathcal{G}'&=ce^{c\rho}\hat{G}+e^{c\rho}\hat{G}'=e^{c\rho}(c\hat{G}+\hat{G}'),\\ \mathcal{G}''&=e^{c\rho}(\hat{G}''+2c\hat{G}'+c^2\hat{G}).\end{aligned} \tag{15}$$

Substituting Eqs. (14) and (15) into Eq. (13) yields

$$\rho\hat{G}''+(2c\rho+s+1)\hat{G}'+\left[\left(c^2-\tfrac{1}{4}\right)\rho+c(s+1)+\tfrac{1}{4}a\right]\hat{G}=0. \tag{16}$$

The term containing ρ will disappear from the bracketed terms if $c=\pm\frac{1}{2}$; then

$$\rho\hat{G}''+(s+1\pm\rho)\hat{G}'+\tfrac{1}{4}[a\pm2(s+1)]\hat{G}=0. \tag{17}$$

For the lower sign this is the *associated Laguerre differential equation*,

$$\rho\hat{G}''+(s+1-\rho)\hat{G}'+\tfrac{1}{4}(a-2s-2)\hat{G}=0. \tag{18}$$

With $p\equiv\frac{1}{4}(a-2s-2)=0,1,2,\ldots,$ the solutions are known as *associated Laguerre polyonomials* $L_p{}^{(k)}(\rho)$,

$$\hat{G}=L_p{}^{(k)}(\rho)=\Gamma(p+k+1)\sum_{m=0}^{n}\frac{(-\rho)^m}{\Gamma(k+m+1)(n-m)!m!}, \tag{19}$$

where $k\equiv s+1$.

The solution of Eq. (2) is then given by

$$\hat{G}=e^{-c\rho}\mathcal{G}=e^{a(\sqrt{k/m}\,r)^2/2}\mathcal{G}=e^{\omega^2\mu_0\hat{\epsilon}r^2/2}Gr^{-s}$$

or

$$G=r^s\exp\left(-\frac{\omega r^2}{2}\sqrt{\frac{\mu_0\hat{\epsilon}k}{m}}\right)\hat{G}. \tag{20}$$

For further detail on this subject, the reader should consult the original study of square-law media, by Kurtz and Streifer [1].

10.3. Dispersion

Especially useful in optical fiber communication are the effects of dispersion. Dispersion can be measured by the broadening of a temporal pulse as it propagates along a length of the guide. If a train of pulses is transmitted along a guiding system and dispersion is appreciable, then the pulses will broaden such that they overlap at the output. Consequently, the information-carrying capacity of a system will be higher the smaller its dispersion.

A pure harmonic wave at the input end of a guide, where $z=0$, is given by

$$U^{\mathrm{h}}(x,y,z=0,t)=\hat{u}(x,y)e^{j\omega_1 t}=\sum_k a_k u_k(x,y)e^{j\omega_1 t}, \tag{1}$$

where the superscript h indicates the function of a harmonic wave of a single frequency ω_1 and the index of summation is $k\equiv k(m,n)$. Where $z\neq 0$, this wave function is modified by a factor of $e^{-j\beta z}$; thus

$$U(x,y,z,t)=\sum_k a_k u_k(x,y)e^{j(\omega_1 t-\beta z)} \tag{2}$$

is a traveling wave propagating in the positive z direction.

The field of an arbitrary temporal pulse type can be similarly expressed. At the input, $z=0$, this is given by

$$U(x,y,z=0,t)=\hat{u}(x,y)g(t), \tag{3}$$

where $g(t)$ is the distribution of the pulse in the time domain. The Fourier components are given by the Fourier transform of $g(t)$,

$$G(\omega)=\mathcal{F}[g(t)]=\frac{1}{\sqrt{2\pi}}\int_{-\infty}^{\infty} g(t)e^{-j\omega t}\,dt, \tag{4}$$

and

$$g(t)=\mathcal{F}^{-1}[G(\omega)]=\frac{1}{\sqrt{2\pi}}\int_{-\infty}^{\infty} G(\omega)e^{j\omega t}\,d\omega. \tag{5}$$

Substituting Eq. (5) into Eq. (3) yields

$$U(x,y,z=0,t)=\frac{1}{\sqrt{2\pi}}\int_{-\infty}^{\infty}\hat{u}(x,y)G(\omega)e^{j\omega t}\,d\omega. \tag{6}$$

The function $\hat{u}(x,y)$ can be expressed in terms of orthogonal mode functions as

$$\hat{u}(x,y)=\sum_k a_k(\omega)u_k(x,y,\omega). \tag{7}$$

Then

$$U(x,y,z=0,t)=\frac{1}{\sqrt{2\pi}}\sum_k\int_{-\infty}^{\infty} a_k(\omega)u_k(x,y,\omega)G(\omega)e^{j\omega t}\,d\omega. \tag{8}$$

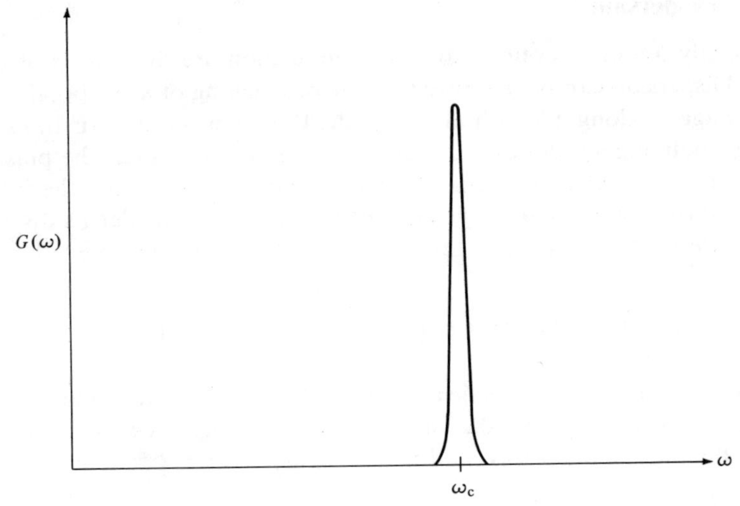

Figure 15. Frequency spectrum of a Gaussian pulse.

At locations other than the input (i.e., for $z \neq 0$), the field is modified by the factor $e^{-j\beta(\omega)z}$:

$$U(x,y,z,t)=\frac{1}{\sqrt{2\pi}}\sum_k \int_{-\infty}^{\infty} a_k(\omega)u_k(x,y,\omega)G(\omega)e^{j[\omega t-\beta(\omega)z]}\,d\omega. \tag{9}$$

To investigate the propagation of a pulse in dispersive medium, consider a temporal Gaussian pulse

$$g(t)=e^{-t^2/2\tau^2+j\omega_c t} \tag{10}$$

where ω_c is the carrier frequency. The Fourier components are then given by

$$\begin{aligned} G(\omega) &= \frac{1}{\sqrt{2\pi}}\int_{-\infty}^{\infty} e^{-t^2/2\tau^2-j(\omega-\omega_c)t}\,dt \\ &= \frac{1}{\sqrt{2\pi}}e^{-\tau^2(\omega-\omega_c)^2/2} \\ &\quad \times \int_{-\infty}^{\infty} e^{-[t+j\tau^2(\omega-\omega_c)]^2/2\tau^2}\,d\left[t+j\tau^2(\omega-\omega_c)\right] \\ &= \tau e^{-\tau^2(\omega-\omega_c)^2/2}. \end{aligned} \tag{11}$$

For a typical nanosecond optical pulse, the product $\omega_c\tau$ is of the order 10^6—that is, $\tau \gg 1/\omega_c$—and the Fourier spectrum $G(\omega)$ is rapidly attenuated as ω is moved away from the value of ω_c (Figure 15).

The functions $a_k(\omega)$ and $u_k(x, y, \omega)$ are rather slowly varying functions of ω as compared to $G(\omega)$. It is therefore permissible to set

$$a_k(\omega)u_k(x, y, z) \simeq a_k(\omega_c)u_k(x, y, \omega_c)$$

and remove them from the integral. Then

$$U(x, y, z, t) = \sum_k a_k(\omega_c)u_k(x, y, \omega_c)\hat{U}(t) \tag{12}$$

and

$$\begin{aligned}\hat{U}(t) &\equiv \frac{1}{\sqrt{2\pi}}\int_{-\infty}^{\infty} G(\omega)e^{j[\omega t-\beta(\omega)z]}\,d\omega \\ &= \frac{\tau}{\sqrt{2\pi}}\int_{-\infty}^{\infty} e^{-\tau^2(\omega-\omega_c)^2/2+j[\omega t-\beta(\omega)z]}\,d\omega.\end{aligned} \tag{13}$$

The function $\beta(\omega)$ can be approximated by the first three terms of its Taylor's expansion.

$$\begin{aligned}\beta(\omega) &= \beta(\omega_c) + (\omega-\omega_c)\frac{\partial\beta}{\partial\omega}\bigg|_{\omega=\omega_c} + \frac{1}{2!}(\omega-\omega_c)^2\frac{\partial^2\beta}{\partial\omega^2}\bigg|_{\omega=\omega_c} + \cdots \\ &\simeq \beta_c + (\omega-\omega_c)\dot{\beta}_c + \frac{1}{2!}(\omega-\omega_c)^2\ddot{\beta}_c,\end{aligned} \tag{14}$$

where

$$\beta_c \equiv \beta(\omega_c), \qquad \dot{\beta}_c \equiv \frac{\partial\beta}{\partial\omega}\bigg|_{\omega=\omega_c}, \qquad \ddot{\beta}_c \equiv \frac{\partial^2\beta}{\partial\omega^2}\bigg|_{\omega=\omega_c}.$$

Substituting Eq. (14) into Eq. (13) yields

$$\begin{aligned}\hat{U}(t) &= \frac{\tau}{\sqrt{2\pi}}e^{j(\omega_c t-\beta_c z)}\int_{-\infty}^{\infty} e^{[-(\omega-\omega_c)^2(\tau^2+j\ddot{\beta}_c z)/2+j(\omega-\omega_c)(t-\dot{\beta}_c z)]}\,d\omega \\ &= \frac{\tau}{\sqrt{2\pi}}\exp\left[j(\omega_c t-\beta_c z) - \frac{1}{2}\frac{\left(t-\dot{\beta}_c z\right)^2}{\tau^2+j\ddot{\beta}_c z}\right] \\ &\quad\times\int_{-\infty}^{\infty}\exp\left\{-\tfrac{1}{2}\left(\tau^2+j\ddot{\beta}_c z\right)\left[(\omega-\omega_c)-j\frac{t-\dot{\beta}_c z}{\tau^2+j\ddot{\beta}_c z}\right]^2\right\} \\ &\quad\times d\left[(\omega-\omega_c)-j\frac{t-\dot{\beta}_c z}{\tau^2+j\ddot{\beta}_c z}\right] \\ &= \left(1+j\frac{\ddot{\beta}_c z}{\tau^2}\right)^{-1/2}\exp\left[-\frac{\left(t-\dot{\beta}_c z\right)^2}{2\left(\tau^2+j\ddot{\beta}_c z\right)} + j(\omega_c t-\beta_c z)\right].\end{aligned} \tag{15}$$

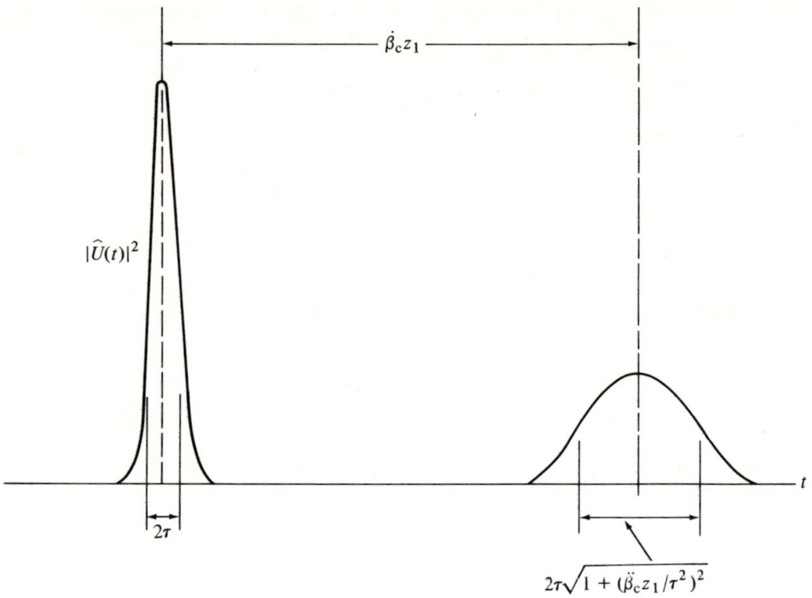

Figure 16. Pulse broadening.

The intensity of the field will be proportional to $|\hat{U}(t)|^2$:

$$|\hat{U}(t)| = \left[\sqrt{1+\left(\frac{\ddot{\beta}_c z}{\tau^2}\right)^2}\,\right]^{-1/2} \exp\left\{-\frac{(t-\dot{\beta}_c z)^2}{2\tau^2\left[1+(\ddot{\beta}_c z/\tau^2)^2\right]}\right\},$$

$$|\hat{U}(t)|^2 = \left[1+\left(\frac{\ddot{\beta}_c z}{\tau^2}\right)^2\right]^{-1/2} \exp\left\{\frac{-(t-\dot{\beta}_c z)^2}{\tau^2\left[1+(\ddot{\beta}_c z/\tau^2)^2\right]}\right\}. \tag{16}$$

At the input $z=0$

$$|\hat{U}(t)|^2|_{z=0} = e^{-t^2/\tau^2}. \tag{17}$$

The maximum intensity occurs at $t=0$ and the pulse width is 2τ (Figure 16). At the output $(z=z_1)$,

$$|\hat{U}(t)|^2|_{z=z_1} = \left[1+\sqrt{1+\left(\frac{\ddot{\beta}_c z_1}{\tau^2}\right)^2}\,\right]^{1/2} \exp\left\{\frac{-(t-\beta_c z_1)^2}{\tau^2\left[1+(\ddot{\beta}_c z_1/\tau^2)^2\right]}\right\}. \tag{18}$$

The maximum intensity occurs at $t=\dot{\beta}_c z_1$ and the pulse width is

$$2\tau\sqrt{1+(\ddot{\beta}_c z_1/\tau^2)^2} \qquad \text{for} \quad (\ddot{\beta}_c z_1/\tau^2)^2 \ll 1.$$

The increase in pulse width is determined by the function

$$\ddot{\beta}_c \equiv \left.\frac{\partial^2\beta}{\partial\omega^2}\right|_{\omega=\omega_c}.$$

References

1. C. N. Kurtz and W. Streifer, Guided Waves in Inhomogeneous Focusing Media, Part I, *IEEE Trans*. MTT-7(1), 11–15 (1969).
2. D. Marcuse, *Theory of Dielectric Optical Waveguides*, Academic, New York, 1974.
3. M. S. Sodha and A. K. Ghatak, *Inhomogeneous Optical Waveguides*, Plenum, New York, 1977.
4. H. G. Unger, *Planar Optical Waveguides and Fibres*, Clarendon, Oxford, 1977.

Appendix 1

Vector Analysis

A1.1. Formulas from Vector Analysis

$$\mathbf{A}\cdot(\mathbf{B}\times\mathbf{C})=\mathbf{B}\cdot(\mathbf{C}\times\mathbf{A})=\mathbf{C}\cdot(\mathbf{A}\times\mathbf{B}), \tag{1}$$

$$\mathbf{A}\times(\mathbf{B}\times\mathbf{C})=\mathbf{B}(\mathbf{A}\cdot\mathbf{C})-\mathbf{C}(\mathbf{A}\cdot\mathbf{B}). \tag{2}$$

$$\nabla(fg)=f\nabla g+g\nabla f, \tag{3}$$

$$\nabla\cdot(f\mathbf{A})=\mathbf{A}\cdot\nabla f+f\nabla\cdot\mathbf{A}, \tag{4}$$

$$\nabla\times(f\mathbf{A})=\nabla f\times\mathbf{A}+f\nabla\times\mathbf{A}; \tag{5}$$

$$\nabla\cdot(\mathbf{A}\times\mathbf{B})=\mathbf{B}\cdot\nabla\times\mathbf{A}-\mathbf{A}\cdot\nabla\times\mathbf{B}, \tag{6}$$

$$\begin{aligned}\nabla\times(\mathbf{A}\times\mathbf{B})&=\nabla_{\mathbf{A}}\times(\mathbf{A}\times\mathbf{B})+\nabla_{\mathbf{B}}\times(\mathbf{A}\times\mathbf{B}),\\ &=[(\mathbf{B}\cdot\nabla)\mathbf{A}-\mathbf{B}(\nabla\cdot\mathbf{A})]+[\mathbf{A}(\nabla\cdot\mathbf{B})-(\mathbf{A}\cdot\nabla)\mathbf{B}].\end{aligned} \tag{7}$$

$$\nabla\times\nabla\times\mathbf{A}=\nabla\nabla\cdot\mathbf{A}-\nabla^2\mathbf{A}, \tag{8}$$

$$\nabla\times\nabla f=0, \tag{9}$$

$$\nabla\cdot(\nabla\times\mathbf{A})=0. \tag{10}$$

$$\int_V \nabla f\,dv=\oint_S f\hat{\mathbf{n}}\,da, \tag{11}$$

$$\int_V \nabla\cdot\mathbf{A}\,dv=\oint_S \mathbf{A}\cdot\hat{\mathbf{n}}\,da, \tag{12}$$

$$\int_V \nabla\times\mathbf{A}\,dv=\oint_S \hat{\mathbf{n}}\times\mathbf{A}\,da, \tag{13}$$

$$\int_S \hat{\mathbf{n}}\times\nabla f\,da=\oint_C f\,d\mathbf{l}. \tag{14}$$

$$\int_S \nabla\times\mathbf{A}\cdot\hat{\mathbf{n}}\,da=\oint_C \mathbf{A}\cdot d\mathbf{l}. \tag{15}$$

A1.2. Green's Theorem

Green's theorem is the result of direct application of the divergence theorem,

$$\int_V \nabla\cdot\mathbf{A}\,dv=\oint_S \mathbf{A}\cdot\hat{\mathbf{n}}\,da. \tag{1}$$

Let $\mathbf{A}\equiv f\nabla g$, where $f\equiv f(\mathbf{r})$ and $g\equiv g(\mathbf{r})$ are arbitrary scalar fields. Now

$$\nabla\cdot\mathbf{A}=\nabla\cdot(f\nabla g)=f\nabla^2 g+\nabla f\cdot\nabla g \tag{2}$$

and

$$\mathbf{A}\cdot\hat{\mathbf{n}}=(f\nabla g)\cdot\hat{\mathbf{n}}=f\frac{\partial g}{\partial n}, \tag{3}$$

This is known as *Green's first identity*.

If **A** is defined with f and g interchanged, as $\mathbf{A}=g\nabla f$, then one obtains

$$\int_V (g\nabla^2 f+\nabla g\cdot\nabla f)\,dv=\oint_S g\frac{\partial f}{\partial n}\,da. \tag{5}$$

This is known as *Green's first identity*.

If **A** is defined with f and g interchanged, as $\mathbf{A}=g\nabla f$, then one obtains

$$\int_V (g\nabla^2 f+\nabla g\cdot\nabla f)\,dv=\oint_S g\frac{\partial f}{\partial n}\,da. \tag{5}$$

Subtracting Eq. (5) from Eq. (4) yields

$$\int_V (f\nabla^2 g-g\nabla^2 f)\,dv=\oint_S\left(f\frac{\partial g}{\partial n}-g\frac{\partial f}{\partial n}\right)da. \tag{6}$$

This is *Green's second identity* or *Green's theorem*.

A1.3. Two-Dimensional Divergence Theorem

Consider a two-dimensional vector $\mathbf{g}^{\mathrm{t}}$ that is a function of variables transverse to the axial direction z:

$$\begin{aligned}\mathbf{g}^{\mathrm{t}}&=\mathbf{g}^{\mathrm{t}}(x,y)\\&=\hat{\mathbf{x}}g_x(x,y)+\hat{y}g_y(x,y)\\&=\hat{\mathbf{x}}g_x+\hat{\mathbf{y}}g_y.\end{aligned} \tag{1}$$

Apply the divergence theorem to a volume V having a uniform cross-sectional area S_0 transverse to the axial direction, so that

$$\int_V \nabla\cdot\mathbf{g}^{\mathrm{t}}\,dv=\oint_S \mathbf{g}^{\mathrm{t}}\cdot\mathbf{n}\,da, \tag{2}$$

where the surface S encloses the volume V. The left side of Eq. (2) is

$$\int_V \nabla \cdot \mathbf{g}^{\mathrm{t}}\, dv = \int_{S_0}\int_{z=0}^{L} \nabla_{\mathrm{t}} \cdot \mathbf{g}^{\mathrm{t}}\, da\, dz \tag{3}$$

$$= L\int_{S_0} \nabla_{\mathrm{t}} \cdot \mathbf{g}^{\mathrm{t}}\, da. \tag{4}$$

The relation $\nabla \cdot \mathbf{g}^{\mathrm{t}} = \nabla_{\mathrm{t}} \cdot \mathbf{g}^{\mathrm{t}}$ (since $\mathbf{g}^{\mathrm{t}}$ is independent of z), is used to obtain Eq. (3), and L is the length of the volume V in the z direction (Figure 17). Equation (4) is possible since the integrand is independent of z. The right side of Eq. (2) is equal to

$$\oint_S \mathbf{g}^{\mathrm{t}} \cdot \hat{\mathbf{n}}\, da = \oint_{S_{\mathrm{t}}+S_z} \mathbf{g}^{\mathrm{t}} \cdot \hat{\mathbf{n}}\, da \tag{5}$$

$$= \oint_{S_{\mathrm{t}}} \mathbf{g}^{\mathrm{t}} \cdot \hat{\mathbf{n}}\, da \tag{6}$$

$$= \oint_C \int_{z=0}^{L} \mathbf{g}^{\mathrm{t}} \cdot \hat{\mathbf{n}}\, dl\, dz \tag{7}$$

$$= L\oint_C \mathbf{g}^{\mathrm{t}} \cdot \hat{\mathbf{n}}\, dl, \tag{8}$$

Figure 17. Volume with uniform cross section.

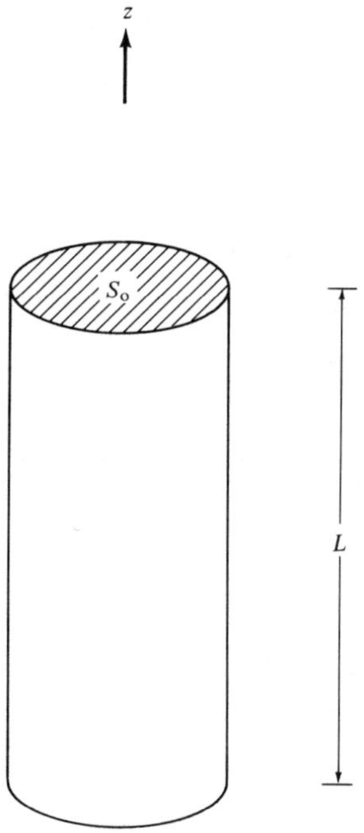

where S_t and S_z are surfaces with normals transverse and parallel to the z axis, respectively, and C is the contour of the cross-sectional area. The surface integral S_z over the end surface vanishes because by definition $\mathbf{g}^t$ has no z component, which leads to Eq. (6). Equation (8) follows since $\mathbf{g}^t$ is independent of z.

Substituting Eqs. (4) and (8) into Eq. (2) yields

$$\int_S \nabla_t \cdot \mathbf{g}^t \, da = \oint_C \mathbf{g}^t \cdot \hat{\mathbf{n}} \, dl. \tag{9}$$

This is known as the *two-dimensional divergence theorem*.

Appendix 2

Delta Function

The *delta function*, or *unit impulse function*, $\delta(x-x_0)$ is defined by the following equations:

$$\begin{aligned} \delta(x-x_0)&=0 \qquad \text{if} \quad x-x_0\neq 0,\\ \delta(x-x_0)&=\infty \qquad \text{if} \quad x-x_0=0;\\ \int_{-\infty}^{\infty}\delta(x-x_0)\,dx&=1, \end{aligned} \tag{1}$$

where x_0 is some reference point. Equation (1) defines $\delta(x)$ as a function which is zero everywhere except at the reference point, where its magnitude is infinite; the value of its integral over all space is unity.

A physical representation of the delta function may be obtained by considering a narrow pulse $f(x)$, as shown in Figure 18. The pulse has a width d and a height $h=1/d$, so that its area is unity. The pulse is symmetrically located about $x=x_0$ and is described by the following relations:

$$\begin{aligned} f(x-x_0)&=0, \qquad |x-x_0|>d/2,\\ f(x-x_0)&=1/d, \qquad |x-x_0|\leq d/2;\\ \int_{-\infty}^{\infty}f(x-x_0)\,dx&=1. \end{aligned} \tag{2}$$

In the limit as the width d of the pulse decreases to zero, its height approaches infinity. The pulse $f(t)$ thus becomes a delta function in the limit as $d\to 0$.

$$\begin{aligned} f(x-x_0)&=0, \qquad |x-x_0|>0,\\ f(x-x_0)&=\infty, \qquad |x-x_0|=0;\\ \int_{-\infty}^{\infty}f(x-x_0)\,dx&=1. \end{aligned} \tag{3}$$

Equation (3) is identical to the definition of a delta function.

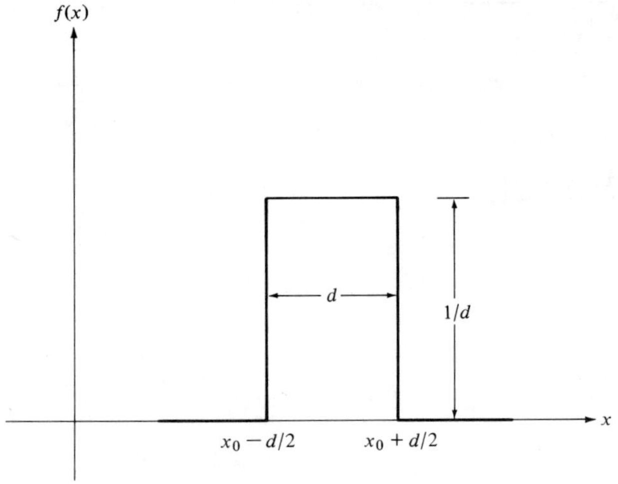

Figure 18. Rectangular pulse.

The Fourier transform of $f(x)$ is

$$\mathcal{F}[f(x)] \equiv F(j\omega) = \int_{-\infty}^{\infty} f(x) e^{-j\omega x}\, dx$$

$$= \int_{x_0 - d/2}^{x_0 + d/2} \frac{1}{d} e^{-j\omega x}\, dx$$

$$= e^{-j\omega x_0} \frac{\sin \omega d/2}{\omega d/2}. \tag{4}$$

Since

$$\lim_{d \to 0} f(x - x_0) = \delta(x - x_0), \tag{5}$$

it follows that

$$\Delta(j\omega) \equiv \mathcal{F}[\delta(x - x_0)] = \lim_{d \to 0} F(j\omega) = e^{-j\omega x_0}. \tag{6}$$

The inverse Fourier transform of $\Delta(j\omega)$ is

$$\delta(x - x_0) = \mathcal{F}^{-1}[\Delta(j\omega)] = \frac{1}{2\pi} \int_{-\infty}^{\infty} \Delta(j\omega) e^{j\omega x}\, d\omega$$

$$= \frac{1}{2\pi} \int_{-\infty}^{\infty} e^{j\omega(x - x_0)}\, d\omega$$

or

$$\int_{-\infty}^{\infty} e^{j\omega(x - x_0)}\, d\omega = 2\pi\delta(x - x_0). \tag{7}$$

A change of variable in Eq. (7), let $\omega=-\omega'$, yields

$$2\pi\delta(x-x_0)=\int_{\infty}^{-\infty} e^{-j\omega'(x-x_0)}(-d\omega)'$$
$$=\int_{-\infty}^{\infty} e^{-j\omega'(x-x_0)}\,d\omega'$$
$$=\int_{-\infty}^{\infty} e^{-j\omega(x-x_0)}\,d\omega. \tag{8}$$

The change of notation in the last form is permissible since the integral is independent of the dummy variable ω'.

The sum of Eqs. (7) and (8) yields

$$\int_{-\infty}^{\infty}\cos\omega(x-x_0)\,d\omega=2\pi\delta(x-x_0). \tag{9}$$

Another useful relation can be obtained as follows.

$$\int_0^{\infty}\cos\omega x\cos\Omega x\,dx=\int_0^{\infty}\tfrac{1}{4}(e^{j\omega x}+e^{-j\omega x})(e^{j\Omega x}+e^{-j\Omega x})\,dx$$
$$=\tfrac{1}{4}\int_{-\infty}^{\infty}\cos(\omega+\Omega)x\,dx+\tfrac{1}{4}\int_{-\infty}^{\infty}\cos(\omega-\Omega)x\,dx$$
$$=\tfrac{1}{4}[2\pi\delta(\omega+\Omega)+2\pi\delta(\omega-\Omega)]$$
$$=(\pi/2)\delta(\omega-\Omega)\qquad\text{for}\quad\omega\geq 0\quad\text{and}\quad\Omega>0. \tag{10}$$

Equation (9) is used to obtain the last expression, and $\delta(\omega+\Omega)=0$ since $\omega+\Omega\neq 0$.

Since $\sin\theta=\cos(\theta-\pi/2)$, one has

$$\int_0^{\infty}\sin\omega x\sin\Omega x\,dx=\int_0^{\infty}\cos(\omega x-\pi/2)\cos(\Omega x-\pi/2)\,dx$$
$$=(\pi/2)\delta(\omega-\Omega). \tag{11}$$

Appendix 3

Expansion of an Arbitrary Function in Eigenfunctions

A3.1. Linear Vector Space

It will now be convenient to consider functions defined in an abstract function space, known as a *linear vector space*. As any vector in physical space can be represented by a linear combination of mutually orthogonal components, any function in this suitably defined function space can be expressed in terms of independent *orthogonal functions*. It will be seen that the number of these orthogonal functions is unbounded.

A3.2. Orthogonality

The concept of orthogonality originated in vector analysis. Two vectors **A** and **B** in three-dimensional space are said to be orthogonal if their scalar product vanishes.

$$\mathbf{A} \cdot \mathbf{B} = \sum_{k=1}^{3} A_k B_k = 0. \tag{1}$$

In Cartesian coordinates, this is

$$\mathbf{A} \cdot \mathbf{B} = A_x B_x + A_y B_y + A_z B_z = 0. \tag{2}$$

Similarly, vectors in n dimensions having components A_k and B_k, respectfully, $k = 1, 2, \ldots, n$, are said to be orthogonal if

$$\mathbf{A} \cdot \mathbf{B} = \sum_{k=1}^{n} A_k B_k = 0. \tag{3}$$

This can be extended to an infinite-dimensional space by letting $n \to \infty$. The components A_k and B_k then become continuously distributed quantities and the index, rather than a discrete quantity k, becomes a continuous variable, say x. The scalar product then becomes an integral,

$$\lim_{n\to\infty} \sum_{k=1}^{n} A_k B_k = \int_0^\infty A(x)B(x)\,dx. \tag{4}$$

In linear vector space, the scalar product of two functions $f(\mathbf{r})$ and $g(\mathbf{r})$ with respect to a positive weight function $W(\mathbf{r})$ is given by

$$f(\mathbf{r})\cdot g(\mathbf{r}) \equiv \int_V f(\mathbf{r})g(\mathbf{r})W(\mathbf{r})\,dv. \tag{5}$$

A3.3. Cauchy–Schwarz Inequality

Consider a vector $\mathbf{c}$ in the spatial domain,

$$\mathbf{c} \equiv \mathbf{a} - k\mathbf{b}, \tag{1}$$

where k is a positive real constant and $\mathbf{a}$ and $\mathbf{b}$ are unspecified vectors. The scalar product

$$\mathbf{c}\cdot\mathbf{c} = (\mathbf{a}-k\mathbf{b})\cdot(\mathbf{a}-k\mathbf{b}) = \mathbf{a}\cdot\mathbf{a} - 2\mathbf{a}\cdot\mathbf{b}k + \mathbf{b}\cdot\mathbf{b}k^2$$

or

$$(\mathbf{b}\cdot\mathbf{b})k^2 - (2\mathbf{a}\cdot\mathbf{b})k + \mathbf{a}\cdot\mathbf{a} - \mathbf{c}\cdot\mathbf{c} = 0. \tag{2}$$

This is a quadratic equation in k with real coefficients. Since k is single valued and, by definition, positive real, the solution of Eq. (2) must be a double root. In other words, the quadratic equation should have a vanishing discriminant.

$$(2\mathbf{a}\cdot\mathbf{b})^2 - 4(\mathbf{b}\cdot\mathbf{b})(\mathbf{a}\cdot\mathbf{a} - \mathbf{c}\cdot\mathbf{c}) = 0,$$

$$(\mathbf{a}\cdot\mathbf{b})^2 - (\mathbf{b}\cdot\mathbf{b})(\mathbf{a}\cdot\mathbf{a}) + (\mathbf{b}\cdot\mathbf{b})(\mathbf{c}\cdot\mathbf{c}) = 0,$$

$$(\mathbf{a}\cdot\mathbf{b})^2 - (\mathbf{b}\cdot\mathbf{b})(\mathbf{a}\cdot\mathbf{a}) \le 0,$$

or

$$(\mathbf{a}\cdot\mathbf{b})^2 \le (\mathbf{a}\cdot\mathbf{a})(\mathbf{b}\cdot\mathbf{b}). \tag{3}$$

This relation is known as Cauchy–Schwarz inequality.

The Cauchy–Schwarz inequality for functions in linear vector space takes the following form.

$$\left[\int_V f(\mathbf{r})g(\mathbf{r})W(\mathbf{r})\,dv\right]^2 \le \left[\int_V f^2(\mathbf{r})W(\mathbf{r})\,dv\right]\left[\int_V g^2(\mathbf{r})W(\mathbf{r})\,dv\right]. \tag{4}$$

A3.4. Orthogonal Expansions

The claim of Section A3.1 can now be stated more fully. Any sufficiently smooth function can be expanded in an infinite series whose terms are constant multiples of an orthogonal sequence, where "sufficiently smooth" is taken to mean square integrable. A function $f(\mathbf{r})$ is *square integrable relative to a given weight function* $W(\mathbf{r})>0$ in the region $\mathbf{a}<\mathbf{r}<\mathbf{b}$ if

$$\int_{\mathbf{a}}^{\mathbf{b}} f^2(\mathbf{r})W(\mathbf{r})\,dx_1\cdots dx_n=\int_V f^2(\mathbf{r})W(\mathbf{r})\,dv<+\infty, \tag{1}$$

where $\mathbf{r}=\hat{\mathbf{x}}_1x_1+\hat{\mathbf{x}}_2x_2+\cdots+\hat{\mathbf{x}}_nx_n=\sum_{k=1}^{n}\hat{\mathbf{x}}_kx_k$ is a position vector in n-dimensional function space, and $\hat{\mathbf{x}}_k$ is the unit vector in the x_k coordinate. When the weight function $W(\mathbf{r})$ is unity, $W(\mathbf{r})=1$, then one states simply that the function $f(\mathbf{r})$ is *square integrable* in the region $\mathbf{a}<\mathbf{r}<\mathbf{b}$.

Two functions $g_j(\mathbf{r})$ and $g_k(\mathbf{r})$ are *orthogonal with respect to a positive weight function* $W(\mathbf{r})$ in the region $\mathbf{a}<\mathbf{r}<\mathbf{b}$ if

$$\int_V g_j(\mathbf{r})g_k(\mathbf{r})W(\mathbf{r})\,dv=\int_{\mathbf{a}}^{\mathbf{b}} g_j(\mathbf{r})g_k(\mathbf{r})W(\mathbf{r})\,dv=0, \qquad j\neq k, \tag{2}$$

where $dv\equiv dx_1\,dx_2\cdots dx_n$ in linear vector space (or $dv=dx\,dy\,dz$ in Cartesian coordinates). If $W(\mathbf{r})=1$, then one has simply that $g_j(\mathbf{r})$ and $g_k(\mathbf{r})$ are *orthogonal* in the region $\mathbf{a}<\mathbf{r}<\mathbf{b}$.

Let a sequence $\{g_m(\mathbf{r})\}$ consist of bounded functions $g_1(\mathbf{r}), g_2(\mathbf{r}),\ldots, g_n(\mathbf{r})$ in the region $\mathbf{a}<\mathbf{r}<\mathbf{b}$, that are square integrable and orthogonal to each other with respect to a positive weight function $W(\mathbf{r})$. [One writes $g(\mathbf{r})\equiv g(x_1,\ldots,x_n)$ in linear vector space or $g(\mathbf{r})=g(x,y,z)$ in Cartesian coordinates.]

A series

$$S_n(\mathbf{r})\equiv s_1(\mathbf{r})+s_2(\mathbf{r})+\cdots+s_n(\mathbf{r})=\sum_{k=1}^{n} s_k(\mathbf{r}) \tag{3}$$

is said to *converge* to the sum $S(\mathbf{r})$ in a region $\mathbf{a}<\mathbf{r}<\mathbf{b}$ for all values of $\mathbf{r}$ within this region if

$$S(\mathbf{r})=\lim_{n\to\infty}\sum_{k=1}^{n} s_k(\mathbf{r}) \tag{4}$$

or

$$\lim_{n\to\infty}\left[S(\mathbf{r})-\sum_{k=1}^{n} s_k(\mathbf{r})\right]=0. \tag{5}$$

This implies that the remainder of a convergent series should vanish as n becomes infinite.

The series $S_n(\mathbf{r})$ is said to *converge uniformly* to the function $S(\mathbf{r})$ in the region $\mathbf{a}<\mathbf{r}<\mathbf{b}$ if, corresponding to an arbitrary quantity $\epsilon>0$, there exists a positive integer N depending on ϵ but not on $\mathbf{r}$ such that, for every value of $\mathbf{r}$ within the interval,

$$|S(\mathbf{r})-S_n(\mathbf{r})|<\epsilon \qquad \text{for all} \quad n>N. \tag{6}$$

A power series is uniformly convergent for all points on and within the sphere of convergence. The sphere of convergence of a power series is the largest sphere about the origin such that the series converges at each point on and within the sphere.

Assume that a given function $f(\mathbf{r})$ can be expressed as the limit of a uniformly convergent series in n orthogonal functions belonging to the sequence $\{g_m(r)\}$.

$$f(\mathbf{r})=c_1g_1(\mathbf{r})+\cdots+c_ng_n(\mathbf{r})=\sum_{k=1}^{n}c_kg_k(\mathbf{r}). \tag{7}$$

Multiply Eq. (7) by $g_j(\mathbf{r})W(\mathbf{r})$ and integrate over the region $\mathbf{a}<\mathbf{r}<\mathbf{b}$ term by term, which is permissible for uniformly convergent series.

$$\int_V f(\mathbf{r})g_j(\mathbf{r})W(\mathbf{r})\,dv=\int_V g_j(\mathbf{r})W(\mathbf{r})\sum_{K=1}^{n}c_kg_k(\mathbf{r})\,dv$$

$$=c_j\int_V g_j{}^2(\mathbf{r})W(\mathbf{r})\,dv. \tag{8}$$

The orthogonality property, Eq. (2), has been used to obtain Eq. (8). The coefficients c_k of the series can be evaluated as

$$c_k=\int_V f(\mathbf{r})g_k(\mathbf{r})W(\mathbf{r})\,dv\Big/\int_V g_k{}^2(\mathbf{r})W(\mathbf{r})\,dv. \tag{9}$$

The c_k are known as the generalized Fourier coefficients of the function $f(\mathbf{r})$ relative to the orthogonal sequence $\{g_m(\mathbf{r})\}$

The coefficient c_k exists if

$$\int_V g_k{}^2(\mathbf{r})W(\mathbf{r})\,dv<+\infty, \tag{10}$$

which is the defining equation for a square integrable function, Eq. (A3.1.1). The function $g_k(\mathbf{r})$ is known to be orthonormal with respect to the weight function $W(\mathbf{r})$ if

$$\int_V g_k{}^2(\mathbf{r})W(\mathbf{r})\,dv=1. \tag{11}$$

Thus, for orthonormal functions, the coefficients of expansion are given by

$$c_k=\int_V f(\mathbf{r})g_k(\mathbf{r})W(\mathbf{r})\,dv. \tag{12}$$

It has been shown that an arbitrary function $f(\mathbf{r})$ can be expressed as a uniformly convergent series, Eq. (7), in terms of bounded square-integrable functions $g_k(\mathbf{r})$ that satisfy the orthogonality relation Eq. (2). The coefficients of expansion c_k are given by Eq. (9).

A3.5. Mean-Square Approximation

In the preceding uniformly convergent series were chosen simply because these can be integrated term by term. However, the concept of convergence most suitable for orthogonal expansions is not uniform convergence but mean-square convergence.

Consider an arbitrary function $h(\mathbf{r})$ and the sequence $\{h_n(\mathbf{r})\}$ whose elements are defined by

$$h_n(\mathbf{r}) \equiv A_1 g_1(\mathbf{r}) + A_2 g_2(\mathbf{r}) + \cdots + A_n g_n(\mathbf{r}) = \sum_{k=1}^{n} A_k g_k(\mathbf{r}), \tag{1}$$

where the $g_k(\mathbf{r})$ are orthogonal functions. The functions $h(\mathbf{r})$ and $g_k(\mathbf{r})$ are assumed to be square integrable. The sequence $\{h_n(\mathbf{r})\}$ converges in the mean to the function $h(\mathbf{r})$ with respect to the positive weight function $W(\mathbf{r})$ in the region $\mathbf{a}<\mathbf{r}<\mathbf{b}$ if

$$\lim_{n\to\infty} \int_V [h(\mathbf{r}) - h_n(\mathbf{r})]^2 W(\mathbf{r})\, dv \to 0. \tag{2}$$

The integral of Eq. (2) may be interpreted as the mean-square error [with respect to $W(\mathbf{r})$] of the representation of $h(\mathbf{r})$ by the partial sum $h_n(\mathbf{r})$. In this expression the *error* is

$$\mathcal{E}(r) = h(\mathbf{r}) - h_n(\mathbf{r}) = h(\mathbf{r}) - \sum_{k=1}^{n} A_k g_k(\mathbf{r}). \tag{3}$$

and the *mean-square error* is

$$\begin{aligned}
\mathcal{E}_n(A_k) &\equiv \int_V [h(\mathbf{r}) - h_n(\mathbf{r})]^2 W(\mathbf{r})\, dv \\
&= \int_V [h^2(\mathbf{r}) - 2h(\mathbf{r})h_n(\mathbf{r}) + h_n{}^2(\mathbf{r})] W(\mathbf{r})\, dv \\
&= \int_V h^2(\mathbf{r}) W(\mathbf{r})\, dv - 2\int_V h(\mathbf{r}) W(\mathbf{r}) \sum_{k=1}^{n} A_k g_k(\mathbf{r})\, dv \\
&\quad + \int_V \left[\sum_{k=1}^{n} A_k g_k(\mathbf{r}) \right]^2 W(\mathbf{r})\, dv \\
&= \int_V h^2(\mathbf{r}) W(\mathbf{r})\, dv - 2 \sum_{k=1}^{n} A_k \int_V h(\mathbf{r}) g_k(\mathbf{r}) W(\mathbf{r})\, dv \\
&\quad + \sum_{k=1}^{n} A_k{}^2 \int_V g_k{}^2(\mathbf{r}) W(\mathbf{r})\, dv.
\end{aligned} \tag{4}$$

The orthogonality property is used to obtain the final form. The mean-square error may be minimized by the best choice of the coefficients A_k

that satisfy the following condition:

$$\frac{\partial \mathcal{E}_n(A_k)}{\partial A_k} = 0 = -2\int_V h(\mathbf{r})g_k(\mathbf{r})W(\mathbf{r})\,dv + 2A_k\int_V g_k{}^2(\mathbf{r})W(\mathbf{r})\,dv$$

or

$$A_k = \int_V h(\mathbf{r})g_k(\mathbf{r})W(\mathbf{r})\,dv \Big/ \int_V g_k{}^2(\mathbf{r})W(\mathbf{r})\,dv. \tag{5}$$

The coefficient A_k as given by Eq. (5) is identical to Eq. (A3.4.9) for the coefficient c_k but in a different notation. In other words, the expansion coefficients c_k for an arbitrary function $h(\mathbf{r})$ in terms of a uniformly convergent series of orthogonal functions $g_k(\mathbf{r})$ will minimize the mean-square error of approximating the same function $h(\mathbf{r})$ by a sequence $h_n(\mathbf{r})$ [Eq. (1)] that converges in the mean to $h(\mathbf{r})$. This can also be verified by direct substitution of c_k, from Eq. (A3.4.9), into Eq. (4), that is,

$$c_k = \int_V h_k(\mathbf{r})g_k(\mathbf{r})W(\mathbf{r})\,dv \Big/ \int_V g_k{}^2(\mathbf{r})W(\mathbf{r})\,dv$$

or

$$\int_V h(\mathbf{r})g_k(\mathbf{r})W(\mathbf{r})\,dv = c_k\int_V g_k{}^2(\mathbf{r})W(\mathbf{r})\,dv. \tag{6}$$

Substituting Eq. (6) into Eq. (4) yields

$$\begin{aligned}
\mathcal{E}_n(A_k) &= \int_V h^2(\mathbf{r})W(\mathbf{r})\,dv - 2\sum_{k=1}^{n} A_k c_k \int_V g_k{}^2(\mathbf{r})W(\mathbf{r})\,dv \\
&\quad + \sum_{k=1}^{n} A_k{}^2 \int_V g_k{}^2(\mathbf{r})W(\mathbf{r})\,dv \\
&\quad + \left[\sum_{k=1}^{n} c_k{}^2 \int_V g_k{}^2(\mathbf{r})W(\mathbf{r})\,dv - \sum_{k=1}^{n} c_k{}^2 \int_V g_k{}^2(\mathbf{r})W(\mathbf{r})\,dv\right] \\
&= \int_V h^2(\mathbf{r})W(\mathbf{r})\,dv + \sum_{k=1}^{n}\left[(A_k - c_k)^2 - c_k{}^2\right]\int_V g_k{}^2(\mathbf{r})W(\mathbf{r})\,dv.
\end{aligned} \tag{7}$$

The last term indicates that the mean-square error will be a minimum when $A_k = c_k$ for any region.

The best mean-square approximation of $h(\mathbf{r})$ by the partial sum $h_n(\mathbf{r}) \equiv \sum_{k=1}^{n} A_k g_k(\mathbf{r})$ is known as the *least-square approximation* to $h(\mathbf{r})$; the least-square approximation is obtained when the mean-square error of the approximation, Eq. (4), is minimized. The coefficient of expansion A_k for the least-square approximation of $h(\mathbf{r})$ is the same for all $n \geq k$. The error of the least-square approximation is given by Eq.(7) with A_k set equal to

c_k; that is,

$$\mathcal{E}_n(A_k)=\int_V h^2(\mathbf{r})W(\mathbf{r})\,dv-\sum_{k=1}^{n} c_k{}^2\int_V g_k{}^2(\mathbf{r})W(\mathbf{r})\,dv. \tag{8}$$

The mean-square error is also given by

$$\begin{aligned}\mathcal{E}_n(A_k)&=\int_V [h(\mathbf{r})-h_n(\mathbf{r})]^2W(\mathbf{r})\,dv\\ &=\int_V\left[h(\mathbf{r})-\sum_{k=1}^{n} c_k g_k(\mathbf{r})\right]^2 W(\mathbf{r})\,dv.\end{aligned} \tag{9}$$

setting Eq. (8) equal to Eq. (9) yields

$$\int_V\left[h^2(\mathbf{r})-\sum_{k=1}^{n} c_k{}^2 g_k{}^2(\mathbf{r})\right]W(\mathbf{r})\,dv=\int_V\left[h(\mathbf{r})-\sum_{k=1}^{n} c_k g_k(\mathbf{r})\right]^2 W(\mathbf{r})\,dv. \tag{10}$$

This is known as *Bessel's identity*. Since the integrand of the right-hand side is always positive, one has

$$\int_V\left[h^2(\mathbf{r})-\sum_{k=1}^{n} c_k{}^2 g_k{}^2(\mathbf{r})\right]W(\mathbf{r})\,dv\geq 0$$

or

$$\int_V h^2(\mathbf{r})W(\mathbf{r})\,dv>\sum_{k=1}^{n} c_k{}^2\int_V g_k{}^2(\mathbf{r})W(\mathbf{r})\,dv. \tag{11}$$

This is known as *Bessel's inequality*. In view of the fact that the left-hand side of Eq. (11) is independent of the index n, Eq. (11) will still be valid when n tends to infinity.

$$\int_V h^2(\mathbf{r})W(\mathbf{r})\,dv\geq\sum_{k=1}^{\infty} c_k{}^2\int_V g_k{}^2(\mathbf{r})W(\mathbf{r})\,dv. \tag{12}$$

A3.6. Orthonormal Functions

Many of the formulas will be simplified if orthonormal functions are used. The orthogonal functions $\bar{g}_k(\mathbf{r})$ are said to be orthonormal with respect to the weight function $W(\mathbf{r})$ if

$$\int_V \bar{g}_k{}^2(\mathbf{r})W(\mathbf{r})\,dv=1. \tag{1}$$

A function $f(\mathbf{r})$ can be expressed as a series of the orthonormal functions $\bar{g}_k(\mathbf{r})$,

$$f(\mathbf{r}) = \sum_{k=1}^{\infty} \bar{c}_k \bar{g}_k(\mathbf{r}). \tag{2}$$

The expansion coefficients $\bar{c}_k$ are given by Eq. (A3.4.12),

$$\bar{c}_k = \int_V f(\mathbf{r}) \bar{g}_k(\mathbf{r}) W(\mathbf{r})\, dv. \tag{3}$$

When $f(\mathbf{r})$ is approximated by a finite sum,

$$f(\mathbf{r}) = \sum_{k=1}^{n} \bar{c}_k \bar{g}_k(\mathbf{r}), \tag{4}$$

then the error of such an approximation is

$$\mathcal{E}(\bar{c}_k) = f(\mathbf{r}) - \sum_{k=1}^{n} \bar{c}_k \bar{g}_k(\mathbf{r}). \tag{5}$$

The mean-square error with respect to the weight function $W(\mathbf{r})$ is evaluated by

$$\mathcal{E}_n(\bar{c}_k) = \int_V \left[f(\mathbf{r}) - \sum_{k=1}^{n} \bar{c}_k \bar{g}_k(\mathbf{r}) \right]^2 W(\mathbf{r})\, dv$$

$$= \int_V f^2(\mathbf{r}) W(\mathbf{r})\, dv - \sum_{k=1}^{n} \bar{c}_k{}^2. \tag{6}$$

The orthonormal relation, Eq. (1), is used to arrive at the last expression. Bessel's identity, Eq. (A3.5.10), for orthonormal functions is

$$\int_V \left[f(\mathbf{r}) - \sum_{k=1}^{n} \bar{c}_k \bar{g}_k(\mathbf{r}) \right]^2 W(r)\, dv = \int_V f^2(\mathbf{r}) W(\mathbf{r})\, dv - \sum_{k=1}^{n} \bar{c}_k{}^2. \tag{7}$$

Using the fact that the left-hand side of Eq. (7) is never negative, one obtains Bessel's inequality for orthonormal functions,

$$\int_V f^2(\mathbf{r}) W(\mathbf{r})\, dv \geq \sum_{k=1}^{n} \bar{c}_k{}^2. \tag{8}$$

This also implies that if $f(\mathbf{r})$ is expanded into a convergent series, $f(\mathbf{r})$ must be square integrable with respect to the weight function $W(\mathbf{r})$, or

$$\int_V f^2(\mathbf{r}) W(\mathbf{r})\, dv < +\infty. \tag{9}$$

A3.7. Completeness

It has been shown that an arbitrary function $f(\mathbf{r})$ can be expressed as a polynomial of n orthogonal functions $g_k(\mathbf{r})$ belonging to the set $\{g_m(\mathbf{r})\}$. Here the function $f(\mathbf{r})$ is continuous and square integrable with respect to the positive weight function $W(\mathbf{r})$. The functions $g_k(\mathbf{r})$ are orthogonal with respect to the same weight function $W(\mathbf{r})$. The mean-square error of such a representation is

$$\mathcal{E}_n(A_k)=\int_V\left[f(\mathbf{r})-\sum_{k=1}^{n}A_k g_k(\mathbf{r})\right]^2 W(\mathbf{r})\,dv. \tag{1}$$

The set $\{g_m(\mathbf{r})\}$ is said to be *complete* if the mean-square error $E_n(A_k)$ approaches zero as the index n tends to infinity for any function $f(\mathbf{r})$,

$$\lim_{n\to\infty}\int_V\left[f(\mathbf{r})-\sum_{k=1}^{n}A_k g_k(\mathbf{r})\right]^2 W(\mathbf{r})\,dv=0,\qquad \mathbf{a}<\mathbf{r}<\mathbf{b}. \tag{2}$$

From Eq. (A3.5.7), this is equivalent to imposing

$$\lim_{n\to\infty}\left\{\left[\int_V f^2(\mathbf{r})W(\mathbf{r})\,dv-\sum_{k=1}^{n}c_k{}^2\int_V g_k{}^2(\mathbf{r})W(\mathbf{r})\,dv\right]\right.$$

$$\left.+\sum_{k=1}^{n}(A_k-c_k)^2\int_V g_k{}^2(\mathbf{r})W(\mathbf{r})\,dv\right\}=0. \tag{3}$$

By virtue of Bessel's inequality (A3.5.11) the quantity within the square bracket is never negative. Furthermore, the last integral is always greater than zero for nontrivial functions $g_k(\mathbf{r})$ and positive weight function $W(\mathbf{r})$. Condition (3) is fulfilled if and only if

$$A_k=c_k \tag{4}$$

and

$$\lim_{n\to\infty}\left[\int_V f^2(\mathbf{r})W(\mathbf{r})\,dv-\sum_{k=1}^{n}c_k{}^2\int_V g_k{}^2(\mathbf{r})W(\mathbf{r})\,dv\right]=0. \tag{5}$$

Equation (5) is satisfied when equality holds in Bessel's inequality (A3.5.11).

Thus, a sequence $\{g_m(\mathbf{r})\}$ of orthogonal and square-integrable functions $g_k(\mathbf{r})$ is complete with respect to a positive weight function $W(\mathbf{r})$ within the space V, $\mathbf{a}<\mathbf{r}<\mathbf{b}$, if and only if

$$\int_V f^2(\mathbf{r})W(\mathbf{r})\,dv=\sum_{k=1}^{n}\left[\int_V f(\mathbf{r})g_k(\mathbf{r})W(\mathbf{r})\,dv\right]^2\Big/\int_V g_k{}^2(\mathbf{r})W(\mathbf{r})\,dv \tag{6}$$

for all continuous square-integrable functions $f(\mathbf{r})$. Equation (6) is identical to Eq. (5) with c_k expressed by Eq. (A3.4.9).

The criterion for completeness of a sequence of orthogonal functions may be stated in another way. Consider a sequence $\{g_m(\mathbf{r})\}$ of orthogonal square-integrable functions in a space V with respect to a positive weight function $W(\mathbf{r})$. The sequence is complete if and only if all continuous square-integrable functions within a space V, $\mathbf{a}<\mathbf{r}<\mathbf{b}$, can be approximated arbitrarily closely in the sense of least squares by a polynomial of

$g_k(\mathbf{r})$ for a finite number of terms, that is, if

$$\int_V \left[f(\mathbf{r}) - \sum_{k=1}^{n} c_k g_k(\mathbf{r}) \right]^2 W(\mathbf{r})\, dv < \epsilon, \tag{7}$$

where ϵ is some positive constant and c_k is the Fourier coefficient from Eq. (A3.4.9).

Appendix 4

Decomposition of Fields

In many problems, it is easiest to find one component of the field in terms of which all other components can be expressed. For this purpose, it is convenient to introduce the following notation:

$$\mathbf{E} \equiv \mathbf{E}_{\parallel} + \mathbf{E}_{\perp}, \qquad \mathbf{H} = \mathbf{H}_{\parallel} + \mathbf{H}_{\perp}, \qquad \nabla = \nabla_{\parallel} + \nabla_{\perp}, \tag{1}$$

where $\mathbf{E}_{\parallel}$, $\mathbf{H}_{\parallel}$, and $\nabla_{\parallel}$ are the components parallel to the desired direction, and $\mathbf{E}_{\perp}$, $\mathbf{H}_{\perp}$, and $\nabla_{\perp}$ are the components transverse (perpendicular) to the desired direction. For example, if z is chosen to be the desired direction in rectangular coordinates, then

$$\nabla = \hat{\mathbf{x}}\frac{\partial}{\partial x} + \hat{\mathbf{y}}\frac{\partial}{\partial y} + \hat{\mathbf{z}}\frac{\partial}{\partial z} = \nabla_{\perp} + \nabla_{\parallel},$$

$$\nabla_{\perp} = \hat{\mathbf{x}}\frac{\partial}{\partial x} + \hat{\mathbf{y}}\frac{\partial}{\partial y}, \quad \nabla_{\parallel} = \hat{z}\frac{\partial}{\partial z}.$$

With the notation defined in Eq. (1), Maxwell's equations become

$$\nabla \times \mathbf{E} = -j\omega\mu\mathbf{H},$$

$$(\nabla_{\parallel} + \nabla_{\perp}) \times (\mathbf{E}_{\parallel} + \mathbf{E}_{\perp}) = -j\omega\mu(\mathbf{H}_{\parallel} + \mathbf{H}_{\perp}),$$

$$\nabla_{\parallel} \times \mathbf{E}_{\perp} + \nabla_{\perp} \times \mathbf{E}_{\parallel} + \nabla_{\perp} \times \mathbf{E}_{\perp} = -j\omega\mu(\mathbf{H}_{\parallel} + \mathbf{H}_{\perp}). \tag{2}$$

This is a vector equation and the corresponding components of each side must be equal. Thus

$$\nabla_{\perp} \times \mathbf{E}_{\perp} = -j\omega\mu\mathbf{H}_{\parallel}, \tag{3}$$

$$\nabla_{\parallel} \times \mathbf{E}_{\perp} + \nabla_{\perp} \times \mathbf{E}_{\parallel} = -j\omega\mu\mathbf{H}_{\perp}. \tag{4}$$

Similarly, for $\varepsilon_r \equiv \varepsilon_r(s_{\|})$, where $s_{\|}$ is the variable in the parallel direction,

$$\nabla \times \mathbf{H} = j\omega\epsilon_0\varepsilon_r\mathbf{E},$$

$$\nabla_{\perp} \times \mathbf{H}_{\perp} = j\omega\epsilon_0\varepsilon_r\mathbf{E}_{\|}, \tag{5}$$

$$\nabla_{\|} \times \mathbf{H}_{\perp} + \nabla_{\perp} \times \mathbf{H}_{\|} = j\omega\epsilon_0\varepsilon_r\mathbf{E}_{\perp}. \tag{6}$$

To express $\mathbf{H}_{\perp}$ in terms of fields in the parallel direction, $\mathbf{E}_{\perp}$ may be eliminated between Eqs. (4) and (6). From Eq. (6),

$$\mathbf{E}_{\perp} = \frac{1}{j\omega\epsilon_0\varepsilon_r}(\nabla_{\|} \times \mathbf{H}_{\perp} + \nabla_{\perp} \times \mathbf{H}_{\|}). \tag{7}$$

Substituting Eq. (7) into Eq. (4) yields

$$-\gamma_0^2\mathbf{H}_{\perp} = \nabla_{\|} \times \left[\frac{1}{\varepsilon_r}(\nabla_{\|} \times \mathbf{H}_{\perp} + \nabla_{\perp} \times \mathbf{H}_{\|})\right]$$

$$+ j\omega\epsilon_0\nabla_{\perp} \times \mathbf{E}_{\|}, \qquad \gamma_0^2 \equiv -\omega^2\mu\epsilon_0,$$

$$-\gamma_0^2\mathbf{H}_{\perp} = \nabla_{\|}\left(\frac{1}{\varepsilon_r}\right) \times (\nabla_{\|} \times \mathbf{H}_{\perp} + \nabla_{\perp} \times \mathbf{H}_{\|})$$

$$+ \frac{1}{\varepsilon_r}[\nabla_{\|} \times \nabla_{\|} \times \mathbf{H}_{\perp} + \nabla_{\|} \times \nabla_{\perp} \times \mathbf{H}_{\|}] + j\omega\epsilon_0\nabla_{\perp} \times \mathbf{E}_{\|},$$

$$-\gamma_0^2\mathbf{H}_{\perp} = \frac{-1}{\varepsilon_r{}^2}\nabla_{\|}\varepsilon_r \times (\nabla_{\|} \times \mathbf{H}_{\perp} + \nabla_{\perp} \times \mathbf{H}_{\|})$$

$$+ \frac{1}{\varepsilon_r}[-(\nabla_{\|}\cdot\nabla_{\|})\mathbf{H}_{\perp} + \nabla_{\perp}(\nabla_{\|}\cdot\mathbf{H}_{\|})] + j\omega\epsilon_0\nabla_{\perp} \times \mathbf{E}_{\|},$$

$$-\gamma_0^2\varepsilon_r\mathbf{H}_{\perp} = (-1/\varepsilon_r)[-(\nabla_{\|}\varepsilon_r\cdot\nabla_{\|})\mathbf{H}_{\perp} + \nabla_{\perp}^H(\nabla_{\|}\varepsilon_r\cdot\mathbf{H}_{\|})]$$

$$-\nabla_{\|}^2\mathbf{H}_{\perp} + \nabla_{\perp}(\nabla_{\|}\cdot\mathbf{H}_{\|}) + j\omega\epsilon_0\varepsilon_r\nabla_{\perp} \times \mathbf{E}_{\|},$$

$$\nabla_{\|}^2\mathbf{H}_{\perp} - \gamma_0^2\varepsilon_r\mathbf{H}_{\perp} - \frac{1}{\varepsilon_r}(\nabla_{\|}\varepsilon_r\cdot\nabla_{\|})\mathbf{H}_{\perp}$$

$$= j\omega\epsilon_0\varepsilon_r\nabla_{\perp} \times \mathbf{E}_{\|} + \nabla_{\perp}(\nabla_{\|}\cdot\mathbf{H}_{\|})$$

$$-\frac{1}{\varepsilon_r}\nabla_{\perp}^H(\nabla_{\|}\varepsilon_r\cdot\mathbf{H}_{\|}). \tag{8}$$

where the operator $\nabla_{\perp}^H$ operates on the function $\mathbf{H}_{\|}$ only.

The fields $\mathbf{E}_{\perp}$ can be expressed in terms of components in the parallel direction by eliminating $\mathbf{H}_{\perp}$ between Eqs. (4) and (6). From Eq. (4),

$$\mathbf{H}_{\perp} = \frac{1}{-j\omega\mu}(\nabla_{\|} \times \mathbf{E}_{\perp} + \nabla_{\perp} \times \mathbf{E}_{\|}). \tag{9}$$

Substituting Eq. (9) into Eq. (6) yields

$$j\omega\epsilon_0\varepsilon_r\mathbf{E}_\perp = \nabla_\parallel\times\left[\frac{1}{-j\omega\mu}(\nabla_\parallel\times\mathbf{E}_\perp+\nabla_\perp\times\mathbf{E}_\parallel)\right]+\nabla_\perp\times\mathbf{H}_\parallel,$$

$$\nabla_\parallel^2\mathbf{E}_\perp-\gamma_0^2\varepsilon_r\mathbf{E}_\perp = \nabla_\perp(\nabla_\parallel\cdot\mathbf{E}_\parallel)-j\omega\mu\nabla_\perp\times\mathbf{H}_\parallel. \tag{10}$$

The Helmholtz equations in an inhomogeneous dielectric medium characterized by $\varepsilon_r(s_\parallel)$ are [Eqs. (7.1.1) and (7.1.2)]

$$\nabla^2\mathbf{E}+\nabla\left(\frac{1}{\varepsilon_r}\nabla_\parallel\varepsilon_r\cdot\mathbf{E}_\parallel\right)-\gamma_0^2\varepsilon r\mathbf{E}=0, \tag{11}$$

$$\nabla^2\mathbf{H}+\frac{1}{\varepsilon_r}\nabla_\parallel\varepsilon_r\times(\nabla\times\mathbf{H})-\gamma_0^2\varepsilon_r\mathbf{H}=0. \tag{12}$$

Equations (11) and (12) may be decomposed into their parallel and transverse components by using the notation defined by Eq. (1):

$$\nabla^2(\mathbf{E}_\parallel+\mathbf{E}_\perp)+(\nabla_\parallel+\nabla_\perp)\left(\frac{1}{\varepsilon_r}\nabla_\parallel\varepsilon_r\cdot\mathbf{E}_\parallel\right)-\gamma_0^2\varepsilon_r(\mathbf{E}_\parallel+\mathbf{E}_\perp)=0$$

or

$$\nabla^2\mathbf{E}_\parallel+\nabla_\parallel\left(\frac{1}{\varepsilon_r}\nabla_\parallel\varepsilon_r\cdot\mathbf{E}_\parallel\right)-\gamma_0^2\varepsilon_r\mathbf{E}_\parallel=0, \tag{13}$$

$$\nabla^2\mathbf{E}_\perp+\nabla_\perp\left(\frac{1}{\varepsilon_r}\nabla_\parallel\varepsilon_r\cdot\mathbf{E}_\parallel\right)-\gamma_0^2\epsilon_r\mathbf{E}_\perp=0, \tag{14}$$

$$\nabla^2(\mathbf{H}_\parallel+\mathbf{H}_\perp)+\frac{1}{\varepsilon_r}\nabla_\parallel\varepsilon_r\times\left[(\nabla_\parallel+\nabla_\perp)\times(\mathbf{H}_\parallel\times\mathbf{H}_\perp)\right]$$
$$-\gamma_0^2\varepsilon_r(\mathbf{H}_\parallel+\mathbf{H}_\perp)=0$$

or

$$\nabla^2\mathbf{H}_\parallel-\gamma_0^2\varepsilon_r\mathbf{H}_\parallel=0, \tag{15}$$

$$\nabla^2\mathbf{H}_\perp-\frac{1}{\varepsilon_r}(\nabla_\parallel\varepsilon_r\cdot\nabla_\parallel)\mathbf{H}_\perp-\gamma_0^2\varepsilon_r\mathbf{H}_\perp=\frac{-1}{\varepsilon_r}\nabla_\perp^H(\nabla_\parallel\varepsilon_r\cdot\mathbf{H}_\parallel). \tag{16}$$

Subtracting Eq. (16) from Eq. (8) yields

$$\nabla_\parallel^2\mathbf{H}_\perp-\nabla^2\mathbf{H}_\perp=j\omega\epsilon_0\varepsilon_r\nabla_\perp\times\mathbf{E}_\parallel+\nabla_\perp(\nabla_\parallel\cdot\mathbf{H}_\parallel),$$

but

$$\nabla^2\mathbf{H}_\perp=\nabla_\parallel^2\mathbf{H}_\perp+\nabla_\perp^2\mathbf{H}_\perp.$$

Therefore,

$$-\nabla_\perp^2\mathbf{H}_\perp=j\omega\epsilon_0\varepsilon_r\nabla_\perp\times\mathbf{E}_\parallel+\nabla_\perp(\nabla_\parallel\cdot\mathbf{H}_\parallel). \tag{17}$$

Subtracting Eq. (14) from Eq. (10) yields

$$-\nabla_{\perp}^{2}\mathbf{E}_{\perp}=\boldsymbol{\nabla}_{\perp}\left(\frac{1}{\varepsilon_r}\boldsymbol{\nabla}_{\|}\varepsilon_r\cdot\mathbf{E}_{\|}\right)+\boldsymbol{\nabla}_{\perp}(\boldsymbol{\nabla}_{\|}\cdot\mathbf{E}_{\|})-j\omega\mu\boldsymbol{\nabla}_{\perp}\times\mathbf{H}_{\|}. \qquad (18)$$

Equations (17) and (18) are the desired relations expressing the transverse field components in terms of the parallel components.

For example, if the field's dependency on the variables transverse to z (the parallel direction) is of the form $e^{-\gamma_x x-\gamma_y y}$, then

$$\nabla_{\perp}^{2}\mathbf{H}_{\perp}=\left(\frac{\partial^2}{\partial x^2}+\frac{\partial^2}{\partial y^2}\right)\mathbf{H}_{\perp}=({\gamma_x}^2+{\gamma_y}^2)\mathbf{H}_{\perp},$$

and Eq. (17) becomes

$$\mathbf{H}_{\perp}=\frac{-1}{{\gamma_x}^2+{\gamma_y}^2}\left[j\omega\epsilon_0\varepsilon_r\boldsymbol{\nabla}_{\perp}\times\mathbf{E}_{\|}+\boldsymbol{\nabla}_{\perp}(\boldsymbol{\nabla}_{\|}\cdot\mathbf{H}_{\|})\right].$$

Similarly,

$$\mathbf{E}_{\perp}=\frac{-1}{{\gamma_x}^2+{\gamma_y}^2}\left[\boldsymbol{\nabla}_{\perp}\left(\frac{1}{\varepsilon_r}\boldsymbol{\nabla}_{\|}\varepsilon_r\cdot\mathbf{E}_{\|}\right)+\boldsymbol{\nabla}_{\perp}(\boldsymbol{\nabla}_{\|}\cdot\mathbf{E}_{\|})-j\omega\mu\boldsymbol{\nabla}_{\perp}\times\mathbf{H}_{\|}\right].$$

Appendix 5

Curvilinear Coordinates

A5.1. Curvilinear Coordinates

The equation

$$f(x, y, z) = C, \tag{1}$$

where C is a constant, determines a family of surfaces in x, y, z space. Each surface is characterized by a specific value of the parameter C. When $C = x$, then

$$f(x, y, z) = x \tag{2}$$

determines a surface of constant x.

Consider three families of surfaces defined by

$$\begin{aligned} f_1(x, y, z) &= u_1, \\ f_2(x, y, z) &= u_2, \\ f_3(x, y, z) &= u_3. \end{aligned} \tag{3}$$

This set of equations is chosen so that when u_1, u_2, and u_3 are constant, the three families of surfaces are orthogonal to each other.

A point in space can be defined as the intersection of three of these surfaces, one from each family. The point is then specified by the parameters of these surfaces,

$$P = P(u_1, u_2, u_3). \tag{4}$$

The variables u_1, u_2, and u_3 are known as the *curvilinear coordinates* of the point.

Let dl_k be an incremental length normal to the $u_k = \text{const}$ plane. The element of length dl_k between u_k and $u_k + du_k$ is given by

$$dl_k = h_k\, du_k, \qquad k = 1, 2, \text{ or } 3, \tag{5}$$

where h_k is, in general, a function of u_1, u_2, and u_3.

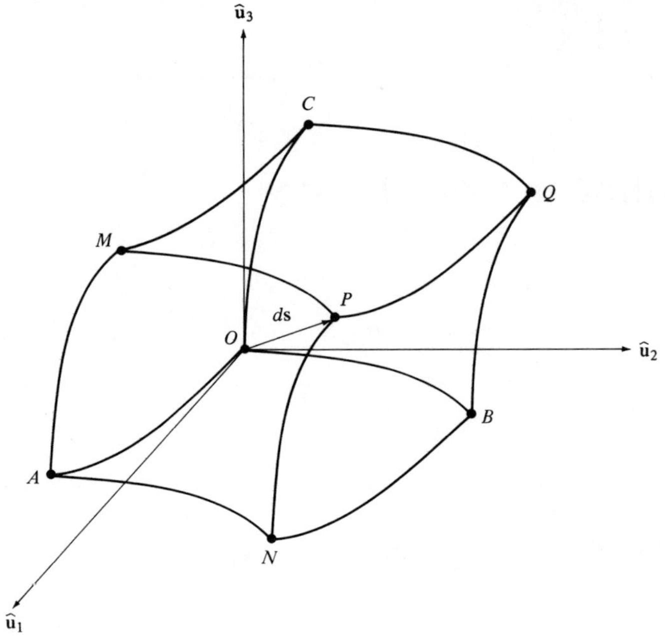

Figure 19. Curvilinear coordinates.

The unit vector $\hat{\mathbf{u}}_k$ is a vector of unit length normal to the $u_k = \text{const}$ plane and points in the direction of increasing values of u_k.

The right-handed coordinate system will be used, that is, the system satisfying the relation

$$\hat{\mathbf{u}}_1 \times \hat{\mathbf{u}}_2 = \hat{\mathbf{u}}_3. \tag{6}$$

The incremental vector $d\mathbf{s}$ (Figure 19) is

$$d\mathbf{s} = \hat{\mathbf{u}}_1\, dl_1 + \hat{\mathbf{u}}_2\, dl_2 + \hat{\mathbf{u}}_3\, dl_3 \tag{7}$$

$$= \hat{\mathbf{u}}_1 h_1\, du_1 + \hat{\mathbf{u}}_2 h_2\, du_2 + \hat{\mathbf{u}}_3 h_3\, du_3, \tag{8}$$

where dl_i is the arclength measured along the $\hat{\mathbf{u}}_i$ direction; in Eq. (8) the arclengths are given in terms of the coordinates du_i. The h_i are determined as follows.

Cartesian Coordinates.

$$d\mathbf{s} = \hat{\mathbf{x}}\, dx + \hat{\mathbf{y}}\, dy + \hat{\mathbf{z}}\, dz = \hat{\mathbf{u}}_1 h_1\, du_1 + \hat{\mathbf{u}}_2 h_2\, du_2 + \hat{\mathbf{u}}_2 h_3\, du_3.$$

If $x = u_1$, $y = u_2$, $z = u_3$, then $dx = du_1$, $dy = du_2$, and $dz = du_3$. Therefore,

$$\hat{\mathbf{x}}\, dx = \hat{\mathbf{u}}_1 h_1\, du_1 = \hat{\mathbf{x}} h_1\, dx. \tag{9}$$

Therefore $h_1 = 1$. Similarly, $h_2 = 1$ and $h_3 = 1$.

Cylindrical Coordinates.

$$d\mathbf{s}=\hat{\mathbf{r}}\,dr+\hat{\boldsymbol{\phi}}\,r\,d\phi+\hat{\mathbf{z}}\,dz=\hat{\mathbf{u}}_1 h_1\,du_1+\hat{u}_2 h_2\,du_2+\hat{\mathbf{u}}_3 h_3\,du_3.$$

With $u_1=r$, $u_2=\phi$, $u_3=z$,

$$\begin{aligned} \hat{\mathbf{r}}\,dr&=\hat{\mathbf{u}}_1 h_1\,du_1=\hat{\mathbf{r}} h_1\,dr, && h_1=1,\\ \hat{\boldsymbol{\phi}}\,r\,d\phi&=\hat{\mathbf{u}}_2 h_2\,du_2=\hat{\mathbf{r}} h_2\,d\phi, && h_2=r,\\ \hat{\mathbf{z}}\,dz&=\hat{\mathbf{u}}_3 h_3\,du_3=\hat{\mathbf{z}} h_3\,dz, && h_3=1. \end{aligned} \tag{10}$$

Spherical Coordinates.

$$d\mathbf{s}=\hat{\mathbf{r}}\,dr+\hat{\boldsymbol{\theta}}\,r\,d\theta+\hat{\boldsymbol{\phi}}\,r\sin\theta\,d\phi=\hat{\mathbf{u}}_1 h_1\,du_1+\hat{\mathbf{u}}_2 h_2\,du_2+\hat{\mathbf{u}}_3 h_3\,du_3.$$

With $r=u_1$, $\theta=u_2$, and $\phi=u_3$,

$$\begin{aligned} \hat{\mathbf{r}}\,dr&=\hat{\mathbf{r}} h_1\,du_1=\hat{\mathbf{r}} h_1\,dr, && h_1=1,\\ \hat{\boldsymbol{\theta}}\,r\,d\theta&=\hat{\boldsymbol{\theta}} h_2\,d\theta, && h_2=r,\\ \hat{\boldsymbol{\phi}}\,r\sin\theta\,d\phi&=\hat{\boldsymbol{\phi}} h_3\,d\phi, && h_3=r\sin\theta. \end{aligned} \tag{11}$$

A5.2. Gradient

The gradient of a function $f(u_1, u_2, u_3)$ is given by

$$\begin{aligned} \nabla f&=\hat{\mathbf{u}}_1\frac{\partial f}{\partial l_1}+\hat{\mathbf{u}}_2\frac{\partial f}{\partial l_2}+\hat{\mathbf{u}}_3\frac{\partial f}{\partial l_3}\\ &=\hat{\mathbf{u}}_1\frac{1}{h_1}\frac{\partial f}{\partial u_1}+\hat{\mathbf{u}}_2\frac{1}{h_2}\frac{\partial f}{\partial u_2}+\hat{\mathbf{u}}_3\frac{1}{h_3}\frac{\partial f}{\partial u_3}, \end{aligned} \tag{1}$$

where the relation

$$dl_i=h_i\,du_i, \qquad i=1, 2, \text{ or } 3, \tag{2}$$

was used to obtain the second form.

Cartesian Coordinates. $h_1=h_2=h_3=1$; $u_1=x$, $u_2=y$, and $u_3=z$.

$$\nabla f=\hat{\mathbf{x}}\frac{\partial f}{\partial x}+\hat{\mathbf{y}}\frac{\partial f}{\partial y}+\hat{\mathbf{z}}\frac{\partial f}{\partial z}. \tag{3}$$

Cylindrical Coordinates. $h_1=h_3=1$, $h_2=r$; $u_1=r$, $u_2=\phi$, and $u_3=z$.

$$\nabla f=\hat{\mathbf{r}}\frac{\partial f}{\partial r}+\hat{\boldsymbol{\phi}}\frac{1}{r}\frac{\partial f}{\partial \phi}+\hat{\mathbf{z}}\frac{\partial f}{\partial z}. \tag{4}$$

Spherical Coordinates. $h_1=1$, $h_2=r$, $h_3=r\sin\theta$; $u_1=r$, $u_2=\theta$, and $u_3=\phi$.

$$\nabla f=\hat{\mathbf{r}}\frac{\partial f}{\partial r}+\hat{\boldsymbol{\theta}}\frac{1}{r}\frac{\partial f}{\partial \theta}+\hat{\boldsymbol{\phi}}\frac{1}{r\sin\theta}\frac{\partial f}{\partial \phi}. \tag{5}$$

A5.3. Divergence

The divergence of a vector

$$\mathbf{A}(u_1, u_2, u_3) = \hat{\mathbf{u}}_1 A_1 + \hat{\mathbf{u}}_2 A_2 + \hat{\mathbf{u}}_3 A_3 \tag{1}$$

is evaluated from the divergence theorem,

$$\int_V \nabla \cdot \mathbf{A}\, dv = \oint_S \mathbf{A} \cdot \hat{\mathbf{n}}\, da. \tag{2}$$

Equation (2) is valid for any volume V, including an infinitesimal volume.

$$\begin{aligned} \lim_{V \to 0} \int_V \nabla \cdot \mathbf{A}\, dv &= \nabla \cdot \mathbf{A}\, dv = \nabla \cdot \mathbf{A}\, dl_1\, dl_2\, dl_3 \\ &= \nabla \cdot \mathbf{A}\, h_1 h_2 h_3\, du_1\, du_2\, du_3, \end{aligned} \tag{3}$$

where $dl_i = h_i\, du_i$ is the length of the side of volume dv. The contribution to the closed surface integral of Eq. (2) through the area $S_1(u_1 = u_{10}) = \mathrm{OBQC}$ (see Figure 19) is

$$\lim_{S_1 \to 0} \int_{S_1(u_{10})} \mathbf{A} \cdot \hat{\mathbf{n}}\, da = \mathbf{A}(u_{10}) \cdot (-\hat{\mathbf{u}}_1\, dl_2\, dl_3) = -A_1(u_{10}) h_2 h_3\, du_2\, du_3. \tag{4}$$

The contribution through the surface $S_1(u_{10} + du_1) = \mathrm{AMPN}$ is

$$\begin{aligned} \lim_{S_1 \to 0} \int_{S_1(u_{10} + du_1)} \mathbf{A} \cdot \hat{\mathbf{n}}\, da &= \mathbf{A}(u_{10} + du_1) \cdot \hat{\mathbf{u}}_1\, da_1(u_{10} + du_1) \\ &= \left(\mathbf{A}_1 \Big|_{u_{10}} + \frac{\partial \mathbf{A}_1}{\partial u_1} \Big|_{u_{10}} du_1 + \cdots \right) \\ &\quad \cdot \hat{\mathbf{u}}_1 \left[da_1 \Big|_{u_{10}} + \frac{\partial (da_1)}{\partial u_1} \Big|_{u_{10}} du_1 + \cdots \right] \\ &= A_{10}\, da_{10} + \frac{\partial A_1}{\partial u_1} \Big|_{u_{10}} du_1\, da_1 + A_1 \frac{\partial (da_1)}{\partial u_1} \Big|_{u_{10}} du_1 + \text{h.o.t.} \end{aligned} \tag{5}$$

where $A_{10} \equiv A_1(u = u_{10})$, $da_{10} \equiv da_1(u = u_{10})$, and h.o.t. stands for higher-order terms. The second and third terms may be combined to give

$$\begin{aligned} \frac{\partial A_{10}}{\partial u_1} du_1\, da_1 + A_1 \frac{\partial (da_{10})}{\partial u_1} du_1 &= \frac{\partial A_{10}}{\partial u_1} du_1 (dl_2\, dl_3) + A_1 \frac{\partial (dl_2\, dl_3)}{\partial u_1} du_1 \\ &= \frac{\partial A_{10}}{\partial u_1} du_1 (h_2 h_3\, du_2\, du_3) \end{aligned}$$

$$+A_1\left[\frac{\partial}{\partial u_1}(h_2h_3\,du_2\,du_3)\right]du_1$$

$$=\left[h_2h_3\frac{\partial A_{10}}{\partial u_1}+A_1\frac{\partial(h_2h_3)}{\partial u_1}\right]du_1\,du_2\,du_3$$

$$=\frac{\partial(A_1h_2h_3)}{\partial u_1}\bigg|_{u_{10}}du_1\,du_2\,du_3 \qquad \left(\frac{\partial A_{10}}{\partial u_1}\equiv\frac{\partial A_1}{\partial u_1}\bigg|_{u_{10}}\right). \tag{6}$$

Substituting Eq. (6) into Eq. (5) yields

$$\lim_{S_1\to 0}\int_{S_1(u_{10}+du_1)}\mathbf{A}\cdot\hat{\mathbf{n}}\,da=A_1\,da_1+\frac{\partial(A_1h_2h_3)}{\partial u_1}du_1\,du_2\,du_3. \tag{7}$$

The higher-order terms have been omitted since they vanish in the limit. The contribution from both $S_1(u_{10})$ and $S_1(u_{10}+du_1)$ is given by the sum of Eqs. (4) and (7):

$$\lim_{S_1\to 0}\int_{S_1(u_{10})+S_1(u_{10}+du_1)}\mathbf{A}\cdot\hat{\mathbf{n}}\,da=\frac{\partial(A_1h_2h_3)}{\partial u_1}\bigg|_{u_{10}}du_1\,du_2\,du_3. \tag{8}$$

Corresponding expressions for the other two pairs of surfaces may be obtained similarly.

$$\lim_{S_2\to 0}\int_{S_2(u_{20})+S_2(u_{20}+du_2)}\mathbf{A}\cdot\hat{\mathbf{n}}\,da=\frac{\partial(A_2h_1h_3)}{\partial u_2}\bigg|_{u_{20}}du_1\,du_2\,du_3, \tag{9}$$

$$\lim_{S_3\to 0}\int_{S_3(u_{30})+S_3(u_{30}+du_3)}\mathbf{A}\cdot\hat{\mathbf{n}}\,da=\frac{\partial(A_3h_1h_2)}{\partial u_3}\bigg|_{u_{30}}du_1\,du_2\,du_3. \tag{10}$$

Next one substitutes of Eqs. (3) and (8)–(10) into Eq. (2). (Since u_{10}, u_{20}, u_{30} can be any reference position, they are omitted in the following expressions.)

$$\nabla\cdot\mathbf{A}\,h_1h_2h_3\,du_1du_2\,du_3=\left[\frac{\partial(A_1h_2h_3)}{\partial u_1}+\frac{\partial(A_2h_1h_3)}{\partial u_2}+\frac{\partial(A_3h_1h_2)}{\partial u_3}\right]\times du_1\,du_2\,du_3.$$

Therefore,

$$\nabla\cdot\mathbf{A}=\frac{1}{h_1h_2h_3}\left[\frac{\partial(A_1h_2h_3)}{\partial u_1}+\frac{\partial(A_2h_1h_3)}{\partial u_2}+\frac{\partial(A_3h_1h_2)}{\partial u_3}\right]. \tag{11}$$

Cartesian Coordinates. $h_1=h_2=h_3=1$; $u_1=x$, $u_2=y$, and $u_3=z$.

$$\nabla\cdot\mathbf{A}=\frac{\partial A_x}{\partial x}+\frac{\partial A_y}{\partial y}+\frac{\partial A_z}{\partial z}. \tag{12}$$

Cylindrical Coordinates. $h_1 = h_3 = 1$, $h_2 = r$; $u_1 = r$, $u_2 = \phi$, and $u_3 = z$.

$$\nabla \cdot \mathbf{A} = \frac{1}{r}\frac{\partial(rA_r)}{\partial r} + \frac{1}{r}\frac{\partial A_\phi}{\partial \phi} + \frac{\partial A_z}{\partial z}. \tag{13}$$

Spherical Coordinates. $h_1 = 1$, $h_2 = r$, $h_3 = r\sin\theta$; $u_1 = r$, $u_2 = \theta$, and $u_3 = \phi$.

$$\nabla \cdot \mathbf{A} = \frac{1}{r^2\sin\theta}\left[\frac{\partial(A_r r^2\sin\theta)}{\partial r} + \frac{\partial(A_\theta r\sin\theta)}{\partial\theta} + \frac{\partial(A_\phi r)}{\partial\phi}\right]$$

$$= \frac{1}{r^2}\frac{\partial(r^2 A_r)}{\partial r} + \frac{1}{r\sin\theta}\frac{\partial(A_\theta\sin\theta)}{\partial\theta} + \frac{1}{r\sin\theta}\frac{\partial A_\phi}{\partial\phi}. \tag{14}$$

A5.4. Curl

The curl of a vector function can be evaluated by Stokes' theorem,

$$\int_S (\nabla\times\mathbf{A})\cdot\hat{\mathbf{n}}\,da = \oint_C \mathbf{A}\cdot d\mathbf{l}. \tag{1}$$

This relation is valid for any surface S, including an infinitesimal area.

$$\lim_{S\to 0}\int_S (\nabla\times\mathbf{A})\cdot\hat{\mathbf{n}}\,da = \lim_{C\to 0}\oint_C \mathbf{A}\cdot d\mathbf{l}. \tag{2}$$

The $\hat{\mathbf{u}}_1$ component of $\nabla\times\mathbf{A}$ is obtained by applying Eq. (2) to the surface OBQC.

$$\lim_{C\to 0}\oint_{\mathrm{OBQC}} \mathbf{A}\cdot d\mathbf{l} = \lim_{C\to 0}\left(\int_0^{\mathrm{B}}\mathbf{A}\cdot d\mathbf{l} + \int_{\mathrm{B}}^{\mathrm{Q}}\mathbf{A}\cdot d\mathbf{l} + \int_{\mathrm{Q}}^{\mathrm{C}}\mathbf{A}\cdot d\mathbf{l} + \int_{\mathrm{C}}^{\mathrm{O}}\mathbf{A}\cdot d\mathbf{l}\right)$$

$$= A_2(u_{30})\,dl_2(u_{30}) - A_2(u_{30}+du_3)\,dl_2(u_{30}+du_3)$$

$$+ A_3(u_{20}+du_2)\,dl_3(u_{20}+du_2) - A_3(u_{20})\,dl_3(u_{20})$$

$$= \lim_{C\to 0}\left\{A_{20}\,dl_2 - \left(A_2 + \left.\frac{\partial A_2}{\partial u_3}\right|_{u_{30}} du_3 + \cdots\right)\right.$$

$$\times\left[dl_{20} + \left.\frac{\partial(dl_2)}{\partial u_3}\right|_{u_{30}} du_3 + \cdots\right]$$

$$+\left(A_{30} + \left.\frac{\partial A_3}{\partial u_2}\right|_{u_{20}} du_2 + \cdots\right)$$

$$\left.\times\left[dl_{30} + \left.\frac{\partial(dl_3)}{\partial u_2}\right|_{u_{20}} du_2 + \cdots\right] - A_{30}\,dl_{30}\right\}$$

$$= \lim_{C\to 0} \left\{ A_{20}\,dl_{20} - \left[A_{20}\,dl_{20} + A_{20} \frac{\partial(dl_2)}{\partial u_3}\bigg|_{u_{30}} du_3 + \frac{\partial A_2}{\partial u_3}\bigg|_{u_{30}} du_3\,dl_{20} + \cdots \right] + \left[A_{30}\,dl_{30} + A_{30} \frac{\partial(dl_2)}{\partial u_2}\bigg|_{u_{20}} du_2 + \frac{\partial A_3}{\partial u_2}\bigg|_{u_{20}} du_2\,dl_3 + \cdots \right] - A_{30}\,dl_{30} \right.$$

$$= \left[-A_{20} \frac{\partial(dl_2)}{\partial u_3}\bigg|_{u_{30}} du_3 - \frac{\partial A_2}{\partial u_3}\bigg|_{u_{30}} du_3\,dl_2 \right] + \left[A_{30} \frac{\partial(dl_3)}{\partial u_2}\bigg|_{u_{20}} du_2 + \frac{\partial A_3}{\partial u_2}\bigg|_{u_{20}} du_2\,dl_3 \right], \quad (3)$$

where $A_{20} \equiv A(u = u_{30})$, $A_{30} \equiv A(u = u_{20})$.

$$dl_{20} \equiv dl(u = u_{30}) = h_2\,du_2, \qquad dl_{30} \equiv dl(u = u_{20}) = h_3\,du_3;$$

$$\lim_{C\to 0} \oint_{\text{OBQC}} \mathbf{A}\cdot d\mathbf{l} = -A_{20} \frac{\partial(h_2\,du_2)}{\partial u_3}\bigg|_{u_{30}} du_3 - \frac{\partial A_2}{\partial u_3}\bigg|_{u_{30}} du_3\,h_2\,du_2 + A_{30} \frac{\partial(h_3\,du_3)}{\partial u_2}\bigg|_{u_{20}} du_2 + \frac{\partial A_3}{\partial u_2}\bigg|_{u_{20}} du_2\,h_3\,du_3$$

$$= \left[-\frac{\partial(A_2 h_2)}{\partial u_3}\bigg|_{u_{30}} + \frac{\partial(h_3 A_2)}{\partial u_2}\bigg|_{u_{20}} \right] du_2\,du_3. \quad (4)$$

The left-hand side of Eq. (1) for the $\hat{\mathbf{u}}_1$ component is

$$\lim_{S\to 0} \int_S (\nabla\times\mathbf{A})\cdot\hat{\mathbf{u}}_1\,da = (\nabla\times\mathbf{A})_{u_1}\,dl_2\,dl_3 = (\nabla\times\mathbf{A})_{u_1} h_2 h_3\,du_2\,du_3. \quad (5)$$

Substituting Eqs. (4) and (5) into Eq. (1) yields

$$(\nabla\times\mathbf{A})_{u_1} h_2 h_3\,du_2\,du_3 = \left[\frac{\partial(h_3 A_3)}{\partial u_2} - \frac{\partial(h_2 A_2)}{\partial u_3} \right] du_2\,du_3,$$

$$(\nabla\times\mathbf{A})_{u_1} = \frac{1}{h_2 h_3} \left[\frac{\partial(h_3 A_3)}{\partial u_2} - \frac{\partial(h_2 A_2)}{\partial u_3} \right]. \quad (6)$$

Since u_{30} and u_{20} can be any reference point, they are dropped from Eq. (6). The other two components may be obtained similarly.

$$(\nabla\times\mathbf{A})_{u_2}=\frac{1}{h_1h_3}\left[\frac{\partial(h_1A_1)}{\partial u_3}-\frac{\partial(h_3A_3)}{\partial u_1}\right], \tag{7}$$

$$(\nabla\times\mathbf{A})_{u_3}=\frac{1}{h_1h_2}\left[\frac{\partial(h_2A_2)}{\partial u_1}-\frac{\partial(h_1A_1)}{\partial u_2}\right]. \tag{8}$$

Equations (6)–(8) may be combined in matrix form as

$$\nabla\times\mathbf{A}=\frac{1}{h_1h_2h_3}\begin{vmatrix} h_1\hat{\mathbf{u}}_1 & h_2\hat{\mathbf{u}}_2 & h_3\hat{\mathbf{u}}_3 \\ \dfrac{\partial}{\partial u_1} & \dfrac{\partial}{\partial u_2} & \dfrac{\partial}{\partial u_3} \\ h_1A_1 & h_2A_2 & h_3A_3 \end{vmatrix}. \tag{9}$$

Cartesian Coordinates. $h_1=h_2=h_3=1$; $u_1=x$, $u_2=y$, and $u_3=z$.

$$\nabla\times\mathbf{A}=\begin{vmatrix} \hat{\mathbf{x}} & \hat{\mathbf{y}} & \hat{\mathbf{z}} \\ \dfrac{\partial}{\partial x} & \dfrac{\partial}{\partial y} & \dfrac{\partial}{\partial z} \\ A_x & A_y & A_z \end{vmatrix}. \tag{10}$$

Cylindrical Coordinates. $h_1=h_3=1$, $h_2=r$; $u_1=r$, $u_2=\phi$, and $u_3=z$.

$$\nabla\times\mathbf{A}=\frac{1}{r}\begin{vmatrix} \hat{\mathbf{r}} & r\hat{\boldsymbol{\phi}} & \hat{\mathbf{z}} \\ \dfrac{\partial}{\partial r} & \dfrac{\partial}{\partial \phi} & \dfrac{\partial}{\partial z} \\ A_r & rA_\phi & A_z \end{vmatrix}. \tag{11}$$

Spherical Coordinates. $h_1=1$, $h_2=r$, $h_3=r\sin\theta$; $u_1=r$, $u_2=\theta$, and $u_3=\phi$.

$$\nabla\times\mathbf{A}=\frac{1}{r^2\sin\theta}\begin{vmatrix} \hat{\mathbf{r}} & r\hat{\boldsymbol{\theta}} & r\sin\theta\hat{\boldsymbol{\phi}} \\ \dfrac{\partial}{\partial r} & \dfrac{\partial}{\partial \theta} & \dfrac{\partial}{\partial \phi} \\ A_r & rA_\theta & r\sin\theta A_\phi \end{vmatrix}. \tag{12}$$

Index